Lecture Notes in Computer Science 12968

More information about this subseries at http://www.springer.com/series/7412

Shadi Albarqouni · M. Jorge Cardoso · Qi Dou ·
Konstantinos Kamnitsas · Bishesh Khanal ·
Islem Rekik · Nicola Rieke · Debdoot Sheet ·
Sotirios Tsaftaris · Daguang Xu · Ziyue Xu (Eds.)

Domain Adaptation and Representation Transfer, and Affordable Healthcare and AI for Resource Diverse Global Health

Third MICCAI Workshop, DART 2021
and First MICCAI Workshop, FAIR 2021
Held in Conjunction with MICCAI 2021
Strasbourg, France, September 27 and October 1, 2021
Proceedings

Springer

Editors
Shadi Albarqouni ⓘ
Helmholtz Zentrum München
Neuherberg, Germany

Qi Dou ⓘ
Chinese University of Hong Kong
Hong Kong, Hong Kong

Bishesh Khanal ⓘ
NepAl Applied Mathematics and Informatics
Institute for Research (NAAMII)
Kathmandu, Nepal

Nicola Rieke ⓘ
Nvidia GmbH
München, Germany

Sotirios Tsaftaris ⓘ
University of Edinburgh
Edinburgh, UK

Ziyue Xu ⓘ
Nvidia Corporation
Santa Clara, CA, USA

M. Jorge Cardoso ⓘ
King's College London
London, UK

Konstantinos Kamnitsas ⓘ
Imperial College London
London, UK

Islem Rekik ⓘ
Istanbul Technical University
Istanbul, Turkey

Debdoot Sheet ⓘ
Indian Institute of Technology Kharagpur
Kharagpur, India

Daguang Xu ⓘ
Nvidia Corporation
Santa Clara, CA, USA

ISSN 0302-9743 ISSN 1611-3349 (electronic)
Lecture Notes in Computer Science
ISBN 978-3-030-87721-7 ISBN 978-3-030-87722-4 (eBook)
https://doi.org/10.1007/978-3-030-87722-4

LNCS Sublibrary: SL6 – Image Processing, Computer Vision, Pattern Recognition, and Graphics

This Springer imprint is published by the registered company Springer Nature Switzerland AG
The registered company address is: Gewerbestrasse 11, 6330 Cham, Switzerland

Preface DART 2021

Computer vision and medical imaging have been revolutionized by the introduction of advanced machine learning and deep learning methodologies. Recent approaches have shown unprecedented performance gains in tasks such as segmentation, classification, detection, and registration. Although these results (obtained mainly on public datasets) represent important milestones for the MICCAI community, most methods lack generalization capabilities when presented with previously unseen situations (corner cases) or different input data domains. This limits clinical applicability of these innovative approaches and therefore diminishes their impact. Transfer learning, representation learning, and domain adaptation techniques have been used to tackle problems such as model training using small datasets while obtaining generalizable representations; performing domain adaptation via few-shot learning; obtaining interpretable representations that are understood by humans; and leveraging knowledge learned from a particular domain to solve problems in another.

The third MICCAI workshop on Domain Adaptation and Representation Transfer (DART 2021) aimed at creating a discussion forum to compare, evaluate, and discuss methodological advancements and ideas that can improve the applicability of machine learning (ML)/deep learning (DL) approaches to clinical settings by making them robust and consistent across different domains.

During the third edition of DART, 23 papers were submitted for consideration and, after peer review, 13 full papers were accepted for presentation. Each paper was rigorously reviewed by three reviewers in a double-blind review process. The papers were automatically assigned to reviewers, taking into account and avoiding potential conflicts of interest and recent work collaborations between peers. Reviewers were selected from among the most prominent experts in the field from all over the world. Once the reviews were obtained the area chairs formulated final decisions over acceptance or rejection of each manuscript. These decisions were always taken according to the reviews and were unappealable.

Additionally, the workshop organization committee granted the Best Paper Award to the best submission presented at DART 2021. The Best Paper Award was assigned as a result of a secret voting procedure where each member of the committee indicated two papers worthy of consideration for the award. The paper collecting the majority of votes was then chosen by the committee.

We believe that the paper selection process implemented during DART 2021 as well as the quality of the submissions resulted in scientifically validated and interesting contributions to the MICCAI community and in particular to researchers working on domain adaptation and representation transfer.

We would therefore like to thank the authors for their contributions, the reviewers for their dedication and professionality in delivering expert opinions about the submissions, and NVIDIA Corporation, which sponsored DART 2021, for the support, resources, and help in organizing the workshop. NVIDIA Corporation also sponsored the award for the best paper at DART 2021, which consisted of a NVIDIA RTX 3090.

August 2021

Shadi Albarqouni
M. Jorge Cardoso
Qi Dou
Konstantinos Kamnitsas
Nicola Rieke
Sotirios Tsaftaris
Daguang Xu
Ziyue Xu

Preface FAIR 2021

As we witness a technological revolution that is spinning diverse research fields including healthcare at an unprecedented rate, we face bigger challenges ranging from the high cost of computational resources to the reproducible design of affordable and innovative solutions. While AI applications have been recently deployed in the healthcare system of high-income countries, their adoption in developing and emerging countries remains limited.

Given the breadth of challenges faced particularly in the field of healthcare and medical data analysis, we presented the first workshop on affordable AI and healthcare (FAIR 2021) aiming to i) raise awareness about the global challenges in healthcare, ii) strengthen the participation of underrepresented communities at MICCAI, and iii) build a community around affordable AI and healthcare in low resource settings. Our workshop stands out from other MICCAI workshops as it prioritizes and focuses on developed AI solutions and research suited to low infrastructure, point-of-care-testing, and edge devices. Examples include, but are not limited to, AI deployments in conjunction with conventional X-rays, ultrasound, microscopic imaging, retinal scans, fundus imaging, and skin lesions. Moreover, we encouraged works that identify often neglected diseases prevalent in low resource countries and propose affordable AI solutions for such diseases using medical images. In particular, we were looking for contributions on (a) making AI affordable for healthcare, (b) making healthcare affordable with AI, or (c) pushing the frontiers of AI in healthcare that enables (a) or (b).

In the first edition of the MICCAI FAIR workshop 23 papers, 17 regular and 6 white papers, were submitted for consideration, and after the peer review process, only 10 regular papers were accepted for publication (an acceptance rate of 59%). The topics of the accepted submissions include image-to-image translation, particularly for low-dose or low-resolution settings; model compactness and compression; domain adaptation and transfer learning; and active, continual and meta-learning. We followed the same review process used in the main MICCAI conference: a double-blind review process with three reviewers per submission. Reviewers were selected from a pool of excellent researchers in the field, who have published at top-tier conferences, and manually assigned to the papers avoiding potential conflict of interest. Submissions were ranked based on the overall scores. Final decisions about acceptance/rejection and oral presentations were made by the Program Chairs according to ranking, quality, and the total number of submissions. Springer's editorial policy was shared with the authors to allow for camera-ready contributions.

NVIDIA Corporation sponsored the prize for the best paper award at FAIR 2021, which consisted of an NVIDIA RTX 3090 GPU card. The best paper was selected based on three criteria: reviewers' scores, Program Chairs recommendation, and public voting during the workshop.

We would like to thank the authors for their contributions, the reviewers for their commitment, patience, and constructive feedback, and NVIDIA Corporation for sponsoring the best paper award. Also, we would like to thank the publicity committee and the advisory committee for their endless support. Special thanks go to Sophia Bano, Linda Marrakchi-Kacem, and Yenisil Plasencia Calaa for their great contributions to the workshop organization.

August 2021   Shadi Albarqouni
                                                           Bishesh Khanal
                                                           Islem Rekik
                                                           Nicola Rieke
                                                           Debdoot Sheet

Organization

Organization Committee DART 2021

Shadi Albarqouni	Helmholtz Center Munich, Germany
M. Jorge Cardoso	King's College London, UK
Qi Dou	The Chinese University of Hong Kong, Hong Kong
Konstantinos Kamnitsas	Imperial College London, UK
Nicola Rieke	NVIDIA GmbH, Germany
Sotirios Tsaftaris	University of Edinburgh, UK
Daguang Xu	NVIDIA Corporation, USA
Ziyue Xu	NVIDIA Corporation, USA

Program Committee DART 2021

Carole Sudre	University College London, UK
Cheng Chen	The Chinese University of Hong Kong, Hong Kong
Cheng Ouyang	Imperial College London, UK
David Zimmerer	German Cancer Research Center (DKFZ), Germany
Dong Nie	University of North Carolina, USA
Dwarikanath Mahapatra	Inception Institute of Artificial Intelligence, Abu Dhabi, UAE
Enzo Ferrante	CONICET, Universidad Nacional del Litoral, Argentina
Ertunc Erdil	ETH Zurich, Switzerland
Fatemeh Haghighi	Arizona State University, USA
Gabriele Valvano	IMT School for Advanced Studies Lucca, Italy
Georgios Kaissis	Technische Universitat Munchen, Germany
Gustav Bredell	ETH Zurich, Switzerland
Hao Guan	University of North Carolina at Chapel Hill, USA
Henry Mauricio Orbes Arteaga	King's College London, UK
Ilja Manakov	Ludwig Maximilian University of Munich, Germany
Jianming Liang	Arizona State University, USA
Junlin Yang	Yale University, USA
Krishna Chaitanya	ETH Zurich, Switzerland
Magdalini Paschali	Technische Universitat Munchen, Germany
Martin Menten	Technische Universitat Munchen, Germany

Organization Committee FAIR 2021

Publicity Committee FAIR 2021

Steering Committee FAIR 2021

Program Committee FAIR 2021

Manoj Acharya	Rochester Institute of Technology, USA
Seyed-Ahmad Ahmadi	NVIDIA GmbH, Germany
Omar Al-Kadi	University of Jordan, Jordan
Ahmed M. Anter	Shenzhen University, China
Alaa Bessadok	University of Sousse, Tunisia
Gustav Bredell	ETH Zurich, Switzerland
Krishna Chaitanya	ETH Zurich, Switzerland
Aayush Chaudhary	Rochester Institute of Technology, USA
Ananda Chowdhury	Jadavpur University, India
Bipul Das	GE Healthcare, India
Mustafa Elattar	Nile University, Egypt
Enzo Feerante	Universidad Nacional del Litoral, Argentina
Anubha Gupta	Indraprastha Institute of Information Technology-Delhi, India
Amelia Jiménez-Sánchez	Pompeu Fabra University, Spain
Fadi Kacem	University of Carthage, Tunisia
Islem Mhiri	University of Sousse, Tunisia
Anirban Mukhopadhyay	Technische Universität Darmstadt, Germany
Arijit Patra	AstraZeneca R&D, UK
Anita Rau	University College London, UK
Mehrdad Salehi	ImFusion, Germany
Ahmed H. Shahin	University College London, UK
Francisco Vasconcelos	University College London, UK

Contents

Affordable AI and Healthcare

Domain Adaptation and Representation Transfer

A Systematic Benchmarking Analysis of Transfer Learning for Medical Image Analysis

Mohammad Reza Hosseinzadeh Taher[1], Fatemeh Haghighi[1], Ruibin Feng[2],
Michael B. Gotway[3], and Jianming Liang[1(✉)]

[1] Arizona State University, Tempe, AZ 85281, USA
{mhossei2,fhaghigh,jianming.liang}@asu.edu
[2] Stanford University, Stanford, CA 94305, USA
ruibin@stanford.edu
[3] Mayo Clinic, Scottsdale, AZ 85259, USA
Gotway.Michael@mayo.edu

Abstract. Transfer learning from supervised ImageNet models has been frequently used in medical image analysis. Yet, no large-scale evaluation has been conducted to benchmark the efficacy of newly-developed pre-training techniques for medical image analysis, leaving several important questions unanswered. As the first step in this direction, we conduct a systematic study on the transferability of models pre-trained on iNat2021, the most recent large-scale fine-grained dataset, and 14 top self-supervised ImageNet models on 7 diverse medical tasks in comparison with the supervised ImageNet model. Furthermore, we present a practical approach to bridge the domain gap between natural and medical images by continually (pre-)training supervised ImageNet models on medical images. Our comprehensive evaluation yields new insights: (1) pre-trained models on fine-grained data yield distinctive local representations that are more suitable for medical segmentation tasks, (2) self-supervised ImageNet models learn holistic features more effectively than supervised ImageNet models, and (3) continual pre-training can bridge the domain gap between natural and medical images. We hope that this large-scale open evaluation of transfer learning can direct the future research of deep learning for medical imaging. As open science, all codes and pre-trained models are available on our GitHub page https://github.com/JLiangLab/BenchmarkTransferLearning.

Keywords: Transfer learning · ImageNet pre-training · Self-supervised learning

---

Electronic supplementary material The online version of this chapter (https://doi.org/10.1007/978-3-030-87722-4_1) contains supplementary material, which is available to authorized users.

S. Albarqouni et al. (Eds.): DART 2021/FAIR 2021, LNCS 12968, pp. 3–13, 2021.
https://doi.org/10.1007/978-3-030-87722-4_1

1   Introduction

To circumvent the challenge of annotation dearth in medical imaging, fine-tuning supervised ImageNet models (i.e., models trained on ImageNet via supervised learning with the human labels) has become the standard practice [9, 10, 23, 24, 31]. As evidenced by [31], nearly all top-performing models in a wide range of representative medical applications, including classifying the common thoracic diseases, detecting pulmonary embolism, identifying skin cancer, and detecting Alzheimer's Disease, are fine-tuned from supervised ImageNet models. However, intuitively, achieving outstanding performance on medical image classification and segmentation would require fine-grained features. For instance, chest X-rays all look similar, therefore, distinguishing diseases and abnormal conditions may rely on some subtle image details. Furthermore, delineating organs and isolating lesions in medical images would demand some fine-detailed features to determine the boundary pixels. In contrast to ImageNet, which was created for coarse-grained object classification, iNat2021 [12], the most recent large-scale fine-grained dataset, has recently been created. It consists of 2.7M training images covering 10K species spanning the entire tree of life. As such, the **first question** this paper seeks to answer is: *What advantages can supervised iNat2021 models offer for medical imaging in comparison with supervised ImageNet models?*

In the meantime, numerous self-supervised learning (SSL) methods have been developed. In the afore-discussed transfer learning, models are pre-trained in a supervised manner using expert-provided labels. By comparison, SSL pre-trained models use machine-generated labels. The recent advancement in SSL has resulted in self-supervised pre-training techniques that surpass gold standard supervised ImageNet models in a number of computer vision tasks [5, 7, 14, 26, 30]. Therefore, the **second question** this paper seeks to answer is: *How generalizable are the self-supervised ImageNet models to medical imaging in comparison with supervised ImageNet models?*

More importantly, there are significant differences between natural and medical images. Medical images are typically monochromic and consistent in anatomical structures [9, 10]. Now, several moderately-sized datasets have been created in medical imaging, for instance, NIH ChestX-Ray14 [25] that contains 112K images; CheXpert [13] that consists of 224K images. Naturally, the **third question** this paper seeks to answer is: *Can these moderately-sized medical image datasets help bridge the domain gap between natural and medical images?*

To answer these questions, we conduct the first extensive benchmarking study to evaluate the efficacy of different pre-training techniques for diverse medical imaging tasks, covering various diseases (e.g., embolism, nodule, tuberculosis, etc.), organs (e.g., lung and fundus), and modalities (e.g., CT, X-ray, and funduscopy). Concretely, (1) we study the impact of pre-training data granularity on transfer learning performance by evaluating the fine-grained pre-trained models on iNat2021 for various medical tasks; (2) we evaluate the transferability of 14 state-of-the-art self-supervised ImageNet models to a diverse set of tasks in medical image classification and segmentation; and (3) we investigate domain-

Table 1. We benchmark transfer learning for seven popular medical imaging tasks, spanning over different label structures (binary/multi-label classification and segmentation), modalities, organs, diseases, and data size.

Code[†]	Application	Modality	Dataset
ECC	Pulmonary embolism detection	CT	RSNA PE Detection [2]
DXC$_{14}$	Fourteen thorax diseases classification	X-ray	NIH ChestX-Ray14 [25]
DXC$_5$	Five thorax diseases classification	X-ray	CheXpert [13]
VFS	Blood vessels segmentation	fundoscopic	DRIVE [4]
PXS	Pneumothorax segmentation	X-ray	SIIM-ACR [1]
LXS	Lung segmentation	X-ray	NIH Montgomery [15]
TXC	Tuberculosis detection	X-ray	NIH Shenzhen CXR [15]

[†] The first letter denotes the object of interest ("E" for embolism, "D" for thorax diseases, etc.); the second letter denotes the modality ("X" for X-ray, "F" for Fundoscopic, etc.); the last letter denotes the task ("C" for classification, "S" for segmentation).

adaptive (continual) pre-training [8] on natural and medical datasets to tailor ImageNet models for target tasks on chest X-rays.

Our extensive empirical study reveals the following important insights: (1) Pre-trained models on fine-grained data yield distinctive local representations that are beneficial for medical segmentation tasks, while pre-trained models on coarser-grained data yield high-level features that prevail in classification target tasks (see Fig. 1). (2) For each target task, in terms of the mean performance, there exist at least three self-supervised ImageNet models that outperform the supervised ImageNet model, an observation that is very encouraging, as migrating from conventional supervised learning to self-supervised learning will dramatically reduce annotation efforts (see Fig. 2). (3) Continual (pre-)training of supervised ImageNet models on medical images can bridge the gap between the natural and medical domains, providing more powerful pre-trained models for medical tasks (see Table 2).

2   Transfer Learning Setup

Tasks and Datasets: Table 1 summarizes the tasks and datasets, with more details in Appendix A. We considered a diverse suite of 7 challenging and popular medical imaging tasks covering various diseases, organs, and modalities. These tasks span many common properties of medical imaging tasks, such as imbalanced classes, limited data, and small-scanning areas of pathologies of interest. We use official data split of these datasets if available; otherwise, we randomly divide the data into 80%/20% for training/testing.

Evaluations: We evaluate various models pre-trained with different methods and datasets. Therefore, we control other influencing factors such as preprocessing, network architecture, and transfer hyperparameters. In all experiments, (1) for the classification target tasks, the standard ResNet-50 backbone [11] followed by a task-specific classification head is used, (2) for the segmentation target tasks, a U-Net network with a ResNet-50 encoder is used, where the encoder is initialized with the pre-trained models, (3) all target model parameters are fine-tuned, (4) AUC (area under the ROC curve) and Dice coefficient are used for evaluating classification and segmentation target tasks, respectively, (5) mean and standard deviation of performance metrics over ten runs are reported, and (6) statistical analyses based on independent two-sample t-test are presented. More implementation details are in Appendix B and project's GitHub page.

Pre-trained Models: We benchmark transfer learning from two large-scale natural datasets, ImageNet and iNat2021, and two in-domain medical datasets, CheXpert [13] and ChestX-Ray14 [25]. We pre-train supervised in-domain models which are either initialized randomly or fine-tuned from the ImageNet model. For all other supervised and self-supervised methods, we use existing official and ready-to-use pre-trained models, ensuring that their configurations have been meticulously assembled to achieve the best results in target tasks.

3   Transfer Learning Benchmarking and Analysis

1) Pre-trained models on fine-grained data are better suited for segmentation tasks, while pre-trained models on coarse-grained data prevail on classification tasks. Medical imaging literature mostly has focused on the pre-training with *coarse-grained* natural image datasets, such as ImageNet [17,19,24,27]. In contrast to previous works, we aim to study the capability of pre-training with *fine-grained* datasets for transfer learning to medical tasks. In fine-grained datasets, visual differences between subordinate classes are often subtle and deeply embedded within local discriminative parts. Therefore, a model has to capture visual details in the local regions for solving a fine-grained recognition task [6,29,32]. We hypothesize that a pre-trained model on a fine-grained dataset derives distinctive local representations that are useful for medical tasks which usually rely upon small, local variations in texture to detect/segment pathologies of interest. To put this hypothesis to the test, we empirically validate how well pre-trained models on large-scale fine-grained datasets can transfer to a range of target medical applications. This study represents the first effort to rigorously evaluate the impact of pre-training data *granularity* on transfer learning to medical imaging tasks.

Experimental Setup: We examine the applicability of iNat2021 as a pre-training source for medical imaging tasks. Our goal is to compare the generalization of the learned features from fine-grained pre-training on iNat2021 with the conventional pre-training on the ImageNet. Given this goal, we use existing official and ready-to-use pre-trained models on these two datasets, and fine-tune them for 7 diverse target tasks, covering multi-label classification, binary classi-

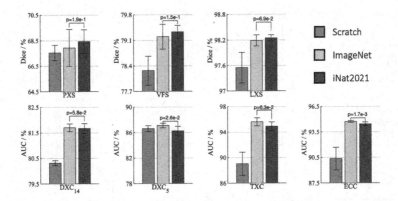

Fig. 1. For segmentation (target) tasks (*i.e.*, PXS, VFS, and LXS), fine-tuning the model pre-trained on iNat2021 outperforms that on ImageNet, while the model pre-trained on ImageNet prevails on classification (target) tasks (*i.e.*, DXC$_{14}$, DXC$_5$, TXC, and ECC), demonstrating the effect of data granularity on transfer learning capability: pre-trained models on the fine-grained data capture subtle features that empowers segmentation target tasks, and pre-trained models on the coarse-grained data encode high-level features that facilitate classification target tasks.

fication, and pixel-wise segmentation (see Table 1). To provide a comprehensive evaluation, we also include results for training target models from scratch.

Observations and Analysis: As evidenced in Fig. 1, fine-tuning from the iNat2021 pre-trained model outperforms the ImageNet counterpart in semantic segmentation tasks, *i.e.*, PXS, VFS, and LXS. This implies that, owing to the finer data granularity of iNat2021, the pre-trained model on this dataset yields a more fine-grained visual feature space, which captures essential pixel-level cues for medical segmentation tasks. This observation gives rise to a natural question of whether this improved performance can be attributed to the larger pre-training data of iNat2021 (2.7M images) compared to ImageNet (1.3M images). In answering this question, we conducted an ablation study on the iNat2021 mini dataset [12] with 500K images to further investigate the impact of data granularity on the learned representations. Our result demonstrates that even with fewer pre-training data, iNat2021 mini pre-trained models can outperform ImageNet counterparts in segmentation tasks (see Appendix C). This demonstrates that recovering discriminative features from iNat2021 dataset should be attributed to fine-grained data rather than the larger training data size.

Despite the success of iNat2021 models in segmentation tasks, fine-tuning of ImageNet pre-trained features outperforms iNat2021 in classification tasks, namely DXC$_{14}$, DXC$_5$, TXC, and ECC (see Fig. 1). Contrary to our intuition (see

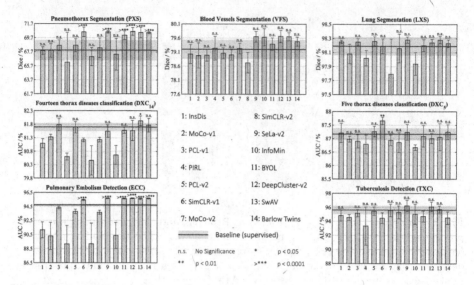

Fig. 2. For each target task, in terms of the mean performance, the supervised ImageNet model can be outperformed by at least three self-supervised ImageNet models, demonstrating the higher transferability of self-supervised representation learning. Recent approaches, SwAV [5], Barlow Twins [28], SeLa-v2 [5], and DeepCluster-v2 [5], stand out as consistently outperforming the supervised ImageNet model in most target tasks. We conduct statistical analysis between the supervised model and each self-supervised model in each target task, and show the results for the methods that significantly outperform the baseline or provide comparable performance. Methods are listed in numerical order from left to right.

Sec. 1), pre-training on a coarser granularity dataset, such as ImageNet, yields high-level semantic features that are more beneficial for classification tasks.

Summary: Fine-grained pre-trained models could be a viable alternative for transfer learning to fine-grained medical tasks, hoping practitioners will find this observation useful in migrating from standard ImageNet checkpoints to reap the benefits we've demonstrated. Regardless of – or perhaps in addition to – other advancements, visually diverse datasets like ImageNet can continue to play a valuable role in building performant medical imaging models.

2) Self-supervised ImageNet models outperform supervised ImageNet models. A recent family of self-supervised ImageNet models has demonstrated superior transferability in an increasing number of computer vision tasks compared to supervised ImageNet models [7,14,30]. Self-supervised models, in particular, capture task-agnostic features that can be easily adapted to different domains [14,26], while high-level features of supervised pre-trained models may be extraneous when the source and target data distributions are far apart [30]. We hypothesize this phenomenon is more pronounced in the medical domain, where there is a remarkable domain shift [7] compared to ImageNet. To test this hypothesis, we dissect the effectiveness of a wide range of recent self-supervised methods, encompassing contrastive learning, clustering, and redundancy-reduction methods, on the broadest benchmark yet of various modalities spanning X-ray, CT, and fundus images. This work represents the first effort to rigorously benchmark SSL techniques to a broader range of medical imaging problems.

Experimental Setup: We evaluate the transferability of 14 popular SSL methods with officially released models, which have been expertly optimized, including contrastive learning (CL) based on instance discrimination (*i.e.,* InsDis, MoCo-v1, MoCo-v2, SimCLR-v1, SimCLR-v2, and BYOL), CL based on JigSaw shuffling (PIRL), clustering (DeepCluster-v2 and SeLa-v2), clustering bridging CL (PCL-v1, PCL-v2, and SwAV), mutual information reduction (InfoMin), and redundancy reduction (Barlow Twins), on 7 diverse medical tasks. All methods are pre-trained on the ImageNet and use ResNet-50 architecture. Details of SSL methods can be found in Appendix F. As the baseline, we consider the standard supervised pre-trained model on ImageNet with a ResNet-50 backbone.

Observations and Analysis: According to Fig. 2, for each target task, there are at least three self-supervised ImageNet models that outperform the supervised ImageNet model on average. Moreover, the top self-supervised ImageNet models remarkably accelerate the training process of target models in comparison with supervised counterpart (see Appendix E). Intuitively, supervised pre-training labels encourage the model to retain more domain-specific high-level information, causing the learned representation to be biased toward the pre-training task/dataset's idiosyncrasies. Self-supervised learners, however, capture low/mid level features that are not attuned to domain-relevant semantics, generalizing better to diverse sorts of target tasks with low-data regimes.

Comparing the classification (DXC_{14}, DXC_5, TXC, and ECC) and segmentation tasks (PXS, VFS, and LXS) in Fig. 2, in the latter, a larger number of SSL methods results in better transfer performance, while supervised pre-training falls short. This suggests that when there are larger domain shifts, self-supervised models can provide more precise localization than supervised models. This is because supervised pre-trained models primarily focus on the smaller discriminative regions of the images, whereas SSL methods attune to larger regions [7,30], which empowers them with deriving richer visual information from the entire image.

Summary: SSL can learn holistic features more effectively than supervised pre-training, resulting in higher transferability to a variety of medical tasks. It's worth noting that no single SSL method dominates in all tasks, implying that universal pre-training remains a mystery. We hope that the results of this benchmarking, resonating with recent studies in the natural image domain [7,14,30], will lead to more effective transfer learning for medical image analysis.

3) Domain-adaptive pre-training bridges the gap between the natural and medical imaging domains. Pre-trained ImageNet models are the

Table 2. Domain-adapted pre-trained models outperform the corresponding ImageNet and in-domain models. For every target task, we performed the independent two sample t-test between the best (bolded) vs. others. Highlighted boxes in green indicate results which have no statistically significant difference at the $p = 0.05$ level. When pre-training and target tasks are the same, transfer learning is not applicable, denoted by "-". The footnotes compare our results with the state-of-the-art performance for each task.

Initialization	Target tasks				
	DXC$_{14}$[a]	DXC$_5$[b]	TXC[c]	PXS[d]	LXS[e]
Scratch	80.31±0.10	86.60±0.17	89.03±1.82	67.54±0.60	97.55±0.36
ImageNet	81.70±0.15	87.10±0.36	95.62±0.63	67.93±1.45	98.19±0.13
ChestX-ray14 [25]	-	**87.40±0.26**	96.32±0.65	68.92±0.98	98.18±0.06
CheXpert [13]	81.99±0.08	-	97.07±0.95	69.30±0.50	98.25±0.04
ImageNet→ChestX-ray14	-	87.09±0.44	**98.47±0.26**	**69.52±0.38**	98.27±0.03
ImageNet→CheXpert	**82.25±0.18**	-	97.33±0.26	69.36±0.49	**98.31±0.05**

[a] [16] holds an AUC of 82.00% vs. 82.25% ± 0.18% (ours)
[b] [18] holds an AUC of 89.40% w/ disease dependencies (DD) vs. 87.40% ± 0.26% (ours w/o DD)
[c] [20] holds an AUC of 95.35% ± 1.86% vs. 98.47% ± 0.26% (ours)
[d] [10] holds a Dice of 68.41% ± 0.14% vs. 69.52% ± 0.38% (ours)
[e] [21] holds a Dice of 96.94% ± 2.67% vs. 98.31% ± 0.05% (ours)

predominant standard for transfer learning as they are free, open-source models which can be used for a variety of tasks [3,9,17,27]. Despite the prevailing use of ImageNet models, the remarkable covariate shift between natural and medical images restrain transfer learning [19]. This constraint motivates us to present a practical approach that tailors ImageNet models to medical applications. Towards this end, we investigate domain-adaptive pre-training on natural and medical datasets to tune ImageNet models for medical tasks.

Experimental Setup: The domain-adaptive paradigm originated from natural language processing [8]. This is a sequential pre-training approach in which a model is first pre-trained on a massive general dataset, such as ImageNet, and then pre-trained on domain-specific datasets, resulting in domain-adapted pre-trained models. For the first pre-training step, we used the supervised ImageNet model. For the second pre-training step, we created two new models that were initialized through the ImageNet model followed by supervised pre-training on CheXpert (ImageNet→CheXpert) and ChestX-ray14 (ImageNet→ChestX-ray14). We compare the domain-adapted models with (1) the ImageNet model, and (2) two supervised pre-trained models on CheXpert and ChestX-ray14, which are randomly initialized. In contrast to previous work [3] which is limited to two classification tasks, we evaluate domain-adapted models on a broader range of five target tasks on chest X-ray scans; these tasks span classification and segmentation, ascertaining the generality of our findings.

Observations and Analysis: We draw the following observations from Table 2. (1) Both ChestX-ray14 and CheXpert models consistently outperform the ImageNet model in all cases. This observation implies that in-domain medical transfer learning, whenever possible, is preferred over ImageNet transfer learning. Our conclusion is opposite to [27], where in-domain pre-trained models outperform ImageNet models in controlled setups but lag far behind the real-world ImageNet models. (2) The overall trend showcases the advantage of domain-adaptive pre-training. Specifically, for DXC_{14}, fine-tuning the ImageNet→CheXpert model surpasses both ImageNet and CheXpert models. Furthermore, the dominance of domain-adapted models (ImageNet→CheXpert and ImageNet→ChestX-ray14) over ImageNet and corresponding in-domain models (CheXpert and ChestX-ray14) is conserved at LXS, TXC, and PXS. This suggests that domain-adapted models leverage the learning experience of the ImageNet model and further refine it with domain-relevant data, resulting in more pronounced representation. (3) In DXC_5, the domain-adapted performance decreases relative to corresponding ImageNet and in-domain models. This is most likely due to the lesser number of images in the in-domain pre-training dataset than the target dataset (75K vs. 200K), suggesting that in-domain pre-training data should be larger than the target data [8,22].

Summary: Continual pre-training can bridge the domain gap between natural and medical images. Concretely, we leverage the readily conducted annotation efforts to produce more performant medical imaging models and reduce future annotation burdens. We hope our findings posit new research directions for developing specialized pre-trained models in medical imaging.

4   Conclusion and Future Work

We provide the first fine-grained and up-to-date study on the transferability of various brand-new pre-training techniques for medical imaging tasks, answering central and timely questions on transfer learning in medical image analysis. Our empirical evaluation suggests that: (1) what truly matters for the segmentation tasks is fine-grained representation rather than high-level semantic features, (2) top self-supervised ImageNet models outperform the supervised ImageNet model, offering a new transfer learning standard for medical imaging, and (3) ImageNet models can be strengthened with continual in-domain pre-training.

Future Work: In this work, we have considered transfer learning from the supervised ImageNet model as the baseline, on which all our evaluations are benchmarked. To compute p-values for statistical analysis, 14 SSL, 5 supervised, and 2 domain-adaptive pre-trained models were run 10 times each on a set of 7 target tasks—leading to a large number of experiments (1,420). Nevertheless, our self-supervised models were all pre-trained on ImageNet with ResNet50 as the backbone. While ImageNet is generally regarded as a strong source for pre-training [12,27], pre-training modern self-supervised models with iNat2021 and in-domain medical image data on various architectures may offer even deeper insights into transfer learning for medical imaging.

Acknowledgments. This research has been supported partially by ASU and Mayo Clinic through a Seed Grant and an Innovation Grant, and partially by the NIH under Award Number R01HL128785. The content is solely the responsibility of the authors and does not necessarily represent the official views of the NIH. This work has utilized the GPUs provided partially by the ASU Research Computing and partially by the Extreme Science and Engineering Discovery Environment (XSEDE) funded by the National Science Foundation (NSF) under grant number ACI-1548562. We thank Nahid Islam for evaluating the self-supervised methods on the PE detection target task. The content of this paper is covered by patents pending.

References

1. SIIM-ACR pneumothorax segmentation (2019). https://www.kaggle.com/c/siim-acr-pneumothorax-segmentation/
2. RSNA STR pulmonary embolism detection (2020). https://www.kaggle.com/c/rsna-str-pulmonary-embolism-detection/overview
3. Azizi, S., et al.: Big self-supervised models advance medical image classification. arXiv:2101.05224 (2021)
4. Budai, A., Bock, R., Maier, A., Hornegger, J., Michelson, G.: Robust vessel segmentation in fundus images. Int. J. Biomed. Imaging **2013** (2013). Article ID 154860
5. Caron, M., Misra, I., Mairal, J., Goyal, P., Bojanowski, P., Joulin, A.: Unsupervised learning of visual features by contrasting cluster assignments. arXiv:2006.09882 (2021)
6. Chang, D., et al.: The devil is in the channels: mutual-channel loss for fine-grained image classification. IEEE Trans. Image Process. **29**, 4683–4695 (2020)
7. Ericsson, L., Gouk, H., Hospedales, T.M.: How well do self-supervised models transfer? In: Proceedings of the IEEE/CVF Conference on Computer Vision and Pattern Recognition (CVPR), pp. 5414–5423, June 2021
8. Gururangan, S., et al.: Don't stop pretraining: adapt language models to domains and tasks. arXiv:2004.10964 (2020)
9. Haghighi, F., Hosseinzadeh Taher, M.R., Zhou, Z., Gotway, M.B., Liang, J.: Learning semantics-enriched representation via self-discovery, self-classification, and self-restoration. In: Martel, A.L., et al. (eds.) MICCAI 2020. LNCS, vol. 12261, pp. 137–147. Springer, Cham (2020). https://doi.org/10.1007/978-3-030-59710-8_14
10. Haghighi, F., Taher, M.R.H., Zhou, Z., Gotway, M.B., Liang, J.: Transferable visual words: exploiting the semantics of anatomical patterns for self-supervised learning. arXiv:2102.10680 (2021)
11. He, K., Zhang, X., Ren, S., Sun, J.: Deep residual learning for image recognition. In: Proceedings of the IEEE Conference on Computer Vision and Pattern Recognition, pp. 770–778 (2016)
12. Horn, G.V., Cole, E., Beery, S., Wilber, K., Belongie, S., Aodha, O.M.: Benchmarking representation learning for natural world image collections. arXiv:2103.16483 (2021)
13. Irvin, J., et al.: CheXpert: a large chest radiograph dataset with uncertainty labels and expert comparison. arXiv:1901.07031 (2019)
14. Islam, A., Chen, C.F., Panda, R., Karlinsky, L., Radke, R., Feris, R.: A broad study on the transferability of visual representations with contrastive learning (2021)

15. Jaeger, S., Candemir, S., Antani, S., Wáng, Y.X.J., Lu, P.X., Thoma, G.: Two public chest X-ray datasets for computer-aided screening of pulmonary diseases. Quant. Imaging Med. Surg. 4(6), 475–477 (2014)

16. Kim, E., Kim, S., Seo, M., Yoon, S.: XProtoNet: diagnosis in chest radiography with global and local explanations. In: Proceedings of the IEEE/CVF Conference on Computer Vision and Pattern Recognition (CVPR), pp. 15719–15728 (2021)

17. Mustafa, B., et al.: Supervised transfer learning at scale for medical imaging. arXiv:2101.05913 (2021)

18. Pham, H.H., Le, T.T., Tran, D.Q., Ngo, D.T., Nguyen, H.Q.: Interpreting chest X-rays via CNNs that exploit hierarchical disease dependencies and uncertainty labels. Neurocomputing **437**, 186–194 (2021)

19. Raghu, M., Zhang, C., Kleinberg, J., Bengio, S.: Transfusion: understanding transfer learning with applications to medical imaging. arXiv:1902.07208 (2019)

20. Rajaraman, S., Zamzmi, G., Folio, L., Alderson, P., Antani, S.: Chest X-ray bone suppression for improving classification of tuberculosis-consistent findings. Diagnostics **11**(5) (2021). Article No. 840

21. Reamaroon, N., et al.: Robust segmentation of lung in chest X-ray: applications in analysis of acute respiratory distress syndrome. BMC Med. Imaging **20**, 116–128 (2020)

22. Reed, C.J., et al.: Self-supervised pretraining improves self-supervised pretraining. arXiv:2103.12718 (2021)

23. Shin, H.C., et al.: Deep convolutional neural networks for computer-aided detection: CNN architectures, dataset characteristics and transfer learning. IEEE Trans. Med. Imaging **35**(5), 1285–1298 (2016)

24. Tajbakhsh, N., et al.: Convolutional neural networks for medical image analysis: full training or fine tuning? IEEE Trans. Med. Imaging **35**(5), 1299–1312 (2016)

25. Wang, X., Peng, Y., Lu, L., Lu, Z., Bagheri, M., Summers, R.M.: ChestX-ray8: hospital-scale chest X-ray database and benchmarks on weakly-supervised classification and localization of common thorax diseases. In: Proceedings of the IEEE Conference on Computer Vision and Pattern Recognition, pp. 2097–2106 (2017)

26. Wei, L., et al.: Can semantic labels assist self-supervised visual representation learning? (2020)

27. Wen, Y., Chen, L., Deng, Y., Zhou, C.: Rethinking pre-training on medical imaging. J. Vis. Commun. Image Represent. **78**, 103145 (2021)

28. Zbontar, J., Jing, L., Misra, I., LeCun, Y., Deny, S.: Barlow twins: self-supervised learning via redundancy reduction. arXiv:2103.03230 (2021)

29. Zhao, J., Peng, Y., He, X.: Attribute hierarchy based multi-task learning for fine-grained image classification. Neurocomputing **395**, 150–159 (2020)

30. Zhao, N., Wu, Z., Lau, R.W.H., Lin, S.: What makes instance discrimination good for transfer learning? arXiv:2006.06606 (2021)

31. Zhou, Z., Sodha, V., Pang, J., Gotway, M.B., Liang, J.: Models genesis. Med. Image Anal. **67**, 101840 (2021)

32. Zhuang, P., Wang, Y., Qiao, Y.: Learning attentive pairwise interaction for fine-grained classification. In: Proceedings of the AAAI Conference on Artificial Intelligence, vol. 34, no. 07, pp. 13130–13137 (2020)

Self-supervised Multi-scale Consistency for Weakly Supervised Segmentation Learning

Gabriele Valvano[1,2(✉)], Andrea Leo[1], and Sotirios A. Tsaftaris[2]

[1] IMT School for Advanced Studies Lucca, 55100 Lucca, LU, Italy
gabriele.valvano@imtlucca.it
[2] School of Engineering, University of Edinburgh, Edinburgh EH9 3FB, UK

Abstract. Collecting large-scale medical datasets with fine-grained annotations is time-consuming and requires experts. For this reason, weakly supervised learning aims at optimising machine learning models using weaker forms of annotations, such as scribbles, which are easier and faster to collect. Unfortunately, training with weak labels is challenging and needs regularisation. Herein, we introduce a novel self-supervised multi-scale consistency loss, which, coupled with an attention mechanism, encourages the segmentor to learn multi-scale relationships between objects and improves performance. We show state-of-the-art performance on several medical and non-medical datasets. The code used for the experiments is available at https://vios-s.github.io/multiscale-pyag.

Keywords: Self-supervised learning · Segmentation · Shape prior

1 Introduction

To lessen the need for large-scale annotated datasets, researchers have recently explored weaker forms of supervision [15,32], consisting of weak annotations that are easier and faster to collect. Unfortunately, weak labels provide lower quality training signals, making it necessary to introduce regularisation to prevent model overfitting. Examples of regularisation are: forcing the model to produce similar predictions for similar inputs [22,31], or using prior knowledge about object shape [14,38], intensity [20], and position [13].

Data-driven shape priors learned by Generative Adversarial Networks (GAN) are popular regularisers [34], exploiting unpaired masks' availability to improve training. Recently, GANs have been used in weakly supervised learning, showing that they can provide training signals for the unlabelled pixels of an image [37]. Moreover, multi-scale GANs also provide information on multi-scale relationships among pixels [32], and can be easily paired with attention mechanisms [32,36] to focus on the specific objects and boost performance. However, GANs can be difficult to optimise, and they require a set of compatible masks for training. Annotated on images from a different data source, these masks must contain

© Springer Nature Switzerland AG 2021
S. Albarqouni et al. (Eds.): DART 2021/FAIR 2021, LNCS 12968, pp. 14–24, 2021.
https://doi.org/10.1007/978-3-030-87722-4_2

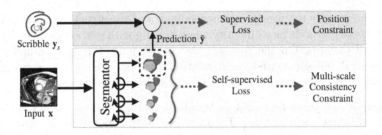

Fig. 1. We train a segmentor to predict masks overlapping with the available scribble annotations (top of the figure). To encourage the segmentor to learn multi-scale relationships between objects (bottom), we use novel attention mechanism (loopy arrows) that we condition with a self-supervised consistency loss.

annotations for the exact same classes used to train the segmentor. Moreover, the structures to segment must be similar across datasets to limit the risk of covariate shift. For example, there are no guarantees that optimising a multi-scale GAN using masks from a paediatric dataset will not introduce biases in a weakly supervised segmentor meant to segment elderly images.

Thus, multi-scale GANs are not always a feasible option. In these cases, it would be helpful to introduce multi-scale relationships without relying on unpaired masks. Herein, we show that it is possible to do so without performance loss. Our **contributions** are: **i)** we present a novel self-supervised method to introduce multi-scale shape consistency *without* relying on unpaired masks for training; **ii)** we train a shape-aware segmentor coupling multi-scale predictions and attention mechanisms through a *mask-free* self-supervised objective; and **iii)** we show comparable performance gains to that of GANs, but without need for unpaired masks. We summarise our idea in Fig. 1.

2 Related Work

Weakly-Supervised Learning for Image Segmentation. Recent research has explored weak annotations to supervise models, including: bounding boxes [15], image-level labels [24], point clouds [25], and scribbles [4,8,17,32]. Although it is possible to extend the proposed approach to other types of weak annotations, herein, we focus on scribbles, which have shown to be convenient to collect in medical imaging, especially when annotating nested structures [4].

A standard way to improve segmentation with scribbles is to post-process model predictions using Conditional Random Fields (CRFs) [4,17]. Recent work avoids the post-processing step and the need of tuning the CRF parameters by including learning constraints during training. For example [2] uses a max-min uncertainty regulariser to limit the segmentor flexibility, while other approaches regularise training using global statistics, such as the size of the target region [14,15,38] or topological priors [15]. Although they increase model performance, the applicability of these constraints is limited to specific assumptions about the objects and usually requires prior knowledge about the structure to segment. As a result, these methods face difficulty when dealing with pathology or uncommon anatomical variants. On the contrary, we do not make strong assumptions: we use a general self-supervised regularisation loss, optimising the segmentor to maintain multi-scale structural consistency in the predicted masks.

Multi-scale Consistency and Attention. Multi-scale consistency is not new to medical image segmentation. For example, deep supervision uses undersampled ground-truth segmentations to supervise a segmentor at multiple resolution levels [9]. Unfortunately, differently from these methods, we cannot afford to undersample the available ground-truth annotations because scribbles, which have thin structures, would risk to disappear at lower scales.

Other methods introduce the shape prior training GAN discriminators with a set of compatible segmentation masks [32,37]. Instead, we remove the need of full masks for training, and we impose multi-scale consistent predictions through an architectural bias localised inside of attention gates within the segmentor.

Attention has been widely adopted in deep learning [11] as it suppresses the irrelevant or ambiguous information in the feature maps. Recently, attention was also successfully used in image segmentation [21,27,28]. While standard approaches do not explicitly constrain the learned attention maps, Valvano et al. [32] have recently shown that conditioning the attention maps to be semantic increases model performance. In particular, they condition the attention maps through an adversarial mask discriminator, which requires a set of unpaired masks to work. Herein, we replace the mask discriminator with a more straightforward and general self-supervised consistency objective, obtaining attention maps coherent with the segmentor predictions at multiple scales.

Self-supervised Learning for Medical Image Segmentation. Self-supervised learning studies how to create supervisory signals from data using pretext tasks: i.e. easy surrogate objectives aimed at reducing human intervention requirements. Several pretext tasks have been proposed in the literature, including image in/out-painting [39], superpixel segmentation [23], coordinate prediction [1], context restoration [6] and contrastive learning [5]. After a self-supervised training phase, these models need a second-stage fine-tuning on the segmentation task. Unfortunately, choosing a proper pretext task is not trivial, and pre-trained features may not generalise well if unrelated to the final objective [35]. Hence, our method is more similar to those using self-supervision to regularise training, such as using transformation consistency [33] and feature prediction [31].

3   Proposed Approach

Notation. We use capital Greek letters to denote functions $\Phi(\cdot)$, and italic lowercase letters for scalars s. Bold lowercase define two-dimensional images $\mathbf{x} \in \mathbb{R}^{h \times w}$, with $h, w \in \mathbb{N}$ natural numbers denoting image height and width. Lastly, we denote tensors $\mathrm{T} \in \mathbb{R}^{n \times m \times o}$ using uppercase letters, with $n, m, o \in \mathbb{N}$.

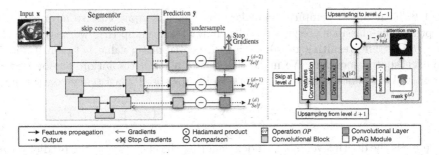

Fig. 2. Left: As a side product of PyAG modules, the segmentor produces segmentation masks at multiple scales. We compare ($\ominus$ symbol) the lower resolution masks (green squares) to those obtained undersampling the full resolution prediction $\tilde{\mathbf{y}}$ (blue squares), computing a self-supervised loss contribution $\mathcal{L}_{Self}^{(\cdot)}$ at each level. To prevent trivial solutions, we stop ($\mathbf{X}$ symbol) gradients (red arrows) from propagating through the highest resolution stream. **Right:** At every depth level d, a convolutional block processes the input features and predicts a low-resolution version of the segmentation mask $\mathbf{y}^{(d)}$ as part of a PyAG attention module (represented in light yellow background). To ensure that the mask $\tilde{\mathbf{y}}^{(d)}$ is consistent with the final prediction $\tilde{\mathbf{y}}$, we use the self-supervised multi-scale loss described in Eq. 1 and graphically represented on the left panel. Using the predicted mask, we compute the probability of pixels belonging to the background and then suppress their activations in the feature map $\mathbf{M}^{(d)}$ according to Eq. 2. (Color figure online)

Method Overview. We assume to have access to pairs of images $\mathbf{x}$ and their weak annotations $\mathbf{y_s}$ (in our case, $\mathbf{y_s}$ are scribbles), which we denote with the tuples $(\mathbf{x}, \mathbf{y_s})$. We present a segmentor incorporating a multi-scale prior learned in a self-supervised manner. We introduce the shape-prior through a specialised attention gate residing at several abstraction levels of the segmentor. These gates produce segmentation masks as an auxiliary task, allowing them to construct semantic attention maps used to suppress background activations in the extracted features. As our model predicts and refines the segmentation at multiple scales, we refer to these attention modules as Pyramid Attention Gates (PyAG).

Model Architecture and Training. The segmentor $\Sigma(\cdot)$ is a modified UNet [26] with batch normalisation [10]. Encoder and decoder of the UNet are interconnected through skip connections, which propagate features across convolutional blocks at multiple depth levels d. We leave the encoder as in the original framework while we modify the decoder at each level, as illustrated in Fig. 2. In particular, we first process the extracted features with two convolutional layers, as in the standard UNet. Next, we refine them with the introduced PyAG module, represented in a light yellow background on the right side of Fig. 2. Each PyAG module consists of: classifier, background extraction, and multiplicative gating operation. As a classifier, we use a convolutional layer with c filters having size $1 \times 1 \times k$, with c the number of segmentation classes including the background, and k the number of input channels. Obtained an input feature map $M^{(d)}$ at depth d, the classifier predicts a multi-channel score map that we pass through a *softmax*. The resulting tensor assigns a probabilistic value between 0 and 1 to each spatial location. We make this tensor a lower-resolution version of the predicted segmentation mask using the self-supervised consistency constraint:

$$\mathcal{L}_{Self} = -\sum_{d=1}^{n}\sum_{i=1}^{c} \tilde{\mathbf{y}}_i^{(0)} \log(\tilde{\mathbf{y}}_i^{(d)}), \tag{1}$$

where d is the depth level, i is an index denoting the class, $\tilde{\mathbf{y}}^{(d)}$ is the prediction at depth d, and $\tilde{\mathbf{y}}^{(0)} = \tilde{\mathbf{y}}$ is the final prediction of the model.[1] Notice that, different from [32], we condition $\tilde{\mathbf{y}}^{(d)}$ with $\mathcal{L}_{Self}$ rather than a multi-scale discriminator.

To prevent affecting the final prediction, we propagate the self-supervised training gradients only through the attention gates and the segmentor encoder, as we graphically show in Fig. 2, left. We further constrain the segmentor to reuse the extracted information by suppressing the activations in the spatial locations of the feature map $M^{(d)}$ which can be associated with the background (Fig. 2, right). This multiplicative gating operation can be formally defined as:

$$M^{(d)} \leftarrow M^{(d)} \cdot \left(1 - \tilde{\mathbf{y}}_{bkd}^{(d)}\right), \tag{2}$$

where $\tilde{\mathbf{y}}_{bkd}^{(d)}$ is the background channel of the predicted mask at the depth level d. The extracted features are finally upsampled to the new resolution level $d-1$ and processed by the next convolutional block.

To supervise the model with scribbles, we use the Partial Cross-Entropy (PCE) loss [30] on the final prediction $\tilde{\mathbf{y}}$. By multiplying the cross-entropy with a labelled pixel identifier $\mathbb{1}(\mathbf{y_s})$, the PCE avoids loss contribution on the unlabelled pixels. The role of the masking function $\mathbb{1}(\mathbf{y_s})$ is to return 1 for annotated pixels, 0 otherwise. Mathematically, we formulate the weakly-supervised loss as:

$$\mathcal{L}_{PCE} = \mathbb{1}(\mathbf{y_s}) \cdot \left[-\sum_{i=1}^{c} \mathbf{y}_{si} \log(\tilde{\mathbf{y}}_i) \right], \tag{3}$$

with $\mathbf{y_s}$ the ground truth scribble annotation.

Considering both weakly-supervised and self-supervised objectives, the overall cost function becomes: $\mathcal{L} = \mathcal{L}_{PCE} + a \cdot \mathcal{L}_{Self}$, where a is a scaling factor that balances training between the two costs. Similar to [32], we find beneficial to use a dynamic value for a, which maintains a fixed ratio between supervised and regularisation cost. In particular, we set $a = a_0 \cdot \frac{\|\mathcal{L}_{Self}\|}{\|\mathcal{L}_{PCE}\|}$, where $a_0 = 0.1$ is meant to give more importance to the supervised objective. We minimise $\mathcal{L}$ using Adam optimiser [16] with a learning rate of 0.0001, and a batch size of 12.

---

[1] Here we assume that the predicted $\tilde{\mathbf{y}}$ is a mask, not a scribble. Intuitively, our hypothesis derives from the observation that unlabelled pixels in the image have an intrinsic uncertainty: thus, the segmentor will look for clues in the image (e.g. anatomical edges and colours) to solve the segmentation task. Since Eq. 3 does not limit model flexibility on the unlabelled pixels, we empirically confirm our hypothesis.

4   Experiments

4.1   Data

ACDC [3] has cardiac MRIs of 100 patients. There are manual segmentations for right ventricle (RV), left ventricle (LV) and myocardium (MYO) at the end-systolic and diastolic cardiac phases. We resample images to the average resolution of $1.51\,\text{mm}^2$, and crop/pad them to 224×224 pixels. We normalise data by removing the patient-specific median and dividing by its interquartile range.

CHAOS [12] contains abdominal images from 20 different patients, with manual segmentation of liver, kidneys, and spleen. We test our method on the available T1 in-phase images. We resample images to $1.89\,\text{mm}^2$ resolution, normalise them in between -1 and 1, and then crop them to 192×192 pixel size.

LVSC [29] has cardiac MRIs of 100 subjects, with manual segmentations of left ventricular myocardium (MYO). We resample images to the average resolution of $1.45\,\text{mm}^2$ and crop/pad them to 224×224 pixels. We normalise data by removing the patient-specific median and dividing by its interquartile range.

PPSS [19] has (non-medical) RGB images of pedestrians with occlusions. Images were obtained from 171 different surveillance videos and cameras. There are manual segmentations for six pedestrian parts: face, hair, arms, legs, upper clothes, and shoes. We resample all the images to the same spatial resolution of the segmentation masks: 80×160; then we normalise images in $[0, 1]$ range.

Scribbles. The above datasets provide fully-annotated masks. To test the advantages of our approach in weakly-supervised learning, we use the manual scribble annotations provided for ACDC in [32]. For the remaining datasets, we follow the guidelines provided by Valvano et al. [32] to emulate synthetic scribbles using binary erosion operations or random walks inside the segmentation masks.

Setup. We divide ACDC, LVSC, and CHAOS data into groups of 70%, 15% and 15% of patients for train, validation, and test set, respectively. In PPSS, we follow recommendations in [19], using images from the first 100 cameras to train (90%) and validate (10%) our model, the remaining 71 cameras for testing it.

4.2   Evaluation Protocol

We compare segmentation performance of our method, termed **UNet$_{\text{PyAG}}$**, to:

- **UNet:** Trained on scribbles using the $\mathcal{L}_{PCE}$ loss [30].
- **UNet$_{\text{Comp.}}$:** UNet segmentor whose training is regularised with the Compactness loss proposed by [18], which models a generic shape compactness prior and prevents the appearance of scattered false positives/negatives in the generated masks. The compactness prior is mathematically defined as: $\mathcal{L}_{\text{Comp.}} = \frac{P^2}{4\pi A}$, where P is the perimeter length and A is the area of the generated mask. As for our method, we dynamically rescale this regularisation term to be 10 times smaller than the supervised cost (Sect. 3).

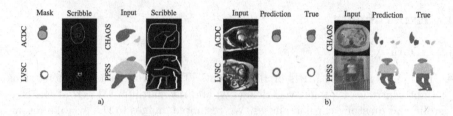

Fig. 3. For each dataset: a) examples of scribbles b) model predictions.

- **UNet$_{CRF}$**: Lastly, we consider post-processing the previous UNet predictions through CRF to better capture the object boundaries [7].[2]

While our method does not need a set of unpaired masks for training, we also compare with methods which learn the shape prior from masks:

- **UNet$_{AAG}$** [32]: The method upon which we build our model by replacing the multi-scale GAN with self-supervision. The subscript AAG stands for Adversarial Attention Gates, which couple adversarial signals and attention.
- **DCGAN**: We consider a standard GAN, learning the shape prior from unpaired masks. This model is the same as UNet$_{AAG}$, but without attention gates and multi-scale connections between segmentor and discriminator.
- **ACCL** [37]: It trains with scribbles using a PatchGAN discriminator to provide adversarial signals, and with the $\mathcal{L}_{PCE}$ [30] on the annotated pixels.

We perform 3-fold cross-validation and measure segmentation quality using Dice and IoU scores, and the Hausdorff Distance. We use Wilcoxon test ($p < 0.01$) to show if improvements w.r.t. the second best model are statistically significant.

4.3 Results

We show examples of predicted masks in Fig. 3 and quantitative results in Fig. 4.

As shown, our method is the best one when we compare it to other approaches that do not require extra masks for training (Fig. 4, top). In particular, a simple UNet has unsatisfying performance, but regularisation considerably helps. Adding the compactness loss aids more with compact shapes, such as those in ACDC, CHAOS and PPSS, while it can be harmful when dealing with non-compact masks, such as that of the myocardium (doughnut-shape) in LVSC.

Post-processing the segmentor predictions with CRF can lead to performance increase when object boundaries are well defined. On the contrary, we could not make the performance increase on CHAOS data, where using CRF made segmentation worse with all the metrics.

---

[2] CRF models the pairwise potentials between pixels using weighted Gaussians, weighting with values ω_1 and ω_2, and parametrising the distributions with the factors $\sigma_\alpha, \sigma_\beta, \sigma_\gamma$. For ACDC and LVSC, we use the cardiac segmentation parameters in [4]: $(\sigma_\alpha, \sigma_\beta, \sigma_\gamma, \omega_1, \omega_2) = (2, 0.1, 5, 5, 10)$. For CHAOS, we manually tune $(\omega_1, \omega_2) = (0.1, 0.2)$. Finally, for PPSS, we tuned them to: $(\sigma_\alpha, \sigma_\beta, \sigma_\gamma, \omega_1, \omega_2) = (80, 3, 3, 3, 3)$.

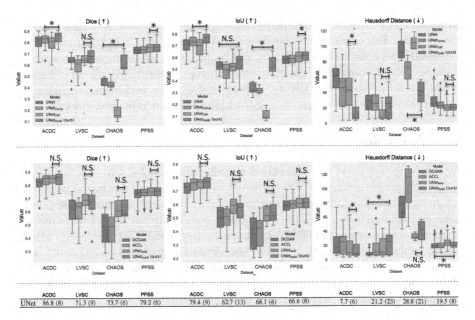

	ACDC	LVSC	CHAOS	PPSS	ACDC	LVSC	CHAOS	PPSS	ACDC	LVSC	CHAOS	PPSS
UNet	86.8 (8)	71.3 (9)	73.7 (6)	79.2 (6)	79.4 (9)	62.7 (13)	68.1 (6)	66.6 (8)	7.7 (6)	21.2 (23)	28.8 (21)	19.5 (8)

Fig. 4. Segmentation performance in terms of Dice ($\uparrow$), IoU ($\uparrow$), Hausdorff distance ($\downarrow$), with arrows showing metric improvement direction. Box plots report median and inter-quartile range (IQR), considering outliers the values outside $2 \times$IQR. For each dataset, we compare the two best performing models (horizontal black lines) and use an asterisk (*) to show if their performance difference is statistically significant (Wilcoxon test, $p < 0.01$) or N.S. otherwise. **Top row:** our method vs baseline (UNet) and methods regularising predictions with Compactness loss (UNet$_{Comp.}$) and CRF as post-processing (UNet$_{CRF}$). Our method is the best across datasets. In this case, we would like to perform better than the benchmarks ($p < 0.01$). **Middle row:** our method vs methods regularising predictions using a shape prior learned from unpaired masks (DCGAN, ACCL, UNet$_{AAG}$). In this case, we would like our method to perform at least as well as methods using masks (i.e. we would like the test to be not statistically significant, N.S.). We observe competitive performance with the best benchmark, while we also do not need masks for training. **Bottom row:** we report performance of a UNet trained with fully-annotated masks. These values can be seen as an upper bound when training with scribbles.

On LVSC, the introduced multi-scale shape consistency prior tends to make the model a bit less conservative on the most apical and basal slices of the cardiac MRI. Unfortunately, whenever there is a predicted mask but the manual segmentation is empty, the Hausdorff distance peaks. In fact, by definition, the distance assumes the maximum possible value (i.e. the image dimension) whenever one of the masks is empty, which makes the performance distribution on the test samples broader (see Hausdorff distance box plots for LVSC, Fig. 4, top).

On CHAOS, Dice and IoU are more skewed for methods not using unpaired masks for training (Fig. 4, top row). This happens because CHAOS is a small dataset, and optimising models using only scribble supervision is challenging. On the contrary, the extra knowledge of unpaired masks may help (bottom row).

Finally, we compare our method with approaches using unpaired masks for training (Fig. 4, bottom). We find competitive performance on all datasets. While, in some cases, the $UNet_{AAG}$ performs slightly better than $UNet_{PyAG}$, we emphasise that our approach can work also without unpaired masks.

5   Conclusion

We introduced a novel self-supervised learning strategy for semantic segmentation. Our approach consists of predicting masks at multiple resolution levels and enforcing multi-scale segmentation consistency. We use these multi-scale predictions as part of attention gating operations, restricting the model to re-use the extracted information on the object shape and position. Our method performs considerably better than other scribble-supervised approaches while having comparable performance to approaches requiring additional unpaired masks to regularise their training. Hoping to inspire future research, we release the code used for the experiments at https://vios-s.github.io/multiscale-pyag.

Acknowledgments. This work was partially supported by the Alan Turing Institute (EPSRC grant EP/N510129/1). S.A. Tsaftaris acknowledges the support of Canon Medical and the Royal Academy of Engineering and the Research Chairs and Senior Research Fellowships scheme (grant RCSRF1819\8\25).

References

1. Bai, W., et al.: Self-supervised learning for cardiac MR image segmentation by anatomical position prediction. In: Shen, D., et al. (eds.) MICCAI 2019. LNCS, vol. 11765, pp. 541–549. Springer, Cham (2019). https://doi.org/10.1007/978-3-030-32245-8_60
2. Belharbi, S., Rony, J., Dolz, J., Ayed, I.B., McCaffrey, L., Granger, E.: Deep interpretable classification and weakly-supervised segmentation of histology images via max-min uncertainty. arXiv preprint arXiv:2011.07221 (2020)
3. Bernard, O.E.A.: Deep learning techniques for automatic MRI cardiac multi-structures segmentation and diagnosis: is the problem solved? IEEE TMI (2018)
4. Can, Y.B., Chaitanya, K., Mustafa, B., Koch, L.M., Konukoglu, E., Baumgartner, C.F.: Learning to segment medical images with scribble-supervision alone. In: Stoyanov, D., et al. (eds.) DLMIA/ML-CDS-2018. LNCS, vol. 11045, pp. 236–244. Springer, Cham (2018). https://doi.org/10.1007/978-3-030-00889-5_27
5. Chaitanya, K., Erdil, E., Karani, N., Konukoglu, E.: Contrastive learning of global and local features for medical image segmentation with limited annotations. NeurIPS 33 (2020)
6. Chen, L., Bentley, P., Mori, K., Misawa, K., Fujiwara, M., Rueckert, D.: Self-supervised learning for medical image analysis using image context restoration. MIA **58**, 101539 (2019)
7. Chen, L.C., Papandreou, G., Kokkinos, I., Murphy, K., Yuille, A.L.: DeepLab: semantic image segmentation with deep convolutional nets, atrous convolution, and fully connected CRFs. IEEE TPAMI **40**(4), 834–848 (2017)

8. Dorent, R., et al.: Scribble-based domain adaptation via co-segmentation. In: Martel, A.L., et al. (eds.) MICCAI 2020. LNCS, vol. 12261, pp. 479–489. Springer, Cham (2020). https://doi.org/10.1007/978-3-030-59710-8_47
9. Dou, Q., et al.: 3D deeply supervised network for automated segmentation of volumetric medical images. MIA **41**, 40–54 (2017)
10. Ioffe, S., Szegedy, C.: Batch normalization: accelerating deep network training by reducing internal covariate shift. In: International Conference on Machine Learning (ICML), pp. 448–456. PMLR (2015)
11. Jetley, S., Lord, N.A., Lee, N., Torr, P.H.S.: Learn to pay attention. ICLR (2018)
12. Kavur, A.E., Selver, M.A., Dicle, O., Bariş, M., Gezer, N.S.: CHAOS - Combined (CT-MR) Healthy Abdominal Organ Segmentation Challenge Data, April 2019
13. Kayhan, O.S., Gemert, J.C.v.: On translation invariance in CNNs: convolutional layers can exploit absolute spatial location. In: CVPR, pp. 14274–14285 (2020)
14. Kervadec, H., Dolz, J., Tang, M., Granger, E., Boykov, Y., Ayed, I.B.: Constrained-CNN losses for weakly supervised segmentation. MIA **54**, 88–99 (2019)
15. Kervadec, H., Dolz, J., Wang, S., Granger, E., Ayed, I.B.: Bounding boxes for weakly supervised segmentation: global constraints get close to full supervision. In: MIDL (2020)
16. Kingma, D.P., Ba, J.: Adam: a method for stochastic optimization. In: ICLR (2015)
17. Lin, D., Dai, J., Jia, J., He, K., Sun, J.: ScribbleSup: scribble-supervised convolutional networks for semantic segmentation. In: CVPR, pp. 3159–3167 (2016)
18. Liu, Q., Dou, Q., Heng, P.-A.: Shape-aware meta-learning for generalizing prostate MRI segmentation to unseen domains. In: Martel, A.L., et al. (eds.) MICCAI 2020. LNCS, vol. 12262, pp. 475–485. Springer, Cham (2020). https://doi.org/10.1007/978-3-030-59713-9_46
19. Luo, P., Wang, X., Tang, X.: Pedestrian parsing via deep decompositional network. In: ICCV, pp. 2648–2655 (2013)
20. Nosrati, M.S., Hamarneh, G.: Incorporating prior knowledge in medical image segmentation: a survey. arXiv preprint arXiv:1607.01092 (2016)
21. Oktay, O., Schlemper, J., Folgoc, L.L., Lee, M., Heinrich, M., et al.: Attention U-net: learning where to look for the pancreas. In: MIDL (2018)
22. Ouali, Y., Hudelot, C., Tami, M.: An overview of deep semi-supervised learning. arXiv preprint arXiv:2006.052 78 (2020)
23. Ouyang, C., Biffi, C., Chen, C., Kart, T., Qiu, H., Rueckert, D.: Self-supervision with superpixels: training few-shot medical image segmentation without annotation. In: Vedaldi, A., Bischof, H., Brox, T., Frahm, J.-M. (eds.) ECCV 2020. LNCS, vol. 12374, pp. 762–780. Springer, Cham (2020). https://doi.org/10.1007/978-3-030-58526-6_45
24. Patel, G., Dolz, J.: Weakly supervised segmentation with cross-modality equivariant constraints. arXiv preprint arXiv: 2104.02488 (2021)
25. Qu, H., et al.: Weakly supervised deep nuclei segmentation using partial points annotation in histopathology images. IEEE TMI **39**(11), 3655–3666 (2020)
26. Ronneberger, O., Fischer, P., Brox, T.: U-Net: convolutional networks for biomedical image segmentation. In: Navab, N., Hornegger, J., Wells, W.M., Frangi, A.F. (eds.) MICCAI 2015. LNCS, vol. 9351, pp. 234–241. Springer, Cham (2015). https://doi.org/10.1007/978-3-319-24574-4_28
27. Schlemper, J., et al.: Attention gated networks: learning to leverage salient regions in medical images. MIA **53**, 197–207 (2019)
28. Sinha, A., Dolz, J.: Multi-scale self-guided attention for medical image segmentation. IEEE J. Biomed. Health Inform. **25**, 121–130 (2020)

29. Suinesiaputra, A., et al.: A collaborative resource to build consensus for automated left ventricular segmentation of cardiac MR images. MIA **18**(1), 50–62 (2014)
30. Tang, M., Djelouah, A., Perazzi, F., Boykov, Y., Schroers, C.: Normalized cut loss for weakly-supervised CNN segmentation. In: CVPR, pp. 1818–1827 (2018)
31. Valvano, G., Chartsias, A., Leo, A., Tsaftaris, S.A.: Temporal consistency objectives regularize the learning of disentangled representations. In: Wang, Q., et al. (eds.) DART/MIL3ID 2019. LNCS, vol. 11795, pp. 11–19. Springer, Cham (2019). https://doi.org/10.1007/978-3-030-33391-1_2
32. Valvano, G., Leo, A., Tsaftaris, S.A.: Learning to segment from scribbles using multi-scale adversarial attention gates. IEEE TMI (2021)
33. Xie, Y., Zhang, J., Liao, Z., Xia, Y., Shen, C.: PGL: prior-guided local self-supervised learning for 3d medical image segmentation. arXiv preprint arXiv: 2011.12640 (2020)
34. Yi, X., Walia, E., Babyn, P.: Generative adversarial network in medical imaging: a review. MIA **58**, 101552 (2019)
35. Zamir, A.R., Sax, A., Shen, W., Guibas, L.J., Malik, J., Savarese, S.: Taskonomy: disentangling Task Transfer Learning. In: CVPR, pp. 3712–3722 (2018)
36. Zhang, H., Goodfellow, I., Metaxas, D., Odena, A.: Self-attention generative adversarial networks. In: ICLR, pp. 7354–7363. PMLR (2019)
37. Zhang, P., Zhong, Y., Li, X.: ACCL: adversarial constrained-CNN loss for weakly supervised medical image segmentation. arXiv:2005.00328 (2020)
38. Zhou, Y., et al.: Prior-aware neural network for partially-supervised multi-organ segmentation. In: ICCV, pp. 10672–10681 (2019)
39. Zhou, Z., et al.: Models genesis: generic autodidactic models for 3D medical image analysis. In: Shen, D., et al. (eds.) MICCAI 2019. LNCS, vol. 11767, pp. 384–393. Springer, Cham (2019). https://doi.org/10.1007/978-3-030-32251-9_42

FDA: Feature Decomposition and Aggregation for Robust Airway Segmentation

Minghui Zhang[1], Xin Yu[3], Hanxiao Zhang[1], Hao Zheng[1], Weihao Yu[1], Hong Pan[2], Xiangran Cai[3], and Yun Gu[1,4(✉)]

[1] Institute of Medical Robotics, Shanghai Jiao Tong University, Shanghai, China
geron762@sjtu.edu.cn
[2] Department of Computer Science and Software Engineering,
Swinburne University of Technology, Victoria, Australia
[3] Medical Image Centre, The First Affiliated Hospital of Jinan University,
Guangzhou, China
[4] Shanghai Center for Brain Science and Brain-Inspired Technology, Shanghai, China

Abstract. 3D Convolutional Neural Networks (CNNs) have been widely adopted for airway segmentation. The performance of 3D CNNs is greatly influenced by the dataset while the public airway datasets are mainly clean CT scans with coarse annotation, thus difficult to be generalized to noisy CT scans (e.g. COVID-19 CT scans). In this work, we proposed a new dual-stream network to address the variability between the clean domain and noisy domain, which utilizes the clean CT scans and a small amount of labeled noisy CT scans for airway segmentation. We designed two different encoders to extract the transferable clean features and the unique noisy features separately, followed by two independent decoders. Further on, the transferable features are refined by the channel-wise feature recalibration and Signed Distance Map (SDM) regression. The feature recalibration module emphasizes critical features and the SDM pays more attention to the bronchi, which is beneficial to extracting the transferable topological features robust to the coarse labels. Extensive experimental results demonstrated the obvious improvement brought by our proposed method. Compared to other state-of-the-art transfer learning methods, our method accurately segmented more bronchi in the noisy CT scans.

Keywords: Clean and noisy domain · Decomposition and aggregation · Airway segmentation

1 Introduction

The novel coronavirus 2019 (COVID-19) has turned into a pandemic, infecting humans all over the world. To relieve the burden of clinicians, many researchers take the advantage of deep learning methods for automated COVID-19 diagnosis and infection measurement from imaging data (e.g., CT scans, Chest X-ray).

S. Albarqouni et al. (Eds.): DART 2021/FAIR 2021, LNCS 12968, pp. 25–34, 2021.
https://doi.org/10.1007/978-3-030-87722-4_3

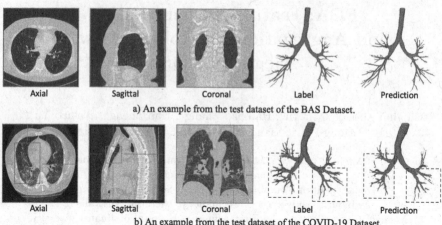

Axial   Sagittal   Coronal   Label   Prediction

a) An example from the test dataset of the BAS Dataset.

Axial   Sagittal   Coronal   Label   Prediction

b) An example from the test dataset of the COVID-19 Dataset.

Fig. 1. Different patterns exist between the BAS dataset and the COVID-19 dataset. One well-trained segmentation model in the clean domain leads to low accuracy when testing in the noisy domain.

Current studies mainly focus on designing a discriminative or robust model to distinguish COVID-19 from other patients with pneumonia [11,17], lesion localization [18], and segmentation [16]. In this work, we tackle another challenging problem, airway segmentation of COVID-19 CT scans. The accurate segmentation enables the quantitative measurements of airway dimensions and wall thickness which can reveal the abnormality of patients with pulmonary disease and the extraction of patient-specific airway model from CT image is required for navigation in bronchoscopic-assisted surgery. It helps the sputum suction for novel COVID-19 patients.

However, due to the fine-grained pulmonary airway structure, manual annotation is time-consuming, error-prone, and highly relies on the expertise of clinicians. Moreover, COVID-19 CT scans share ground-glass opacities in the early stage and pulmonary consolidation in the late stage [3] that adds additional difficulty for annotation. Even though the fully convolutional networks (FCNs) could automatically segment the airway, there remain the following challenges. First, FCNs are data-driven methods, while there are few public airway datasets with annotation and the data size is also limited. The public airway datasets, including EXACT'09 dataset [8] and the Binary Airway Segmentation (BAS) dataset [12], focus on the cases with the abnormality of airway structures mainly caused by chronic obstructive pulmonary disease (COPD). These cases are relatively clean and we term their distribution as **Clean Domain**, on the contrary, we term the distribution of COVID-19 CT scans as **Noisy Domain**. Figure 1 shows that fully convolutional networks (FCNs) methods [5,12] trained on the clean domain cannot be perfectly generalized to the noisy domain. Although this challenge can be addressed via the collection and labeling of new cases, it is impractical for novel diseases, e.g. COVID-19, which cannot guarantee the scale of datasets and the quality of annotation. Second, transfer learning methods (e,g. domain adaptation [2,13], feature alignment [2,14]) can improve the performance on target domains by transferring the knowledge contained in source domains or learning domain-invariant features. However, these methods are inadequate to apply in our scenario because this target noisy domain contains specific features (e.g. patterns of shadow patches) which cannot be learned from the source domain. Third, the annotation of the airway is extremely hard as they are elongated fine structures with plentiful peripheral bronchi of quite different sizes and orientations. The annotation in the EXACT'09 dataset [8] and the BAS dataset [12] are overall coarse and unsatisfactory. However, the deep learning methods are intended to fit the coarse labels, and thereby they are difficult to learn the robust features for airway representation.

To alleviate such challenges, we propose a dual-stream network to extract the robust and transferable features from the clean CT scans (clean domain) and a few labeled COVID-19 CT scans (noisy domain). Our contributions are threefold:

- We hypothesize that the COVID-19 CT scans own the general features and specific features for airway segmentation. The general features (e.g. the topological structure) are likely to learn from the other clean CT scans, while the specific features (e.g. patterns of shadow patches) should be extracted independently. Therefore, we designed a dual-stream network, which takes both the clean CT scans and a few labeled COVID-19 CT scans as input to synergistically learn general features and independently learn specific features for airway segmentation.
- We introduce the feature calibration module and the Signed Distance Map (SDM) for the clean CT scans with coarse labels, and through this way, robust features can be obtained for the extraction of general features.
- With extensive experiments on the clean CT scans and the COVID-19 CT scans, our method revealed the superiority in the extraction of transferable and robust features and achieved improvement compared to other methods under the evaluation of tree length detected rate and the branch detected rate.

2   Method

A new dual-stream network is proposed, which simultaneously processes the clean CT scans and a few noisy COVID-19 CT scans to learn robust and transferable features for airway segmentation. In this section, we detail the architecture of the proposed dual-stream network, which is illustrated in Fig. 2.

2.1   Learning Transferable Features

COVID-19 CT scans share a similar airway topological structure with the clean CT scans, meanwhile introduce unique patterns, e.g., multifocal patchy shadowing and ground-glass opacities, which are not common in clean CT scans. Since

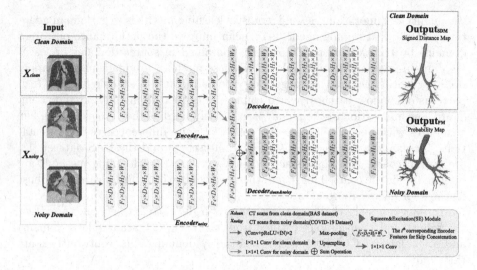

Fig. 2. Detailed structure and workflow of the proposed dual-stream network. $Encoder_{clean}$ aims to synergistically learn transferable features from both X_{clean} and X_{noisy}. $Encoder_{noisy}$ extracts specific features of X_{noisy}. The encoded features of X_{clean} are fed into $Decoder_{clean}$, where SE and SDM modules refine the features. For X_{noisy}, the decomposed features are then aggregated again via the channel-wise summation operation and fed into $Decoder_{clean\&noisy}$.

the number of clean CT scans is relatively large and the airway structure is also clearer, we aim to adapt the knowledge from clean CT scans to improve the airway segmentation of COVID-19 CT scans. Therefore, a dual-stream network is designed to synergistically learn transferable features from both COVID-19 CT scans and clean CT scans, and independently learn specific features only from COVID-19 CT scans. As illustrated in Fig. 2, let $\boldsymbol{X}_{clean}$ denote the input of sub-volume CT scans from the clean CT scans, $\boldsymbol{X}_{noisy}$ from the noisy COVID-19 CT scans. $Encoder_{clean}$ and $Encoder_{noisy}$ are encoder blocks for feature extraction. $Decoder_{clean}$ and $Decoder_{clean\&noisy}$ are decoder blocks to generate the segmentation results based on the features from encoders. The $Output_{SDM}$ represents the output of clean CT scans, and the $Output_{PM}$ represents the output of COVID-19 CT scans, they can be briefly defined as follows:

$$Output_{SDM} = Decoder_{clean}(Encoder_{clean}(\boldsymbol{X}_{clean}))$$
$$Output_{PM} = Decoder_{clean\&noisy}(Encoder_{clean}(\boldsymbol{X}_{noisy}) + Encoder_{noisy}(\boldsymbol{X}_{noisy}))$$

where we omit the detail of $1 \times 1 \times 1$ convolution and the Squeeze&Excitation (SE) module for a straightforward explanation of the overall workflow. In this case, the features of X_{noisy} are decomposed into two parts: $Encoder_{clean}$ aims to extract high-level, semantic and transferable features from both clean CT scans and COVID-19 CT scans; $Encoder_{noisy}$ is designed to obtain the specific features which belongs to the COVID-19 samples. The features of clean CT scans extracted by $Encoder_{clean}$ are fed into $Decoder_{clean}$. For COVID-19 CT images, the decomposed features are then aggregated again via the channel-wise summation operation and fed into $Decoder_{clean\&noisy}$ to reconstruct the volumetric airway structures.

2.2 Refinement of Transferable Features

As is mentioned before, the annotation of the public airway dataset is overall coarse and unsatisfactory. Since we have determined which features to transfer, then the transferable features can be further refined to be more robust through feature recalibration and introducing signed distance map.

Feature Recalibration: 3D Channel SE (cSE) module [20] is designed to investigate the channel-wise attention. We embed this module between $Encoder_{clean}$ and $Decoder_{clean}$, aiming to refine the transferable features. Take $\mathbf{U}$ as input and $\widetilde{\mathbf{U}}$ as output, $\mathbf{U}, \widetilde{\mathbf{U}} \in \mathbb{R}^{F \times D \times H \times W}$ with the number of channels F, depth D, height H, width W. 3D cSE firstly compresses the spatial domain then obtains channel-wise dependencies $\widetilde{\mathbf{Z}}$, which are formulated as follows:

$$Z_i = \frac{1}{D}\frac{1}{H}\frac{1}{W}\sum_{d=1}^{D}\sum_{h=1}^{H}\sum_{w=1}^{W} U_i(d, h, w), \quad i = 1, 2, ..., F \tag{1}$$

$$\widetilde{\mathbf{Z}} = \sigma(\mathbf{W}_2 \delta(\mathbf{W}_1 \mathbf{Z})), \tag{2}$$

where $\delta(\cdot)$ denotes the ReLU function and $\sigma(\cdot)$ refers to sigmoid activation, $\mathbf{W}_1 \in \mathbb{R}^{\frac{F}{r_c} \times F}$, and $\mathbf{W}_2 \in \mathbb{R}^{F \times \frac{F}{r_c}}$. The r_c represents the reduction factor in the channel domain, similar to [4]. The output of 3D cSE is obtained by: $\widetilde{\mathbf{U}} = \mathbf{U} \odot \widetilde{\mathbf{Z}}$.

Signed Distance Map: In recent years, introducing the distance transformed map into CNNs have proven effectivity in medical image segmentation task [6,10,19] due to its superiority of paying attention to the global structural information. The manual annotation of the plentiful tenuous bronchi is error-prone and often be labeled thinner or thicker. The 3D FCNs cooperating with common loss function treat the labeled foreground equally and intend to fit such coarse labels, which are difficult to extract robust features. Even though the thickness of the annotated bronchi is uncertain, the phenomenon of breakage or leakage in the annotation can be avoided by experienced radiologists. Therefore, the overall topology is correctly delineated, and we can use the topological structure instead of the coarse label as a supervised signal. Besides, the intra-class imbalance problem in airway segmentation is severe. Distance transform map is used to rebalance the distribution of trachea, main bronchi, lobar bronchi, and distal segmental bronchi. We use the signed distance map transform as a voxel-wise reweighting method, incorporating with the regression loss that focuses on the relatively small values (such as the lobar bronchi and distal segmental bronchi) by having larger gradient magnitudes.

Given the airway as target structure and each voxel x in the volume set X, we construct the Signed Distance Map (SDM) function termed as $\phi(x)$, defined as:

$$\phi(x) = \begin{cases} 0, & x \in \text{airway and } x \in \mathcal{C}, \\ -\inf_{\forall z \in \mathcal{C}} \|x - z\|_2, & x \in \text{airway and } x \notin \mathcal{C}, \\ +\inf_{\forall z \in \mathcal{C}} \|x - z\|_2, & x \notin \text{airway}, \end{cases} \quad (3)$$

where the $\mathcal{C}$ represents the surface of the airway, we further normalize the SDM into $[-1, +1]$. We then transformed the segmentation task on clean CT scans to an SDM regression problem and introduce the loss function that penalizes the prediction SDM for having the wrong sign and forces the 3D CNNs to learn more robust features that contain topological features for airway. Denote the y_x as the ground truth of SDM and f_x as the prediction of the SDM, the loss function for the regression problem can be defined as follows:

$$L_{reg} = \sum_{\forall x} \|f_x - y_x\|_1 - \sum_{\forall x} \frac{f_x y_x}{f_x y_x + f_x^2 + y_x^2}, \quad (4)$$

where $\|\cdot\|_1$ denotes the L_1 norm.

2.3 Training Loss Functions

The training loss functions consist of two parts, The first part is the L_{reg} for the clean CT scans, and the second part is the L_{seg} for the noisy CT scans, we combine the Dice [9] and Focal loss [7] to construct the L_{seg}:

$$L_{seg} = -\frac{2 \sum_{\forall x} p_x g_x}{\sum_{\forall x} (p_x + g_x)} - \frac{1}{|X|} (\sum_{\forall x} (1 - p_x)^2 log(p_x)), \quad (5)$$

where g_x is the binary ground truth and p_x is the prediction. The total loss is defined as $L_{total} = L_{seg} + L_{reg}$.

3   Experiments and Results

Dataset: We used two datasets to evaluate our method.

- Clean Domain: Binary Airway Segmentation (BAS) dataset [12]. It contains 90 CT scans (70 CT scans from LIDC [1]) and 20 CT scans from the training set of the EXACT'09 dataset [8]. The spatial resolution ranges from 0.5 to 0.82 mm and the slice thickness ranges from 0.5 to 1.0 mm. We randomly split the 90 CT scans into the training set (50 scans), validation set (20 scans), and test set (20 scans).
- Noisy Domain: COVID-19 dataset. We collected 58 COVID-19 patients from three hospitals and the airway ground truth of each COVID-19 CT scan was corrected by three experienced radiologists. The spatial resolution of the COVID-19 dataset ranges from 0.58 to 0.84 mm and slice thickness varies from 0.5 to 1.0 mm. The COVID-19 dataset is randomly divided into 10 scans for training and 48 scans for testing.

Table 1. Results (%) on the test set of the COVID-19 dataset. Values are shown as mean ± standard deviation. 'B' indicates the training set of the BAS dataset and 'C' indicates the training set of the COVID-19 dataset.

Method		Length	Branch	DSC
3D UNet	Train on B only	72.4 ± 4.8	62.1 ± 4.5	93.2 ± 1.5
	Train on B + C	82.8 ± 4.8	83.8 ± 3.8	95.2 ± 1.3
	Train on C only	85.7 ± 5.1	84.9 ± 3.5	95.9 ± 1.2
	Train on B, finetuned on C	86.8 ± 5.3	85.0 ± 4.1	95.7 ± 1.1
3D UNet + cSE (Medical Physics,2019) [20]		86.2 ± 5.3	84.6 ± 4.2	95.8 ± 1.2
Feature Alignment (TMI,2020) [2]		87.9 ± 4.9	85.5 ± 4.8	95.5 ± 1.6
Domain Adaptation (TPAMI,2018) [13]		87.0 ± 4.6	84.9 ± 4.0	96.0 ± 1.3
Proposed w/o cSE&SDM		90.2 ± 5.3	87.6 ± 4.2	96.5 ± 1.2
Proposed w/o cSE		91.1 ± 4.3	86.8 ± 3.7	**96.8 ± 1.0**
Proposed w/o SDM		91.0 ± 4.7	86.3 ± 4.1	96.6 ± 1.0
Proposed		**92.1 ± 4.3**	**87.8 ± 3.7**	**96.8 ± 1.1**

Network Configuration and Implementation Details: As shown in Fig. 2, each block in the encoder or decoder contains two convolutional layers followed by pReLU and Instance Normalization [15]. The initial number of channel is set to 32, thus $\{F_1, F_2, F_3, F_4\} = \{32, 64, 128, 256\}$. During the preprocessing procedure, we clamped the voxel values to $[-1200, 600]$ HU, normalized them into $[0, 255]$, and cropped the lung field to remove unrelated background regions. We adopted a large input size of $128 \times 224 \times 304$ CT cubes densely cropped near airways and chose a batch size of 1 (randomly chose a clean CT scan and noisy COVID-19 CT scan) in the training phase. On-the-fly data augmentation included the random horizontal flipping and random rotation between $[-10°, 10°]$. All models were trained by Adam optimizer with the initial learning rate of 0.002. The total epoch is set to 60 and the learning rate was divide by 10 in the 50th epoch, the hyperparameter of r_c used in 3D cSE module is set to 2. Preliminary experiments confirmed training procedures converged under this setup. In testing phase, we performed the sliding window prediction with stride 48. All the models were implemented in PyTorch framework with a single NVIDIA Geforce RTX 3090 GPU (24 GB graphical memory).

Evaluation Metrics: We adopted three metrics to evaluate methods, including the a) tree length detected rate (Length) [8], b) branch detected rate (Branch) [8], and c) Dice score coefficient (DSC). All metrics are evaluated on the largest component of each airway segmentation result.

Quantitative Results: Experimental results showed that the way of training on the BAS dataset then evaluating on the COVID-19 dataset performed worst, as expected. Training merely on the COVID-19 dataset performed better than training on both the BAS dataset and the COVID-19 dataset, which implied the necessity of transfer learning rather than merely together different datasets.

3D UNet with cSE [20] was trained on the COVID-19 dataset and the results showed no significant improvement. For comparison, three commonly used transfer learning methods, Fine-tuned (pre-trained on BAS dataset, fine-tuned on COVID-19 dataset), Feature Alignment (FA) [2] through adversarial training, and Domain Adaptation (DA) by sharing weights [13] were reimplemented to be applied in our task, the results in Table 1 demonstrated our proposed method is superior to these methods, the proposed method achieved the highest performance on all metrics of Length (92.1%), Branch (87.8%), and DSC (96.8%). We also conducted the ablation study to investigate the effectiveness of each component of the proposed method. In Table 1, we observed that the original dual-stream network had outperformed the other methods, with the achievement of 90.2% Length, 87.6% Branch, and 96.5% DSC. The improvement confirmed the validity of our proposed dual-stream network. Furthermore, cSE module and SDM could boost performance independently and the combination of cSE and SDM brings the highest performance gain, which demonstrated the necessity of refinement for transferable features.

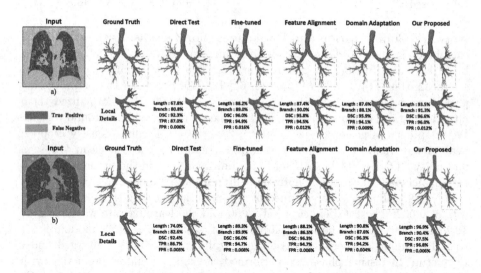

Fig. 3. Visualization of segmentation results. a) is a severe case and b) is a mild case in the test set of the COVID-19 dataset. The blue dotted boxes indicate the local regions full of shadow patches and are zoomed in for better observation.

Qualitative Results: The visualization of segmentation results is presented in Fig. 3. Compared to other methods, the proposed method gains improvement on both the severe and mild cases of the COVID-19 dataset, which accurately detected more bronchi surround by multifocal patchy shadowing of COVID-19.

4 Conclusion

This paper proposed a novel dual-stream network to learn transferable and robust features from clean CT scans to noisy CT for airway segmentation. Our proposed method not only extracted the transferable clean features but also extract unique noisy features separately, transferable features were further refined by the cSE module and SDM. Extensive experimental results showed our proposed method accurately segmented more bronchi than other methods.

Acknowledgements. This work was partly supported by National Natural Science Foundation of China (No. 62003208), National Key R&D Program of China (No. 2019YFB1311503), Shanghai Sailing Program (No. 20YF1420800), Shanghai Municipal of Science and Technology Project (Grant No. 20JC1419500), and Science and Technology Commission of Shanghai Municipality under Grant 20DZ2220400.

References

1. Armato, S.G., II., et al.: The lung image database consortium (LIDC) and image database resource initiative (IDRI): a completed reference database of lung nodules on ct scans. Med. Phys. **38**(2), 915–931 (2011)
2. Chen, C., Dou, Q., Chen, H., Qin, J., Heng, P.A.: Unsupervised bidirectional cross-modality adaptation via deeply synergistic image and feature alignment for medical image segmentation. IEEE Trans. Med. Imag. **39**(7), 2494–2505 (2020)
3. Chung, M., et al.: Ct imaging features of 2019 novel coronavirus (2019-ncov). Radiology **295**(1), 202–207 (2020)
4. Hu, J., Shen, L., Sun, G.: Squeeze-and-excitation networks. In: Proceedings of the IEEE Conference on Computer Vision and Pattern Recognition, pp. 7132–7141 (2018)
5. Juarez, A.G.U., Tiddens, H.A., de Bruijne, M.: Automatic airway segmentation in chest CT using convolutional neural networks. In: Image Analysis for Moving Organ, Breast, and Thoracic Images, pp. 238–250. Springer (2018)
6. Karimi, D., Salcudean, S.E.: Reducing the hausdorff distance in medical image segmentation with convolutional neural networks. IEEE Trans. Med. Imag. **39**(2), 499–513 (2019)
7. Lin, T.Y., Goyal, P., Girshick, R., He, K., Dollár, P.: Focal loss for dense object detection. In: Proceedings of the IEEE International Conference on Computer Vision, pp. 2980–2988 (2017)
8. Lo, P., et al.: Extraction of airways from CT (exact'09). IEEE Trans. Med. Imag. **31**(11), 2093–2107 (2012)
9. Milletari, F., Navab, N., Ahmadi, S.A.: V-net: fully convolutional neural networks for volumetric medical image segmentation. In: 2016 Fourth International Conference on 3D Vision (3DV), pp. 565–571. IEEE (2016)
10. Navarro, F., et al.: Shape-aware complementary-task learning for multi-organ segmentation. In: Suk, H.-I., Liu, M., Yan, P., Lian, C. (eds.) MLMI 2019. LNCS, vol. 11861, pp. 620–627. Springer, Cham (2019). https://doi.org/10.1007/978-3-030-32692-0_71
11. Ouyang, X., et al.: Dual-sampling attention network for diagnosis of Covid-19 from community acquired pneumonia. IEEE Trans. Med. Imag. **39**(8), 2595–2605 (2020)

12. Qin, Y., et al.: Airwaynet: a voxel-connectivity aware approach for accurate airway segmentation using convolutional neural networks. In: International Conference on Medical Image Computing and Computer-Assisted Intervention, pp. 212–220. Springer (2019)
13. Rozantsev, A., Salzmann, M., Fua, P.: Beyond sharing weights for deep domain adaptation. IEEE Trans. Patt. Analy. Mach. Intell. **41**(4), 801–814 (2018)
14. Sun, B., Saenko, K.: Deep CORAL: correlation alignment for deep domain adaptation. In: Hua, G., Jégou, H. (eds.) ECCV 2016. LNCS, vol. 9915, pp. 443–450. Springer, Cham (2016). https://doi.org/10.1007/978-3-319-49409-8_35
15. Ulyanov, D., Vedaldi, A., Lempitsky, V.: Instance normalization: the missing ingredient for fast stylization. arXiv preprint arXiv:1607.08022 (2016)
16. Wang, G., et al.: A noise-robust framework for automatic segmentation of Covid-19 pneumonia lesions from CT images. IEEE Trans. Med. Imag. **39**(8), 2653–2663 (2020)
17. Wang, J., et al.: Prior-attention residual learning for more discriminative Covid-19 screening in CT images. IEEE Trans. Med. Imag. **39**(8), 2572–2583 (2020)
18. Wang, X., et al.: A weakly-supervised framework for Covid-19 classification and lesion localization from chest CT. IEEE Trans. Med. Imag. **39**(8), 2615–2625 (2020)
19. Xue, Y., et al.: Shape-aware organ segmentation by predicting signed distance maps. In: Proceedings of the AAAI Conference on Artificial Intelligence, vol. 34, pp. 12565–12572 (2020)
20. Zhu, W., et al.: Anatomynet: deep learning for fast and fully automated whole-volume segmentation of head and neck anatomy. Med. Phys. **46**(2), 576–589 (2019)

Adversarial Continual Learning for Multi-domain Hippocampal Segmentation

Marius Memmel$^{(\boxtimes)}$, Camila Gonzalez, and Anirban Mukhopadhyay

Technical University of Darmstadt, Karolinenpl. 5, 64289 Darmstadt, Germany
marius.memmel@stud.tu-darmstadt.de

Abstract. Deep learning for medical imaging suffers from temporal and privacy-related restrictions on data availability. To still obtain viable models, continual learning aims to train in sequential order, as and when data is available. The main challenge that continual learning methods face is to prevent catastrophic forgetting, i.e., a decrease in performance on the data encountered earlier. This issue makes continuous training of segmentation models for medical applications extremely difficult. Yet, often, data from at least two different domains is available which we can exploit to train the model in a way that it disregards domain-specific information. We propose an architecture that leverages the simultaneous availability of two or more datasets to learn a disentanglement between the content and domain in an adversarial fashion. The domain-invariant content representation then lays the base for continual semantic segmentation. Our approach takes inspiration from domain adaptation and combines it with continual learning for hippocampal segmentation in brain magnetic resonance imaging (MRI). We showcase that our method reduces catastrophic forgetting and outperforms state-of-the-art continual learning methods.

Keywords: Continual learning · Adversarial training · Feature disentanglement · Hippocampus MRI segmentation

1 Introduction

In medical imaging, privacy regulations and temporal restrictions limit access to data [9]. These limitations inhibit the application of traditional supervised deep learning methods for medical imaging tasks, which require the simultaneous availability of all data during training. Continual learning reframes the problem into a sequential training process, where not all datasets are available at each time step. However, when we evaluate continual learning models, they still experience a significant drop in performance, caused by *catastrophic forgetting*, i.e., the model adapting too strongly to particularities of the last training batch [17].

Electronic supplementary material The online version of this chapter (https://doi.org/10.1007/978-3-030-87722-4_4) contains supplementary material, which is available to authorized users.

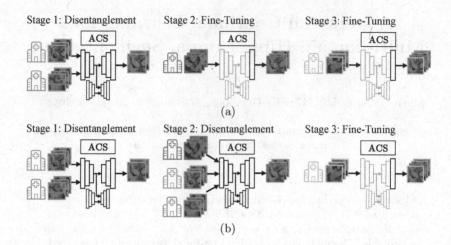

Fig. 1. The Adversarial Continual Segmenter (ACS) can react to different dataset availability through training (black) or freezing (gray) network parts. a) In stage 1, the disentanglement is trained on two datasets from different domains. In stage 2, a new dataset is available, and the previous ones are not. ACS is now fine-tuned on the new one. As a fourth dataset is introduced in stage 3, fine-tuning is repeated. b) If the first two datasets are still available when a third becomes accessible, disentanglement can be repeated in stage 2 with three datasets.

The leading cause of catastrophic forgetting in medical imaging is multi-domain data originating from different domains [25]. These domains result from diverse disease patterns among the examined subjects and divergent technologies and standards used during the acquisition process. In magnetic resonance imaging (MRI) it is common practice for institutions to operate scanners from various vendors and employ disparate protocols [21]. Additionally, MRI datasets frequently contain subjects that are either healthy or suffer from various pathological conditions. To sufficiently solve a task with data from multiple domains, models have to adapt and learn in a domain-invariant fashion.

The limited availability of multi-domain data makes developing a general-purpose model for continual hippocampal segmentation difficult. Commonly, at least two datasets from different domains are accessible simultaneously due to open access, relaxed restrictions, or access to historical data acquired with older scanners and protocols within the institution. This serendipity allows for learning a disentanglement between the domains and the content needed for segmentation, which continual learning methods do not yet exploit. Inspired by image-to-image translation (I2I), we utilize adversarial training to learn a disentanglement between a domain-invariant content representation sufficient for segmentation and a dataset-specific domain representation [11,13,19]. We train an encoder for each representation and share the content encoder with our segmentation module. Finally, we extend our approach to continual learning. In Fig. 1, we describe how our architecture could react to common dataset availability scenarios. We perform experiments on a subset of those hypothetical scenarios.

We contribute the *Adversarial Continual Segmenter* (ACS) for continual semantic segmentation of multi-domain data through adversarial disentanglement and latent space regularization that reduces catastrophic forgetting in hippocampal segmentation of brain MRIs.

2 Related Work

Adversarial Disentanglement: Several Generative Adversarial Networks (GANs) disentangle the feature space to improve interpretability [15]. Chen et al. [6] take a mutual-information-based approach while Karras et al. [14] directly modify the generator to achieve automatic separation of high-level attributes. Adversarial disentanglement shows promising results when applied to segmentation in a multi-domain [12] and multi-modal setting [5]. In domain adaptation, Kamnitsas et al. [13] utilize domain-invariant features for a segmentation task [26] and learn those with adversarial regularization of numerous layer outputs.

Cross-domain Disentanglement: I2I translation extends cross-domain feature disentanglement by splitting the latent space into a content and style encoding to achieve better translation results [22]. The content encoding is assumed only to capture task-specific information. The style encoding holds the domain-specific information. Huang et al. [11] and Lee et al. [19] further assume that the complexity of the content outweighs the domain, which they reflect in different encoder complexities.

Continual Learning: The main problem that continual learning methods face is catastrophic forgetting. For this purpose, regularization-based approaches constrain important parameters from changing [3,8,17,20]. With a similar goal, Memory Aware Synapses (MAS) [2] learn importance weights for each parameter and use those to penalize parameter changes. Knowledge distillation methods try to preserve specific model outputs to retain performance on old data [7,23]. Keeping a subset of training data is also widely used, e.g., in dynamic memory [10] and rehearsal methods [28]. However, keeping parts of the old data is not feasible in most medical imaging scenarios due to privacy concerns [9].

Whereas existing continual learning methods focus solely on a sequential learning process and do not consider the simultaneous availability of datasets and their divergent domains, we specifically exploit these circumstances through adversarial disentanglement.

3 Methods

We first describe how we disentangle input images into content and domain representation and follow up by introducing the adversarial approach. Our domain representation models the heterogeneity of the acquisition modality, e.g., varying protocols and machine vendors, as well as different disease patterns. To learn the domain-invariant content representation, we initially train on two datasets

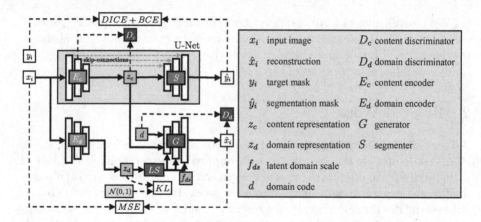

Fig. 2. Detailed visualization of the ACS architecture. We achieve feature disentanglement through two encoders, E_c for content and E_d for domain. The content and domain discriminators D_c and D_d regularize the latent representation z_c, and in the case of D_c, also the skip-connections of the U-Net to be domain-invariant. Segmentation is done by a forward pass of the U-Net.

simultaneously, as is common in I2I translation. This representation then acts as a basis for our U-Net segmenter.

Variational Autoencoder Loss: We model the domain encoder as a variational autoencoder (VAE) [16], which encodes the domain as a distribution z_d parameterized by variance σ^2 and mean μ. Because the complexity of the domain is assumed to be lower than the content complexity, we limit the dimensionality of the domain representation to one float value. We use a combination of a reconstruction loss between the input image x_i and the generator output $\hat{x}_i$ and a Kullback–Leibler regularization term weighted by hyperparameter η to draw the encoding close to a normally distributed Gaussian prior of $\mathcal{N}(0,1)$.

$$\mathcal{L}_{VAE} = \sum_i \|\hat{x}_i - x_i\|_2^2 + \eta \, \mathbb{E}_{x_i \sim X_i, z \sim \mathcal{N}(0,1)} \left[KL \left(E_d \left(x_i \right) \| z \right) \right] \qquad (1)$$

Latent Regression Loss: To prepare the domain information for the generator input, we first sample from the learned domain distribution and then pass the samples into the latent scale layer LS as proposed by Alharbi et al. [1] to produce a latent domain scale f_{ds}. To give G additional information about the domain, we inject a domain code d using central biasing instance normalization [12,29]. We model this domain code with a one-hot vector. As proposed by Jiang and Veeraraghavan [12] and based on Lee et al. [19] and Huang et al. [11], we also define a latent code regression loss $\mathcal{L}_{lr}$, which constrains the generator to produce unique mappings for a latent code z.

GAN Loss: We interpret the encoder of the U-Net as content encoder E_c with the output of the bottom layer treated as content representation z_c. Both z_c and the scaled domain sample f_{ds} are fed into generator G to reconstruct the input sample x_i. To now disentangle the content and domain, we introduce adversarial training. We deploy a domain discriminator D_d that regularizes E_c, E_d, and G by discriminating whether an input image x_i is part of a given domain d. D_d trains on a combination of real and generated images as described in the corresponding discriminator loss in Eq. 2. G generates these images from content z_c of x_i and random domain z. G, E_c, and E_d counter the discriminator by minimizing a negative binary cross-entropy loss shown in Eq. 3. The training forces E_c to produce a domain-invariant output, which we utilize for the segmentation task.

$$\mathcal{L}_{GAN_D} = \sum_i \mathbb{E}_{x_i \sim X_i} \left[\log(D_d(x_i, d_i)) \right]$$
$$+ \mathbb{E}_{x_i \sim X_i} \left[\log(1 - D_d(G(E_c(x_i), LS(z)), d_i)) \right] \tag{2}$$

$$\mathcal{L}_{GAN_G} = -\sum_i \mathbb{E}_{x_i \sim X_i, z \sim \mathcal{N}(0,1)} \left[\log(D_d(G(E_c(x_i), LS(z)), 1)) \right] \tag{3}$$

Content Adversarial Loss: The information about the domain can still flow from E_c to S via the skip-connections of the U-Net. To prevent this, we introduce a content discriminator D_c inspired by Jiang et al. and Kamnitsas et al. [12,13]. D_c regularizes z_c as well as the skip-connections because we want E_c to not leak any domain information to the segmenter S. The discriminator design is similar to a reversed U-Net decoder, i.e., it takes z_c as input and the skip-connections of E_c at each corresponding layer. We train both D_c and E_c using a multi-class cross-entropy loss $\mathcal{L}^c_{adv_{D_c}}$ and $\mathcal{L}^c_{adv_{E_c}}$ respectively. In the case of the generated images, we represent the domain code as a placeholder class. To ensure that the adversarial training is stable, we train the discriminators and the remaining architecture at separate steps [12].

Segmentation: To produce the segmentation mask $\hat{y}_i$, we encode input x_i into z_c through E_c. We then pass z_c and the skip-connections of E_c into S to compute $\hat{y}_i$. We train the U-Net for semantic segmentation through a pixel-wise combination of a Dice and a binary cross-entropy loss between the target mask y_i and the prediction $\hat{y}_i$. After initially training the architecture on two or more datasets, the model has sufficiently learned the disentanglement between the content and the domain. To train on a new dataset, we only fine-tune the last four convolutional layers of S.

4   Datasets and Experiments

Datasets: All datasets are from different domains and contain T1-weighted MRIs. The first dataset was released as part of the *2018 Medical Segmentation Decathlon* challenge [27] and consists of 195 subjects in total, with 90 healthy and 105 with non-affective psychotic disorder. The scans were collected using a Philips Achieva scanner, and the mean size of the volumes is $35 \times 49 \times 35$. The second dataset was published in *Scientific Data* [18] and has a T1-weighted dataset with 25 healthy subjects. All scans were acquired using MRI systems with 3 T units. The mean standard resolution is $48 \times 64 \times 64$. Finally, the third dataset is provided by the *Alzheimer's Disease Neuroimaging Initiative* [4] and consists of 68 subjects that are either part of the control group or suffer from either mild cognitive impairment or Alzheimer's disease. The images were acquired with scanners from Siemens, GE, and Philips with 23, 24, and 21 scans, respectively. The mean volume size is $48 \times 64 \times 64$. Since dataset A only contains regions of the hippocampus, while B and C both comprise scans of the whole brain, we crop each slice of B and C at fixed positions to extract the left and right hippocampus, respectively. All three datasets provide reference segmentation masks for the hippocampus. The masks were annotated manually with the protocols defined in the respective publications. We evaluate our architecture on all three datasets, which we will refer to as A, B, and C, respectively.

Experimental Setup: We split each dataset into 70% train, 20% test, and 10% validation and use the latter to select the hyperparameters. We train slice-by-slice and upsample via bilinear interpolation to achieve uniform slices. We compare ACS with the following baselines. First, just the U-Net block of ACS (U-Net-b) shown in Fig. 2 which is fine-tuned in the same manner as ACS, and second a standard U-Net without fine-tuning (U-Net). Furthermore, we extend the U-Net by knowledge distillation on the output layer (OL-KD) as proposed by Michieli and Zanuttigh [23], and Memory Aware Synapses adapted to brain segmentation (BS-MAS) by Özgün et al. [24]. As suggested in BS-MAS, we divide the surrogate loss by the number of network parameters and normalize the resulting importance values between zero and one. We report the Intersection over Union (IoU) and Dice coefficient on the hippocampus class of the test set. We use a batch size of 40 and train on four Tesla V100 SXM3 GPUs. Each method receives training over 60 epochs. After 30 epochs, the training only continues with the third dataset. We repeat training for every combination of the three datasets (*AB-C*, *AC-B*, *BC-A*), e.g., initial training on datasets A and B, then on C (*AB-C*). Additionally, we jointly train ACS and the U-Net on all datasets simultaneously (*ABC*). To justify the necessity for all mechanisms in our method, as described in Sect. 3, we conduct an ablation study in Table 3. Implementation details, techniques used for hyperparameter tuning, and qualitative results including the disentanglement can be found in the supplementary material and code on github.com/MECLabTUDA/ACS.

5   Result and Discussion

To assess the continual learning performance, we evaluate the results after stages 1 and 2 corresponding to Fig. 1a. An ideal algorithm should perform equally or better on the initial training datasets from epoch 30 to 60 while it should improve on the third dataset added after 30 epochs.

Stage 1: Table 1 shows the results for all methods after training for 30 epochs on the initial two datasets. All baselines observe the same score because they apply the regularization in the second training stage, whereas ACS performs disentanglement during the initial training phase. ACS outperforms them by a Dice of $+0.178 \pm 0.08$ (IoU $+0.190 \pm 0.07$) averaged over all combinations and datasets.

Table 1. Comparison of all baselines to ACS after **30 epochs**.

		Dataset A		Dataset B		Dataset C		**Average**	
		IoU	Dice	IoU	Dice	IoU	Dice	IoU	Dice
AB	Baselines	0.641	0.779	0.779	0.875	0.358	0.512	0.593	0.722
	ACS (ours)	**0.749**	**0.855**	**0.793**	**0.884**	**0.478**	**0.628**	**0.673**	**0.789**
AC	Baselines	0.080	0.147	0.265	0.416	0.380	0.547	0.241	0.370
	ACS (ours)	**0.646**	**0.782**	**0.731**	**0.844**	**0.727**	**0.841**	**0.702**	**0.822**
BC	Baselines	**0.260**	**0.407**	0.749	0.856	0.649	0.784	0.553	0.682
	ACS (ours)	0.239	0.376	**0.798**	**0.887**	**0.710**	**0.829**	**0.582**	**0.698**

Stage 2: To measure overall continual learning performance, i.e., the combination of learning and forgetting, we inspect the average scores over all datasets after 60 epochs in Table 2. While the comparison methods' results fluctuate, our approach achieves a consistently higher performance across all combinations and datasets. This observation manifests in an increase of the average Dice score by $+0.127 \pm 0.10$ over the U-Net, $+0.156 \pm 0.03$ over the U-Net-b, $+0.118 \pm 0.08$ over BS-MAS, and $+0.152 \pm 0.14$ over OL-KD. On combination $AB\text{-}C$, the U-Net drops by an IoU of $>47\%$ (Dice $> 35\%$) on dataset A and by $>26\%$ (Dice $> 17\%$) on dataset B. The remaining methods, including ACS, show a significantly lower decline and effectively reduce catastrophic forgetting.

Combination $AC\text{-}B$ shows the clear advantage of our approach. Dataset A contains four types of disorders recorded by a single scanner, while dataset C holds three disease patterns recorded by three different scanners. The baselines struggle with the diversity of these domains, and our model outperforms them by an IoU of $>91\%$ (Dice $> 53\%$). These observations show that our model learns a sufficient content representation that can deal with diverse cognitive impairments and scans acquired by scanners of various vendors.

We trace back the low performance on dataset A in combination $BC\text{-}A$ to A outnumbering B and C in its variability and number of subjects. The high performance of the U-Net thereby originates from overfitting on A which through its high variability still allows it to perform well on B and C. Because ACS is only fine-tuned on A, it cannot fully exploit this anomaly, but still shows competitive results.

Ablation Study: The conducted ablation study in Table 3 verifies that all losses contribute to the performance of ACS. Only on $AC\text{-}B$, the combination of all losses underperforms slightly, but remains competitive. For more detailed numbers we direct the reader to the supplementary material.

Table 2. Comparison of all baselines and ACS after **60 epochs**. ABC represents joint training on all datasets at once.

		Dataset A		Dataset B		Dataset C		**Average**	
		IoU	Dice	IoU	Dice	IoU	Dice	IoU	Dice
AB-C	U-Net	0.238	0.381	0.501	0.664	0.621	0.764	0.454	0.603
	U-Net-b	0.339	0.503	0.570	0.724	0.473	0.631	0.461	0.619
	BS-MAS	0.304	0.464	0.568	0.722	**0.624**	**0.766**	0.499	0.651
	OL-KD	0.578	0.729	0.727	0.841	0.473	0.633	0.593	0.734
	ACS (ours)	**0.640**	**0.779**	**0.760**	**0.863**	0.572	0.718	**0.657**	**0.787**
AC-B	U-Net	0.295	0.451	0.718	0.836	0.387	0.547	0.467	0.611
	U-Net-b	0.273	0.425	0.567	0.723	0.419	0.584	0.419	0.577
	BS-MAS	0.307	0.466	0.702	0.825	0.364	0.523	0.458	0.604
	OL-KD	0.094	0.171	0.381	0.549	0.400	0.571	0.292	0.430
	ACS (ours)	**0.681**	**0.808**	**0.787**	**0.880**	**0.679**	**0.808**	**0.716**	**0.832**
BC-A	U-Net	**0.745**	**0.852**	0.668	0.800	0.418	0.579	**0.610**	**0.743**
	U-Net-b	0.450	0.615	0.591	0.742	0.497	0.661	0.513	0.673
	BS-MAS	0.731	0.847	0.626	0.768	0.409	0.569	0.589	0.728
	OL-KD	0.347	0.511	**0.766**	**0.867**	**0.639**	**0.776**	0.584	0.718
	ACS (ours)	0.600	0.747	0.649	0.786	0.465	0.631	0.571	0.721
ABC (joint)	U-Net	0.277	0.431	0.458	0.627	0.431	0.599	0.388	0.440
	ACS (ours)	**0.737**	**0.847**	**0.760**	**0.863**	**0.724**	**0.839**	**0.740**	**0.849**

Table 3. Snapshot of the ablation study on ACS trained for 60 epochs with losses active (✓) or deactivated (X) during training. Average reported over all test datasets in a configuration.

$\mathcal{L}^c_{adv}$	$\mathcal{L}_{VAE}$	$\mathcal{L}_{GAN}$	$\mathcal{L}_{lr}$	AB-C		AC-B		BC-A		**Average**	
				IoU	Dice	IoU	Dice	IoU	Dice	IoU	Dice
X	✓	✓	✓	0.620	0.759	0.730	0.843	0.526	0.686	0.626	0.763
✓	X	X	X	0.574	0.721	0.709	**0.849**	0.524	0.681	0.603	0.750
✓	✓	X	✓	0.637	0.772	**0.732**	0.843	0.555	0.711	0.642	0.776
✓	X	✓	✓	0.636	0.772	0.721	0.835	0.551	0.707	0.636	0.772
✓	✓	✓	✓	**0.658**	**0.787**	0.716	0.832	**0.572**	**0.721**	**0.649**	**0.780**

The results demonstrate that leveraging the availability of multiple datasets increases multi-domain segmentation performance by sufficiently learning a domain-invariant representation. This assumption is further supported by the joint training results in Table 2 showing the superior capability of ACS in comparison to the U-Net. Additionally, our method outperforms the state-of-the-art on most continual learning setups and effectively reduces catastrophic forgetting.

6   Conclusion

We propose ACS, an architecture for continual semantic segmentation of multi-domain data that leverages the simultaneous availability of datasets. In real clinical practice, multiple datasets are available at the beginning of the continual training process through, among other sources, public or accessible historical data. Unlike current methods, we leverage this serendipity to disentangle MRI images into content and domain representations through adversarial training. We then perform multi-domain hippocampal segmentation directly on the domain-invariant content representation. We demonstrate drastic improvements through domain disentanglement of multi-domain data in the first training stage. In the second training stage, the benefits of our proposal for continual learning become clear by showcasing that using all available data reduces catastrophic forgetting and outperforms current state-of-the-art methods. Our method pushes continual learning closer towards a clinical application where various degrees of variability such as disease patterns, scan vendors, and acquisition protocols exist and further enables the continual usage of deep learning models in clinical practice.

Acknowledgements. This work was supported by the Bundesministerium für Gesundheit (BMG) with grant [ZMVI1- 2520DAT03A].

References

1. Alharbi, Y., Smith, N., Wonka, P.: Latent filter scaling for multimodal unsupervised image-to-image translation. In: Proceedings of the IEEE/CVF Conference on Computer Vision and Pattern Recognition (CVPR), June 2019
2. Aljundi, R., Babiloni, F., Elhoseiny, M., Rohrbach, M., Tuytelaars, T.: Memory aware synapses: learning what (not) to forget. In: Ferrari, V., Hebert, M., Sminchisescu, C., Weiss, Y. (eds.) ECCV 2018. LNCS, vol. 11207, pp. 144–161. Springer, Cham (2018). https://doi.org/10.1007/978-3-030-01219-9_9
3. Baweja, C., Glocker, B., Kamnitsas, K.: Towards continual learning in medical imaging. CoRR abs/1811.02496 (2018)
4. Boccardi, M., et al.: Training labels for hippocampal segmentation based on the EADC-ADNI harmonized hippocampal protocol. Alzheimers Dementia **11**(2), 175–183 (2015). http://adni.loni.usc.edu/
5. Chartsias, A., et al.: Disentangled representation learning in cardiac image analysis. Med. Image Anal. **58**, 101535 (2019)
6. Chen, X., Duan, Y., Houthooft, R., Schulman, J., Sutskever, I., Abbeel, P.: InfoGAN: interpretable representation learning by information maximizing generative adversarial nets. In: Proceedings of the 30th International Conference on Neural Information Processing Systems, NIPS 2016, pp. 2180–2188. Curran Associates Inc., Red Hook (2016)
7. Douillard, A., Chen, Y., Dapogny, A., Cord, M.: PLOP: learning without forgetting for continual semantic segmentation. In: Proceedings of the IEEE/CVF Conference on Computer Vision and Pattern Recognition (CVPR), pp. 4040–4050, June 2021
8. van Garderen, K.A., Voort, S.V.D., Incekara, F., Smits, M., Klein, S.: Towards continuous learning for glioma segmentation with elastic weight consolidation. ArXiv abs/1909.11479 (2019)

9. González, C., Sakas, G., Mukhopadhyay, A.: What is wrong with continual learning in medical image segmentation? CoRR abs/2010.11008 (2020)

10. Hofmanninger, J., Perkonigg, M., Brink, J.A., Pianykh, O., Herold, C., Langs, G.: Dynamic memory to alleviate catastrophic forgetting in continuous learning settings. In: Martel, A.L. (ed.) MICCAI 2020. LNCS, vol. 12262, pp. 359–368. Springer, Cham (2020). https://doi.org/10.1007/978-3-030-59713-9_35

11. Huang, X., Liu, M.-Y., Belongie, S., Kautz, J.: Multimodal unsupervised image-to-image translation. In: Ferrari, V., Hebert, M., Sminchisescu, C., Weiss, Y. (eds.) ECCV 2018. LNCS, vol. 11207, pp. 179–196. Springer, Cham (2018). https://doi.org/10.1007/978-3-030-01219-9_11

12. Jiang, J., Veeraraghavan, H.: Unified cross-modality feature Disentangler for unsupervised multi-domain MRI abdomen organs segmentation. In: Martel, A.L., et al. (eds.) MICCAI 2020. LNCS, vol. 12262, pp. 347–358. Springer, Cham (2020). https://doi.org/10.1007/978-3-030-59713-9_34

13. Kamnitsas, K., et al.: Unsupervised domain adaptation in brain lesion segmentation with adversarial networks. In: Niethammer, M., et al. (eds.) IPMI 2017. LNCS, vol. 10265, pp. 597–609. Springer, Cham (2017). https://doi.org/10.1007/978-3-319-59050-9_47

14. Karras, T., Laine, S., Aila, T.: A style-based generator architecture for generative adversarial networks. In: Proceedings of the IEEE/CVF Conference on Computer Vision and Pattern Recognition (CVPR), June 2019

15. Kazeminia, S., et al.: GANs for medical image analysis. CoRR abs/1809.06222 (2018)

16. Kingma, D.P., Welling, M.: Auto-encoding variational Bayes. In: Bengio, Y., LeCun, Y. (eds.) ICLR (2014)

17. Kirkpatrick, J., et al.: Overcoming catastrophic forgetting in neural networks. Proc. Natl. Acad. Sci. **114**(13), 3521–3526 (2017)

18. Kulaga-Yoskovitz, J., et al.: Multi-contrast submillimetric 3 tesla hippocampal subfield segmentation protocol and dataset. Sci. Data **2**(1), 150059 (2015). https://doi.org/10.5061/dryad.gc72v. https://datadryad.org/stash/dataset/

19. Lee, H.-Y., Tseng, H.-Y., Huang, J.-B., Singh, M., Yang, M.-H.: Diverse image-to-image translation via disentangled representations. In: Ferrari, V., Hebert, M., Sminchisescu, C., Weiss, Y. (eds.) ECCV 2018. LNCS, vol. 11205, pp. 36–52. Springer, Cham (2018). https://doi.org/10.1007/978-3-030-01246-5_3

20. Lenga, M., Schulz, H., Saalbach, A.: Continual learning for domain adaptation in chest x-ray classification. In: Arbel, T., Ben Ayed, I., de Bruijne, M., Descoteaux, M., Lombaert, H., Pal, C. (eds.) Proceedings of the Third Conference on Medical Imaging with Deep Learning. Proceedings of Machine Learning Research, vol. 121, pp. 413–423. PMLR, 06–08 July 2020

21. Li, H., et al.: Denoising scanner effects from multimodal MRI data using linked independent component analysis. NeuroImage **208**, 116388 (2020)

22. Liu, Y.C., Yeh, Y.Y., Fu, T.C., Wang, S.D., Chiu, W.C., Wang, Y.C.F.: Detach and adapt: learning cross-domain disentangled deep representation. In: Proceedings of the IEEE Conference on Computer Vision and Pattern Recognition (CVPR), June 2018

23. Michieli, U., Zanuttigh, P.: Incremental learning techniques for semantic segmentation. In: Proceedings of the IEEE/CVF International Conference on Computer Vision (ICCV) Workshops, October 2019

24. Özgün, S., Rickmann, A.-M., Roy, A.G., Wachinger, C.: Importance driven continual learning for segmentation across domains. In: Liu, M., Yan, P., Lian, C., Cao, X. (eds.) MLMI 2020. LNCS, vol. 12436, pp. 423–433. Springer, Cham (2020). https://doi.org/10.1007/978-3-030-59861-7_43

25. Pianykh, O.S., et al.: Continuous learning AI in radiology: implementation principles and early applications. Radiology **297**(1), 6–14 (2020). pMID: 32840473

26. Prangemeier, T., Wildner, C., Françani, A.O., Reich, C., Koeppl, H.: Multiclass yeast segmentation in microstructured environments with deep learning. In: 2020 IEEE Conference on Computational Intelligence in Bioinformatics and Computational Biology (CIBCB), pp. 1–8 (2020)

27. Simpson, A.L., et al.: A large annotated medical image dataset for the development and evaluation of segmentation algorithms. CoRR abs/1902.09063 (2019). http://medicaldecathlon.com/

28. Sokar, G., Mocanu, D.C., Pechenizkiy, M.: Learning invariant representation for continual learning. CoRR abs/2101.06162 (2021)

29. Yu, X., Ying, Z., Li, G.: Multi-mapping image-to-image translation with central biasing normalization. CoRR abs/1806.10050 (2018)

Self-supervised Multimodal Generalized Zero Shot Learning for Gleason Grading

Dwarikanath Mahapatra[1]([✉]), Behzad Bozorgtabar[2], Shiba Kuanar[3], and Zongyuan Ge[4,5]

[1] Inception Institute of Artificial Intelligence, Abu Dhabi, UAE
dwarikanath.mahapatra@inceptioniai.org
[2] Signal Processing Laboratory 5, EPFL, Lausanne, Switzerland
[3] Yale University, New Haven, USA
[4] Monash University, Melbourne, Australia
[5] Airdoc Research, Melbourne, Australia

Abstract. Gleason grading from histopathology images is essential for accurate prostate cancer (PCa) diagnosis. Since such images are obtained after invasive tissue resection quick diagnosis is challenging under the existing paradigm. We propose a method to predict Gleason grades from magnetic resonance (MR) images which are non-interventional and easily acquired. We solve the problem in a generalized zero-shot learning (GZSL) setting since we may not access training images of every disease grade. Synthetic MRI feature vectors of unseen grades (classes) are generated by exploiting Gleason grades' ordered nature through a conditional variational autoencoder (CVAE) incorporating self-supervised learning. Corresponding histopathology features are generated using cycle GANs, and combined with MR features to predict Gleason grades of test images. Experimental results show our method outperforms competing feature generating approaches for GZSL, and comes close to performance of fully supervised methods.

Keywords: GZSL · CVAE · Gleason grading · Histopathology · MRI

1 Introduction

Early and accurate diagnosis of prostate cancer (PCa) is an important clinical problem. High resolution histopathology images provide the gold standard but involve invasive tissue resection. Non-invasive techniques like magnetic resonance imaging (MRI) are useful for early abnormality detection [15] and easier to acquire, but their low resolution and noise makes it difficult to detect subtle differences between benign conditions and cancer. A combination of MRI and histopathology features can potentially leverage their respective advantages for

Electronic supplementary material The online version of this chapter (https://doi.org/10.1007/978-3-030-87722-4_5) contains supplementary material, which is available to authorized users.

© Springer Nature Switzerland AG 2021
S. Albarqouni et al. (Eds.): DART 2021/FAIR 2021, LNCS 12968, pp. 46–56, 2021.
https://doi.org/10.1007/978-3-030-87722-4_5

improved accuracy in early PCa detection. Accessing annotated multimodal data of same patient from same examination instance is a challenge. Hence a machine learning model to generate one domain's features from the other is beneficial to combine them for PCa detection. Current supervised learning [2], and multiple instance learning [5] approaches for Gleason grading use all class labels in training. Accessing labeled samples of all Gleason grades is challenging, and we also encounter known and unknown cases at test time, making the problem one of generalized zero-shot learning (GZSL). We propose to predict Gleason grades for PCa by generating histopathology features from MRI features and combining them for GZSL based disease classification.

GZSL classifies natural images from seen and unseen classes [6] and uses Generative Dual Adversarial Network (GDAN) [11], overcomplete distributions [12] and domain aware visual bias elimination [26]. But it has not been well explored for medical images. One major reason being the availability of class attribute vectors for natural images that describe characteristics of seen and unseen classes, but are challenging to obtain for medical images. Self-supervised learning (SSL) also addresses labeled data shortage and has found wide use in medical image analysis by using innovative pre-text tasks for active learning [25], anomaly detection [3], and data augmentation [20]. SSL has been applied to histopathology images using domain specific pretext tasks [14], semi-supervised histology classification [17], stain normalization [21], registration [32] and cancer subtyping using visual dictionaries. Wu et al. [33] combine SSL and GZSL for natural images but rely on class attribute vectors and unlabeled target data for training.

Our current work makes the following contributions: 1) We propose a **multimodal framework for seen and unseen Gleason grade prediction from MR images by combining GZSL and SSL.** 2) Unlike previous methods used for natural images we do not require class attribute vectors nor unlabeled target data during training. 3) We propose a **self-supervised learning** (SSL) approach for feature synthesis of seen and unseen classes that exploits the ordered relationship between different Gleason grades to generate new features. Although the method by [2] uses features from MRI and histopathology images: 1) it is a fully supervised method that has no Unseen classes during test time; 2) it also uses MRI and histopathology features from available data while we generate synthetic features to overcome unavailability of one data modality.

2   Method

2.1   Feature Extraction and Transformation

Figure 1 depicts our overall proposed workflow, which has 3 networks for: feature extraction, feature transformation and self supervised feature synthesis. Let the training set consist of *histopathology and MR images* from PCa cases with known Gleason grades. We train a network that can correctly predict seen and unseen Gleason grades using *MR images only*. Let the dataset of 'Seen' and 'Unseen' classes be S, U and their corresponding labels be $\mathscr{Y}_s, \mathscr{Y}_u$. $\mathscr{Y} = \mathscr{Y}_s \cup \mathscr{Y}_u$, and $\mathscr{Y}_s \cap \mathscr{Y}_u = \emptyset$. Previous works [6] show that generating feature vectors, instead of

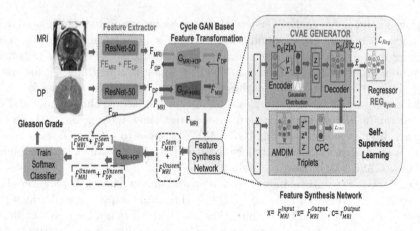

Fig. 1. Training Workflow: Feature extraction from MR and digital pathology images generates respective feature vectors F_{MRI} and F_{DP}. A F_{MRI} is input to a feature synthesis network to generate new MR features for 'Seen' and 'Unseen' classes, which are passed through the feature transformation network to obtain corresponding F_{DP}. Combined MR and histopathology features are used to train a softmax classifier for identifying 'Seen' and 'Unseen' test classes. A self-supervised module contributes to the sample generation step through $\mathscr{L}_{CPC}$.

images, performs better for GZSL classification since output images of generative models can be blurry, especially with multiple objects (e.g., multiple cells in histopathology images). Additionally, generating high-resolution histopathology images from low resolution and noisy MRI is challenging, while finding a transformation between their feature vectors is much more feasible.

We train two separate ResNet-50 networks [9] as feature extractors for MR and DP images. Histopathology image feature extractor (FE_{DP}) is pre-trained on the PANDA dataset [4] that has a large number of whole slide images for PCa classification. FE_{MRI} is pre-trained with the PROMISE12 challenge dataset [15] in a self-supervised manner. Since PROMISE12 is for prostate segmentation from MRI and does not have classification labels, we use a pre-text task of predicting if an image slice has the prostate or not. Gleason grades of MRI are the same as corresponding histopathology images. The images are processed through the convolution blocks and the 1000 dimensional output of the last FC layer is the feature vector for MRI (F_{MRI}) and pathology images (F_{DP}).

Feature Transformation is achieved using CycleGANs [19,22–24,35] to learn mapping functions $G : X \rightarrow Y$ and $F : Y \rightarrow X$, between feature vectors $X = F_{MRI}$ and $Y = F_{DP}$. Adversarial discriminator D_X differentiates between real features F_{DP} and generated features $\widehat{F}_{DP}$, and D_Y distinguishes between F_{MRI} and $\widehat{F}_{MRI}$. The adversarial loss (Eq. 1) and cycle consistency loss (Eq. 2) are,

$$
\begin{aligned}
L_{adv}(G, D_Y) &= \mathbb{E}_y \left[\log D_Y(y)\right] + \mathbb{E}_x \left[\log \left(1 - D_Y(G(x))\right)\right], \\
L_{adv}(F, D_X) &= \mathbb{E}_x \left[\log D_X(x)\right] + \mathbb{E}_y \left[\log \left(1 - D_X(F(y))\right)\right].
\end{aligned}
\tag{1}
$$

$$L_{cyc}(G, F) = E_x \|F(G(x)) - x\|_1 + E_y \|G(F(y)) - y\|_1 . \tag{2}$$

Network Training: is done using $L_{adv}(G, D_Y) + L_{adv}(F, D_X) + L_{cyc}(G, F)$. Generator G is a multi-layer perceptron (MLP) with a hidden layer of 4096 nodes having LeakyReLU [18]. The output layer with 2048 nodes has ReLU activation [28]. G's weights are initialized with a truncated normal of $\mu = 0$, $\sigma = 0.01$, and biases initialized to 0. Discriminator D is an MLP with a hidden layer of 2048 nodes activated by LeakyReLU, and the output layer has no activation. D's initialization is the same as G, and we use Adam optimizer [13].

2.2 CVAE Based Feature Generator Using Self Supervision

The conditional variational autoencoder (CVAE) generator synthesizes feature vectors F_{MRI}^{output} of desired class c_{MRI}^{output}, given input features F_{MRI}^{input} with known class c_{MRI}^{input}. Let $x = F_{MRI}^{input}, z = F_{MRI}^{output}$ and $c = c_{MRI}^{output}$. The CVAE loss is,

$$\min_{\theta_G, \theta_E} \mathscr{L}_{CVAE} + \lambda_c \cdot \mathscr{L}_c + \lambda_{reg} \cdot \mathscr{L}_{reg} + \lambda_E \cdot \mathscr{L}_E + \lambda_{CPC} \cdot \mathscr{L}_{CPC} \tag{3}$$

Denoting CVAE encoder as $p_E(z|x)$ with parameters θ_E, and the regressor output distribution as $p_R(c|x)$, $\mathscr{L}_{CVAE}$ is:

$$\mathscr{L}_{CVAE}(\theta_E, \theta_G) = -E_{p_E(z|x), p(c|x)} [\log p_G(x|z, c)] + KL(p_E(z|x)\|p(z)) \tag{4}$$

$E_{p_E(z|x), p(c|x)}(.)$ is the generator's reconstruction error, and KL divergence, $KL(.)$, encourages CVAE posterior (the encoder) to be close to the prior. Encoder $p_E(z|x)$, conditional decoder/generator $p_G(x|z, c)$, and regressor $p_R(c|x)$ are modeled as Gaussian distributions. The latent code (z, c) is represented by disentangled representations $p_E(z|x)$ and $p_R(c|x)$ to avoid posterior collapse [10].

Regression/Discriminator Module - REG_{Synth}: is a feedforward neural network mapping input feature vector $x \in \mathscr{R}^D$ to its corresponding class-value $c \in \mathscr{R}^1$. REG_{Synth} is a probabilistic model $p_R(c|x)$ with parameters θ_R and is trained using supervised ($\mathscr{L}_{Sup}$) and unsupervised ($\mathscr{L}_{Unsup}$) losses:

$$\min_{\theta_R} \mathscr{L}_R = \mathscr{L}_{Sup} + \lambda_R \cdot \mathscr{L}_{Unsup} \tag{5}$$

$\lambda_R = 0.2$ and $\mathscr{L}_{Sup}(\theta_R) = -E_{\{x_n, c_n\}}[p_R(c_n|x_n)]$ is defined on labeled examples $\{x_n, c_n\}_{n=1}^N$ from the seen class. $\mathscr{L}_{Unsup}(\theta_R) = -E_{p_{\theta_G}(\hat{x}|z, c)p(z)p(c)}[p_R(c|\hat{x})]$ is defined on synthesized examples $\hat{x}$ from the generator. $\mathscr{L}_{Unsup}$ is obtained by sampling z from $p(z)$, and class-value c sampled from $p(c)$ to generate an exemplar for $p_{\theta_G}(\hat{x}|z, c)$), and we calculate the expectation w.r.t. these distributions.

Discriminator-Driven Learning: The error back-propagated from REG_{Synth} improves quality of synthetic samples $\hat{x}$ making them similar to the desired output class-value c. This is achieved using multiple loss functions. The first generates samples whose regressed class value is close to the desired value,

$$\mathscr{L}_c(\theta_G) = -\mathbb{E}_{p_G(\widehat{x}|z,c)p(z)p(c)}\left[\log p_R(c|\widehat{x})\right] \tag{6}$$

The second term draws samples from prior $p(z)$ and combines with class-value from $p(c)$, to ensure the synthesized features are similar to the training data,

$$\mathscr{L}_{Reg}(\theta_G) = -\mathbb{E}_{p(z)p(c)}\left[\log p_G(\widehat{x}|z,c)\right] \tag{7}$$

The third loss term ensures independence (disentanglement) [10] of z from the class-value c. The encoder ensures that the sampling distribution and the one obtained from the generated sample follow the same distribution.

$$\mathscr{L}_E(\theta_G) = -\mathbb{E}_{\widehat{x}\sim p_G(\widehat{x}|z,c)}KL\left[\log p_E(z|\widehat{x})||q(z)\right] \tag{8}$$

The distribution $q(z)$ could be the prior $p(z)$ or the posterior from a labeled sample $p(z|x_n)$.

Self Supervised Loss: Gleason grades have a certain ordering, i.e., grade 3 indicates higher severity than grade 1, and grade 5 is higher than grade 3. Contrastive Predictive Coding (CPC) [30] learns self-supervised representations by predicting future observations from past ones and requires that observations be ordered in some dimension. Inspired by CPC we train our network to predict features of desired Gleason grade from input features of a different grade. From the training data we construct pairs of $\{F_{MRI}^{input}, c_{MRI}^{input}, F_{MRI}^{output}, c_{MRI}^{output}\}$, the input and output features and desired class label value c_{MRI}^{output} of synthesized vector.

Since the semantic gap between F_{MRI}^{input} and F_{MRI}^{output} may be too large, we use random transformations to first generate intermediate representation of positive z^+, anchor z^a and negative z^- features using the mutual information-based AMDIM approach of [1]. AMDIM ensures that the mutual information between similar samples z^+, z^a is high while for dissimilar samples z^-, z^a it is low, where $z = F_{MRI}^{output}$. The CPC objective (Eq. 9) evaluates the quality of the predictions using a contrastive loss where the goal is to correctly recognize the synthetic vector z among a set of randomly sampled feature vectors $z_l = \{z^+, z^a, z^-\}$.

$$\mathscr{L}_{CPC} = -\sum \log \frac{\exp^{(F_{MRI}^{input})^T F_{MRI}^{output}}}{\exp^{(F_{MRI}^{input})^T F_{MRI}^{output}} + \sum \exp^{(F_{MRI}^{input})^T F_{MRI}^{output}}} \tag{9}$$

This loss is the InfoNCE inspired from Noise-Contrastive Estimation [8,27] and maximizes the mutual information between c and z [30]. By specifying the desired class label value of synthesized vector and using it to guide feature generation in the CVAE framework we avoid the pitfalls of unconstrained and unrealistic feature generation common in generative model based GZSL (e.g., [16,34]).

Training and Implementation: Our framework was implemented in PyTorch. The CVAE Encoder has two hidden layers of 2000 and 1000 units respectively while the CVAE Generator is implemented with one hidden layer of 1000 hidden units. The Regressor has only one hidden layer of 800 units. We choose Adam [13] as our optimizer, and the momentum is set to $(0.9, 0.999)$. The learning rate for CVAE and Regressor is 0.0001. First the CVAE loss (Eq. 4) is pre-trained followed by joint training of regressor and encoder-generator pair in optimizing Eq. 7 and Eq. 3 till convergence. Hyperparameter values were chosen using a train-validation split with $\lambda_R = 0.1, \lambda_c = 0.1, \lambda_{reg} = 0.1, \lambda_E = 0.1$, and $\lambda_{CPC} = 0.2$. Training the feature extractor for 50 epochs takes 12 h and the feature synthesis network for 50 epochs takes 17 h, all on a single NVIDIA V100 GPU (32 GB RAM).

Evaluation Protocol: The seen class S has samples from 2 or more Gleason grades, and the unseen class U contains samples from remaining classes. $80 - 20$ split of S is done into S_{Train}/S_{Test}. F_{MRI} is synthesized from $S + U$ using the CVAE and combined with the corresponding synthetic F_{DP} to obtain a single feature vector for training a softmax classifier minimizing the negative log-likelihood loss. Following standard practice for GZSL, average class accuracies are calculated for two settings: 1) *Setting* **A**: training is performed on synthesized samples of $S + U$ classes and test on S_{Test}. 2) *Setting* **B**: training is performed on synthesized samples of $S + U$ classes and test on U. Following GZSL protocol we report the harmonic mean: $H = \frac{2 \times Acc_U \times Acc_S}{Acc_U + Acc_S}$; Acc_S and Acc_U are classification accuracy of images from seen (setting A) and unseen (setting B) classes respectively.

3   Experimental Results

Dataset Details: We use images from 321 patients (48–70 years; mean: 58.5 years) undergone prostate surgery with pre-operative T2-w and Apparent Diffusion Coefficient MRI, and post-operative digitized histopathology images. To ensure prostate tissue is sectioned in same plane as T2w MRI custom 3D printed molds were used. Majority votes among three pathologists provided Gleason grades was the consensus annotation. Pre-operative MRI and histopathology images were registered [31] to enable accurate mapping of cancer labels. We have the following classes: Class 1: *Grade* – 3, 67 patients, Class 2: *Grades* – 3 + 4/4 + 3, 60 patients, Class 3: *Grade* – 4, 74 patients, Class 4: *Grades* – 4 + 5/5 + 4, 57 patients, Class 5: *Grade* – 5, 63 patients. In the supplementary we show results for the Kaggle DR challenge [7] and PANDA challenge [4].

Pre-processing: Histopathology images were smoothed with a Gaussian filter ($\sigma = 0.25$). They were downsampled to 224×224 with X-Y resolution 0.25×0.25 mm^2. The T2w images, prostate masks, and Gleason labels are projected on the corresponding downsampled histopathology images and resampled to the same X-Y resolution. This ensures corresponding pixels in each modality represents the same physical area. MR images were normalized using histogram alignment [29]. The training, validation, and test sets had 193/64/64 patients.

We compare results of our method MM$_{GZSL}$ (Multimodal GZSL) with different feature generation based GZSL methods: 1) CVAE based feature synthesis method of [11]; 2) GZSL using over complete distributions [12]; 3) self-supervised learning GZSL method of [33]; 4) cycle-GAN GZSL method of [6]; 5) FSL-the fully supervised learning method of [2] using the same data split, actual labels of 'Unseen' class and almost equal representation of all classes.

Table 1. GZSL and Ablation Results: Average classification accuracy (%) and harmonic mean accuracy of generalized zero-shot learning when test samples are from Seen (Setting A) or unseen (Setting B) classes. Mean and variance are reported when the number of classes in the 'Seen' set is 3. p values are with respect to **Harmonic Mean of MM$_{GZSL}$**.

	Acc_S	Acc_U	H	p	Sen_S	Spe_S	Sen_U	Spe_U
Comparison methods								
MM$_{GZSL}$	83.6(2.4)	81.7(3.0)	82.6(2.8)	–	84.1(3.1)	82.9(2.6)	81.2(3.0)	80.1(3.3)
[11]	80.3(3.5)	73.4(3.6)	76.7(2.8)	0.002	81.2(3.6)	79.9(3.5)	74.1(3.2)	72.8(3.4)
[12]	80.6(3.4)	72.8(3.0)	76.5(3.2)	0.001	81.1(2.9)	80.0(3.2)	73.5(3.1)	72.1(3.4)
[33]	81.1(2.9)	73.2(3.2)	76.9(3.1)	0.001	81.8(3.5)	80.7(3.1)	74.0(3.5)	72.9(3.7)
[6]	81.2(3.7)	72.8(3.8)	76.7(3.8)	0.004	81.8(3.1)	80.7(3.4)	73.1(4.0)	71.9(4.2)
FSL	83.9(2.2)	83.3(2.5)	83.5(2.3)	0.01	84.9(2.4)	83.5(2.6)	83.7(2.8)	82.5(2.6)
Ablation studies								
MM$_{wCPC}$	79.1(3.1)	72.3(3.8)	75.5(3.4)	0.001	79.8(3.4)	78.3(3.5)	73(3.6)	82.5(3.1)
MM$_{wReg}$	80.1(3.7)	72.2(3.5)	75.9(3.5)	0.0001	80.5(3.4)	78.8(3.5)	72.8(3.8)	71.3(4.0)
MM$_{wC}$	79.2(3.9)	72.9(4.0)	75.9(3.9)	0.009	80.1(3.8)	78.6(4.3)	73.3(4.1)	72.1(4.2)
MM$_{wE}$	80.2(3.7)	73.0(3.9)	76.4(3.7)	0.005	80.8(3.3)	79.5(3.8)	74.1(3.8)	72.4(3.7)
MM$_{MR}$	75.6(4.2)	71.1(4.5)	73.3(4.3)	0.007	76.3(4.5)	75.0(4.3)	71.9(4.1)	70.4(4.3)
MM$_{DP}$	81.2(3.0)	79.1(3.6)	80.1(3.4)	0.008	82.0(3.4)	80.6(3.5)	79.9(3.9)	78.4(4.2)

3.1   Generalized Zero Shot Learning Results

Table 1 summarizes the results of our algorithm and other methods when the Seen set has samples from 3 classes. The numbers are an average of 5 runs. Samples from 3 labeled classes presents the optimum scenario balancing high classification accuracy, and generation of representative synthetic samples. Setting A does better than setting B. Since GZSL is a very challenging problem, it is expected that classification performance on Unseen classes will not match those of Seen classes. MM$_{GZSL}$'s performance is closest to FSL. Since FSL has been trained with all classes in training and test sets it gives the best results.

We use the McNemar test and determine that the difference in Acc_S of MM$_{GZSL}$ and FSL is not significant ($p = 0.062$) although the corresponding values for Acc_U are significant ($p = 0.031$) which is not surprising since real Unseen examples have not been encountered in GZSL. Since MM$_{GZSL}$'s performance is closest to FSL for Unseen classes, it demonstrates MM$_{GZSL}$'s effectiveness in generating realistic features of unseen classes. We refer the reader to the supplementary material for additional results (e.g. using different number of classes in the Seen dataset). The individual Per Class mean accuracy and variance are: **Setting A:-** Class1 = 84.1(2.4), Class2 = 83.8(2.8), Class3 = 84.1(2.3), Class4 = 82.9(2.8), Class5 = 82.9(3.1). **Setting B-** Class1 = 83.1(2.7),

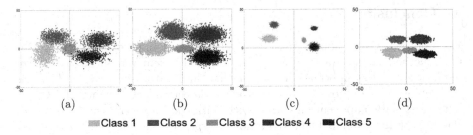

(a)   (b)   (c)   (d)

■Class 1   ■Class 2   ■Class 3   ■Class 4   ■Class 5

Fig. 2. Feature visualizations: (a) Seen+Unseen classes from actual dataset; distribution of synthetic samples generated by b) MM_{GZSL}; (c) MM_{GZSL} without self-supervision; (d) for [33]. Different colours represent different classes. (b) is closer to (a), while (c) and (d) are quite different.

Class2 = 82.9 (2.8), Class3 = 81.1 (2.9), Class4 = 79.0(3.2), Class5 = 78.2(3.9). This shows that our feature generation and classification is not biased to any specific class.

Ablation Studies: Table 1 shows ablation results where each row denotes a specific setting without the particular term in the final objective function in Eq. 3 - e.g., MM_{wCPC} denotes our proposed method MM_{GZSL} without the self-supervised loss $\mathscr{L}_{CPC}$. MM_{wCPC} shows the worst performance indicating $\mathscr{L}_{CPC}$'s correspondingly higher contribution than other terms. The p–values indicate each term's contribution is significant for the overall performance of MM_{GZSL} and excluding any one leads to significant performance degradation.

We also show in Table 1 the result of using: 1) only the synthetic MR features (MM_{MR}); 2) only digital histopathology features (MM_{DP}) obtained by transforming F_{MRI} to get F_{DP}. MM_{DP} gives performance metrics closer to MM_{GZSL}, which indicates that the digital histopathology images provide highly discriminative information compared to MR images. However, MR images also have a notable contribution, as indicated by the p-values. In another set of experiments we redesign the GZSL experiments to generate synthetic histopathology features F_{DP} instead of F_{MRI} and then transforming them to get MR features. We obtain very similar performance ($Acc_S = 83.7, Acc_U = 80.8, H = 82.2$) to the original MM_{GZSL} setting. This demonstrates that cycle GAN based feature transformation network does a good job of learning accurate histopathology features from the corresponding MR features.

Visualization of Synthetic Features: Figure 2(a) shows t-SNE plot of real data features where the classes are spread over a wide area, with overlap amongst consecutive classes. Figures 2(b, c, d) show, respectively, distribution of synthetic features generated by MM_{GZSL}, MM_{wCPC} and [33]. MM_{GZSL} features are the most similar to the original data. MM_{wCPC} and [33] synthesize sub-optimal feature representation of actual features, resulting in poor classification performance on unseen classes.

4   Conclusion

We propose a GZSL approach without relying on class attribute vectors. Our novel method can accurately predict Gleason grades from MR images with lower resolution and less information content than histopathology images. This has the potential to improve accuracy of early detection and staging of PCa. A self-supervised component ensures the semantic gap between Seen and Unseen classes is easily covered. The distribution of synthetic features generated by our method are close to the actual distribution, while removing the self-supervised term results in unrealistic distributions. Results show our method's superior performance and synergy between different loss terms leads to improved GZSL classification. We observe failure cases where the acquired MR images are not of sufficiently good resolution to allow accurate registration of histopathology and MRI. Future work will aim to address this issue.

References

1. Bachman, P., Hjelm, R.D., Buchwalter, W.: Learning representations by maximizing mutual information across views. In: Proceedings of NeurIPS, pp. 15509–15519 (2019)
2. Bhattacharya, I., et al.: CorrSigNet: learning CORRelated prostate cancer SIGnatures from radiology and pathology images for improved computer aided diagnosis. In: Martel, A.L., et al. (eds.) MICCAI 2020. LNCS, vol. 12262, pp. 315–325. Springer, Cham (2020). https://doi.org/10.1007/978-3-030-59713-9_31
3. Bozorgtabar, B., Mahapatra, D., Vray, G., Thiran, J.-P.: SALAD: self-supervised aggregation learning for anomaly detection on X-rays. In: Martel, A.L., et al. (eds.) MICCAI 2020. LNCS, vol. 12261, pp. 468–478. Springer, Cham (2020). https://doi.org/10.1007/978-3-030-59710-8_46
4. Bulten, W., et al.: Automated deep-learning system for Gleason grading of prostate cancer using biopsies: a diagnostic study. Lancet Oncol. 21(2), 233–241 (2020)
5. Campanella, G., Silva, V.M., Fuchs, T.J.: Terabyte-scale deep multiple instance learning for classification and localization in pathology. arXiv preprint arXiv:1805.06983 (2018)
6. Felix, R., Vijay Kumar, B.G., Reid, I., Carneiro, G.: Multi-modal cycle-consistent generalized zero-shot learning. In: Ferrari, V., Hebert, M., Sminchisescu, C., Weiss, Y. (eds.) ECCV 2018. LNCS, vol. 11210, pp. 21–37. Springer, Cham (2018). https://doi.org/10.1007/978-3-030-01231-1_2
7. California Healthcare Foundation: Diabetic retinopathy detection. https://www.kaggle.com/c/diabetic-retinopathy-detection
8. Gutmann, M., Hyvarinen, A.: A new estimation principle for unnormalized statistical models. In: Proceedings of the AISTATS, pp. 297–304 (2010)
9. He, K., Zhang, X., Ren, S., Sun, J.: Deep residual learning for image recognition. In: Proceedings of the CVPR, pp. 770–778 (2016)
10. Hu, Z., Yang, Z., Liang, X., Salakhutdinov, R., Xing, E.P.: Toward controlled generation of text. In: Proceedings of the ICML, pp. 1587–1596 (2017)
11. Huang, H., Wang, C., Yu, P.S., Wang, C.D.: Generative dual adversarial network for generalized zero-shot learning. In: The IEEE Conference on Computer Vision and Pattern Recognition (CVPR), pp. 801–810, June 2019

12. Keshari, R., Singh, R., Vatsa, M.: Generalized zero-shot learning via over-complete distribution. In: The IEEE Conference on Computer Vision and Pattern Recognition (CVPR), pp. 13300–13308, June 2020

13. Kingma, D.P., Ba, J.: Adam: a method for stochastic optimization. arXiv preprint arXiv:1412.6980 (2014)

14. Koohbanani, N.A., Unnikrishnan, B., Khurram, S.A., Krishnaswamy, P., Rajpoot, N.: Self-path: self-supervision for classification of pathology images with limited annotations. arXiv:2008.05571 (2020)

15. Litjens, G., Toth, R., de Ven, W., Hoeks, C., et al.: Evaluation of prostate segmentation algorithms for MRI: the PROMISE12 challenge. Med. Image Anal. **18**(2), 359–373 (2014)

16. Long, Y., Liu, L., Shen, F., Shao, L., Li, X.: Zero-shot learning using synthesised unseen visual data with diffusion regularisation. IEEE Trans. Pattern Anal. Mach. Intell. **40**(10), 2498–2512 (2017)

17. Lu, M.Y., Chen, R.J., Wang, J., Dillon, D., Mahmood, F.: Semi-supervised histology classification using deep multiple instance learning and contrastive predictive coding. arXiv:1910.10825 (2019)

18. Maas, A.L., Hannun, A.Y., Ng, A.Y.: Rectifier nonlinearities improve neural network acoustic models. In: Proceedings of the ICML (2013)

19. Mahapatra, D., Antony, B., Sedai, S., Garnavi, R.: Deformable medical image registration using generative adversarial networks. In: Proceedings of the IEEE ISBI, pp. 1449–1453 (2018)

20. Mahapatra, D., Bozorgtabar, B., Shao, L.: Pathological retinal region segmentation from OCT images using geometric relation based augmentation. In: Proceedings of the IEEE CVPR, pp. 9611–9620 (2020)

21. Mahapatra, D., Bozorgtabar, B., Thiran, J.-P., Shao, L.: Structure preserving stain normalization of histopathology images using self supervised semantic guidance. In: Martel, A.L., et al. (eds.) MICCAI 2020. LNCS, vol. 12265, pp. 309–319. Springer, Cham (2020). https://doi.org/10.1007/978-3-030-59722-1_30

22. Mahapatra, D., Ge, Z.: Training data independent image registration with GANs using transfer learning and segmentation information. In: Proceedings of the IEEE ISBI, pp. 709–713 (2019)

23. Mahapatra, D., Ge, Z.: Training data independent image registration using generative adversarial networks and domain adaptation. Pattern Recogn. **100**, 1–14 (2020)

24. Mahapatra, D., Ge, Z., Sedai, S., Chakravorty, R.: Joint registration and segmentation of Xray images using generative adversarial networks. In: Shi, Y., Suk, H.-I., Liu, M. (eds.) MLMI 2018. LNCS, vol. 11046, pp. 73–80. Springer, Cham (2018). https://doi.org/10.1007/978-3-030-00919-9_9

25. Mahapatra, D., Poellinger, A., Shao, L., Reyes, M.: Interpretability-driven sample selection using self supervised learning for disease classification and segmentation. IEEE TMI 1–15 (2021)

26. Min, S., Yao, H., Xie, H., Wang, C., Zha, Z.J., Zhang, Y.: Domain-aware visual bias eliminating for generalized zero-shot learning. In: The IEEE Conference on Computer Vision and Pattern Recognition (CVPR), pp. 12664–12673, June 2020

27. Mnih, A., Kavukcuoglu, K.: Learning word embeddings efficiently with noise-contrastive estimation. In: Proceedings of the NeurIPS, pp. 2265–2273 (2013)

28. Nair, V., Hinton, G.E.: Rectified linear units improve restricted Boltzmann machines. In: Proceedings of the ICML, pp. 807–814 (2010)

29. Nyul, L., Udupa, J., Zhang, X.: New variants of a method of MRI scale standardization. IEEE Trans. Med. Imaging **19**(2), 143–150 (2000)

30. van den Oord, A., Li, Y., Vinyals, O.: Representation learning with contrastive predictive coding. arXiv:1807.03748 (2018)
31. Rusu, M., Shao, W., Kunder, C.A., Wang, J.B., Soerensen, S.J.C., et al.: Registration of pre-surgical MRI and histopathology images from radical prostatectomy via RAPSODI. Med. Phys. **47**(9), 4177–4188 (2020)
32. Tong, J., Mahapatra, D., Bonnington, P., Drummond, T., Ge, Z.: Registration of histopathology images using self supervised fine grained feature maps. In: Albarqouni, S., et al. (eds.) DART/DCL-2020. LNCS, vol. 12444, pp. 41–51. Springer, Cham (2020). https://doi.org/10.1007/978-3-030-60548-3_5
33. Wu, J., Zhang, T., Zha, Z.J., Luo, J., Zhang, Y., Wu, F.: Self-supervised domain-aware generative network for generalized zero-shot learning. In: The IEEE Conference on Computer Vision and Pattern Recognition (CVPR), pp. 12767–12776, June 2020
34. Xian, Y., Lorenz, T., Schiele, B., Akata, Z.: Feature generating networks for zero-shot learning. In: Proceedings of the IEEE CVPR, pp. 5542–5551 (2018)
35. Zhu, J.Y., Park, T., Isola, P., Efros, A.A.: Unpaired image-to-image translation using cycle-consistent adversarial networks. arXiv preprint arXiv:1703.10593 (2017)

Self-supervised Learning of Inter-label Geometric Relationships for Gleason Grade Segmentation

Dwarikanath Mahapatra[1](✉), Shiba Kuanar[2], Behzad Bozorgtabar[3], and Zongyuan Ge[4,5]

[1] Inception Institute of Artificial Intelligence, Abu Dhabi, UAE
dwarikanath.mahapatra@inceptioniai.org
[2] Yale University, New Haven, USA
[3] Signal Processing Laboratory 5, EPFL, Lausanne, Switzerland
[4] Monash University, Melbourne, Australia
[5] Airdoc Research, Melbourne, Australia

Abstract. Segmentation of Prostate Cancer (PCa) tissues from Gleason graded histopathology images is vital for accurate diagnosis. Although deep learning (DL) based segmentation methods achieve state-of-the-art accuracy, they rely on large datasets with manual annotations. We propose a method to synthesize PCa histopathology images by learning the geometrical relationship between different disease labels using self-supervised learning. Manual segmentation maps from the training set are used to train a Shape Restoration Network (ShaRe-Net) that predicts missing mask segments in a self-supervised manner. Using Dense-UNet as the backbone generator architecture we incorporate latent variable sampling to inject diversity in the image generation process and thus improve robustness. Experimental results demonstrate the superiority of our method over competing image synthesis methods for segmentation tasks. Ablation studies show the benefits of integrating geometry and diversity in generating high-quality images. Our self-supervised approach with limited class-labeled data achieves better performance than fully supervised learning.

Keywords: Self-supervised learning · Geometric modeling · GANs

1 Introduction

Gleason grading for Prostate Cancer (PCa) staging is achieved by analysis of stained biopsy tissue images Since it is subjective due to high level of heterogeneity [29], computer-aided methods can improve consistency, speed, accuracy, and reproducibility of diagnosis. Deep learning (DL) segmentation methods require large image datasets for training. Owing to scarcity of such datasets,

Electronic supplementary material The online version of this chapter (https://doi.org/10.1007/978-3-030-87722-4_6) contains supplementary material, which is available to authorized users.

© Springer Nature Switzerland AG 2021
S. Albarqouni et al. (Eds.): DART 2021/FAIR 2021, LNCS 12968, pp. 57–67, 2021.
https://doi.org/10.1007/978-3-030-87722-4_6

most approaches apply image augmentation to increase dataset size. Traditional augmentations such as image rotations or deformations do not fully represent the underlying data distribution of the training set, and are sensitive to parameter choices. Recent data augmentation methods of [4,9,28] have used generative adversarial networks (GANs) [7], and show success for medical image classification and segmentation. However, *they have limited relevance for segmentation* since they do not model geometric relation between different organs. Hence there is a need for augmentation methods that consider the geometric relation between different anatomical regions (labels) and generate realistic images for diseased and healthy cases.

Existing work on prostate biopsies using conventional data augmentation include semantic segmentation [10], feature-engineering [16], and multiple instance learning for binary classification of clinical specimens [5]. GANs have also been used for few shot learning [1], registration [19,22,24] and medical image generation [8,20]. However, there is no explicit attempt to model attributes such as shape variations that are important to capture different conditions across a population. [26] augmented medical images with simulated anatomical changes but demonstrate inconsistent performance based on transformation functions and parameters.

Self-supervised learning (SSL) also addresses labeled data shortage and has found wide use in medical image analysis by using innovative pre-text tasks such as patients' MR scan recognition to detect vertebra [12], and context restoration for classification, segmentation, and disease localization [6]. SSL has been applied to histopathology images using domain specific pretext tasks [15], semi-supervised histology classification [18], registration [30], stain normalization [21], anomaly detection [3], cancer subtyping using visual dictionaries [27] and active learning based segmentation [25].

Our Contribution: In this paper we propose a self-supervised learning based method to learn inter label geometric relationships by considering the intrinsic relationships between shape and geometry of anatomical structures. We present a **G**eometry-**A**ware **S**hape **GAN** (**GASGAN**) that learns to generate plausible images of different Gleason graded regions while preserving learned relationships between geometry and shape. Generated images are used for segmentation of Gleason graded regions.

2   Method

Our objective is to synthesize informative histopathology patches from a base image patch (and mask) and train a segmentation model. The first stage of synthetic image generation is a spatial transformation network (STN) that transforms the base mask with different location, scale, and orientation attributes. This initially transformed mask is input to the Dense-UNet based 'Generator' that injects diversity, and the discriminator ensures that the final generated mask is representative of the dataset. In addition to the adversarial loss (L_{Adv}) we incorporate losses from an auxiliary classifier (L_{Class} to ensure the image has the desired label) and geometric relationship module (L_{Shape} from ShaRe-Net).

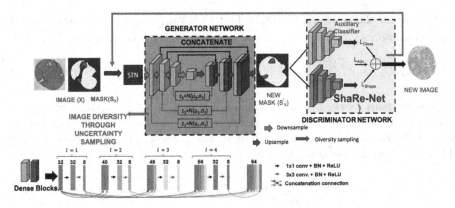

Fig. 1. Overview of training stage. Segmentation masks S_X of image X input to a STN whose output is fed to the generator network. The DenseUNet based Generator network injects diversity at different levels through uncertainty sampling to generate a new mask S'_X. S'_X is fed to the discriminator which evaluates its accuracy based on L_{class}, L_{shape} and L_{adv}. The provided feedback is used for weight updates to obtain the final model. Feature-maps from previous layers are concatenated together as the input to the following layers.

2.1 Geometry Aware Shape Generation

Let us denote an input image as x, the corresponding segmentation masks as s_x and its label (Gleason score) as l_x. The first stage is a spatial transformer network (STN) [11] that transforms the base mask with different attributes of location, scale and orientation. The transformations used to obtain new segmentation mask s'_x are applied to x to get corresponding transformed image x'. Since the primary aim of our approach is to learn contours and other shape specific information of anatomical regions, a modified DenseUNet architecture as the generator network effectively captures hierarchical shape information and easily introduces diversity at multiple levels of image generation.

The generator $\mathbf{G}_g$ takes input $\mathbf{s}_x$ and a desired label vector of output mask c_g to output an affine transformation matrix $\mathbf{A}$ via a STN, i.e., $\mathbf{G}_g(\mathbf{s}_x, c_g) = \mathbf{A}$. $\mathbf{A}$ is used to generate s'_x and x'. The discriminator $\mathbf{D}_{class}$ determines whether output image has the desired label c_g or not. The discriminator $\mathbf{D}_g$ ensures the generated masks are realistic. Let the minimax criteria between $\mathbf{G}_g$ and $\mathbf{D}_g$ be $\min_{\mathbf{G}_g} \max_{\mathbf{D}_g} \mathbf{L}_g(\mathbf{G}_g, \mathbf{D}_g)$. The loss function $\mathbf{L}_g$ is,

$$L_g = L_{adv} + \lambda_1 L_{class} + \lambda_2 L_{shape} \tag{1}$$

where 1) $\mathbf{L}_{adv}$ is an adversarial loss to ensure $\mathbf{G}_g$ outputs realistic deformations; 2) $\mathbf{L}_{class}$ ensures generated image has same label as x and preserves it's Gleason grade; and 3) $\mathbf{L}_{shape}$ ensures new masks have realistic shapes. $\lambda_1 = 0.95, \lambda_2 = 1$ balance each term's contribution.

Adversarial Loss- $L_{adv}(G_g, D_g)$: The STN outputs $\widetilde{A}$, a prediction for $\mathbf{A}$ conditioned on $\mathbf{s}_x$, and a new semantic map $\mathbf{s}_x \oplus \widetilde{A}(\mathbf{s}_x)$ is generated. $\mathbf{L}_{adv}$ is,

$$L_{adv}(G_g, D_g) = \mathbb{E}_x \left[\log D_g(\mathbf{s}_x \oplus \widetilde{A}(\mathbf{s}_x)) \right] + \mathbb{E}_{\mathbf{s}_x} \left[\log(1 - D_g(\mathbf{s}_x \oplus \widetilde{A}(\mathbf{s}_x))) \right], \quad (2)$$

Classification Loss- L_{class}: The affine transformation $\widetilde{A}$ is applied to the base image $\mathbf{x}$ to obtain the generated image $\mathbf{x}'$. We add an auxiliary classifier (to ensure the output image has the desired label) when optimizing both $\mathbf{G}_g$ and $\mathbf{D}_g$ and define the classification loss as,

$$L_{class} = \mathbb{E}_{\mathbf{x}', c_g}[- \log D_{class}(c_g | x')], \quad (3)$$

where the term $D_{class}(c_g | x')$ represents a probability distribution over classification labels computed by D.

Self Supervised Modeling of Inter-label Geometric Relationships: We propose a novel approach to model the relationship between different anatomical labels. Given a dataset of masks $S = \{s_1, s_2, \cdots, s_N\}$ consisting of N label masks, a new dataset $\hat{S} = f(S)$ is generated, where f is a function altering the original label maps by arbitrarily masking labeled regions or swapping patches. A CNN is trained to predict the intensity values of the altered pixels which correspond to the mask's Gleason labels. The task here is to reconstruct the masks when part of it has been altered. By learning to predict missing labels the network implicitly learns the geometrical relationship between different Gleason grades.

We train an Encoder-Decoder style network similar to UNet where the input is the altered mask and the output is the original mask, and use a L_2 loss term. We call this network as the **Shape Restoration Network (ShaRe-Net)**. To compute L_{shape} we obtain the feature maps of the generated mask $(SN(G_g(\mathbf{s}_x))$ and original mask $SN(\mathbf{s}_x)$ using all layers of ShaRe-Net, and calculate the mean square error values between them.

$$L_{shape} = \frac{1}{N} \sum_i^N (SN(\mathbf{s}_x) - SN(G_g(\mathbf{s}_x)))^2 \quad (4)$$

$SN(\cdot)$ denotes the input processed through ShaRe-Net.

2.2 Sample Diversity from Uncertainty Sampling

To inject diversity in the shape generation step we adapt the method of [2] by replacing the UNet generator with a Dense UNet architecture [17]. It enables reuse of information from previous layers and introduces complementary information across layers. This enables the generation step to factor in the dependence across different levels of shape abstraction.

The generated mask s'_x is obtained by fusing L levels of the generator G_g (as shown in Fig. 1), each of which is associated with a latent variable z_l.

(a)   (b)   (c)   (d)   (e)   (f)

Fig. 2. (a) Base image with Gleason graded regions shown as blue outline; Example of generated images using: (b) Our proposed *GASGAN* method; (c) [31]; (d) [8]; (e) *DAGAN* method by [1]; (f) *cGAN* method by [23]. (Color figure online)

We use probabilistic uncertainty sampling to model conditional distribution of segmentation masks and use separate latent variables at multi-resolutions to factor inherent uncertainties. The hierarchical approach introduces diversity at different stages and influences different features (e.g., low level features at the early layers and abstract features in the later layers). Denoting the generated mask as s for simplicity, we obtain conditional distribution $p(s|x)$ for L levels:

$$p(s|x) = \int p(s|z_1, \cdots , z_L)p(z_1|z_2, x) \cdots p(z_{L-1}|z_L, x)p(z_L|x)dz_1 \cdots dz_L. \quad (5)$$

z_l's dimensionality is $r_x 2^{-l+1} \times r_y 2^{-l+1}$, where r_x, r_y are image dimensions. Latent variable z_l models diversity at resolution 2^{-l+1} of the original image (e.g. z_1 and z_3 denote the original and 1/4 image resolution). A variational approximation $q(z|s, x)$ approximates the posterior distribution $p(z|s, x)$ where $z = \{z_1, ..., z_L\}$. $\log p(s|x) = L(s|x) + KL(q(z|s, x)||p(z|s, x))$, where L is the evidence lower bound, and $KL(., .)$ is the Kullback-Leibler divergence. The prior and posterior distributions are parameterized as normal distributions $\mathcal{N}(z|\mu, \sigma)$.

3   Experimental Results

3.1   Dataset Description

We use the public Gleason grading challenge dataset[1] [13]. A total of 333 Tissue Microarrays (TMAs) from 231 patients who had undergone radical prostatectomy The TMAs had been stained in $H\&E$ and scanned at 40x magnification with a SCN400 Slide Scanner (Leica Microsystems, Wetzlar, Germany). Six pathologists having $27, 15, 1, 24, 17$, and 5 years of experience were asked to annotate the TMA images in detail. Four of them annotated all 333 cores. The other two annotated 191 and 92 of the cores. Pixel-wise majority voting was used to construct the "ground truth label". The training set had 200 TMAs while the validation set had 44 TMAs. A separate test set consisting of 87 TMAs from 60 other patients was used. Patients were randomly assigned to training and test set. All images and masks were resized to 512×512 to facilitate training.

---

[1] https://gleason2019.grand-challenge.org/Home.

Figure 2 shows example cases of synthetic images generated by our method and other competing techniques. A sample image generated by our approach is more realistic in appearance compared to the base image, while other images have artifacts, noise, or distorted regions.

3.2 Experimental Setup, Baselines and Metrics

Our method has the following steps: 1) Use the default training, validation, and test folds of the dataset. 2) Use training images to train the image generator. 3) Generate shapes from the training set and train UNet++ segmentation network [32] on the generated images. 4) Use trained UNet++ to segment test images. 5) Repeat the above steps for different data augmentation methods. Our model is implemented in PyTorch, on a NVIDIA V100 GPU having 32 GB RAM. We trained all models using Adam optimiser [14] with a learning rate of 10^{-3}, batch-size of 16 and using batch-normalisation. The values of parameters λ_1 and λ_2 in Eq. 1 were set by a detailed grid search on a separate dataset of 45 TMAs that was not used for training or testing. They were varied between $[0, 1]$ in steps of 0.05 by fixing λ_1 and varying λ_2 for the whole range. This was repeated for all values of λ_1. The best segmentation accuracy was obtained for $\lambda_1 = 0.97$ and $\lambda_2 = 0.93$, which were our final parameter values.

We denote our method as GASGAN (**G**eometry **A**ware **S**hape GANs) and compare it's performance against other methods such as: 1) *DA-* conventional data augmentation consisting of rotation, translation and scaling; 2) *DAGAN* - data augmentation GANs of [1]; 3) *cGAN* - the conditional GAN based method of [23]; 4) *HGAN* - pathology image augmentation method of [31]; 5) the image enrichment method of [8]. Segmentation performance on the training set is evaluated in terms of Dice Metric (DM), Hausdorff Distance (HD) and Mean Absolute error (MAE).

Ablation Experiments: The following variants of our method were used: 1) GASGAN$_{noL_{class}}$- GASGAN without classification loss (Eq. 3); 2) GASGAN$_{noL_{shape}}$ - GASGAN without shape relationship modeling term (Eq. 4); 3) GASGAN$_{NoSamp}$ - GASGAN without uncertainty sampling for injecting diversity.

3.3 Segmentation Results on Gleason Training Data

Table 1 shows the average DM, HD, and MAE for different augmentation methods on the Gleason challenge training dataset, and their p values with respect to GASGAN. Figure 3 shows the segmentation results using a PSPNet trained on images from different image synthesis methods. Figure 3(a) shows the test image and Fig. 3(b) shows the manual mask. Figures 3(c)–(f) show, respectively, the segmentation masks obtained by GASGAN, [8,31], *DAGAN* and *cGAN*. GASGAN's DM is highest and also has the lowest HD and MAE values amongst all other methods. Our results clearly show that modeling geometrical features leads to better performance than state of the art segmentation networks.

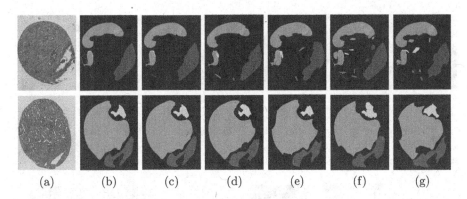

(a)   (b)   (c)   (d)   (e)   (f)   (g)

Fig. 3. Segmentation results on the Gleason training dataset: (a) original images; (b) manual segmentation masks. Segmented masks using data generated by: (c) GASGAN; (d) [31]; (e) [8]; (f) *DAGAN*; (g) *cGAN*.

Table 1. Segmentation results on the training dataset for the Gleason challenge using PSPNet. Mean and standard deviation (in brackets) are shown. The best results per metric are shown in bold. p values are with respect to GASGAN.

	DA	DAGAN	cGAN	[8]	[31]	GASGAN
DM	0.862(0.10)	0.907(0.13)	0.909(0.11)	0.917(0.05)	0.924(0.08)	**0.963(0.05)**
p	0.0003	0.005	0.003	0.0007	0.0002	–
MAE	0.081(0.016)	0.068(0.016)	0.073(0.012)	0.063(0.017)	0.048(0.012)	**0.021(0.009)**
HD	12.1(4.3)	10.7(3.7)	10.9(3.2)	10.1(3.4)	9.0(3.0)	**7.8(2.3)**

We show results for PSPNet since the top ranking method[2] uses this architecture and gives the following values $DM = 0.911(0.05), HD = 8.6(2.9), MAE = 0.049(0.009)$. We outperform this method by a significant margin, thus clearly demonstrating the effectiveness of geometrical modeling in generating informative images. GASGAN's superior segmentation accuracy is attributed to its capacity to learn the geometrical relationship between different labels (through L_{shape}). Thus our attempt to model the intrinsic geometrical relationships between different labels could generate superior quality masks. In supplementary material we show results using UNet++ [32] architecture. The PSPNet and UNet++ results demonstrate that our image augmentation method does equally well for different segmentation architectures.

Ablation Studies: Numbers for ablation experiments are: $GASGAN_{noL_{cls}}$: DM = 0.891(0.07), HD = 9.6(3.1), MAE = 0.071(0.015); $GASGAN_{noL_{shape}}$: 0.899 (0.08), 9.1(3.2), 0.065(0.016); $GASGAN_{noSamp}$:- 0.895(0.06), 9.4(3.4), 0.062(0.016). The segmentation outputs in Fig. 4 are quite different from the ground truth and GASGAN (Fig. 3). In some cases, the normal regions are included as pathological areas, while parts of the diseased regions are not

---

[2] https://github.com/hubutui/Gleason.

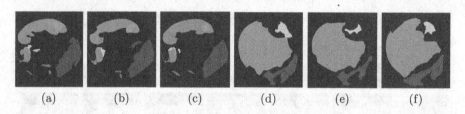

(a)   (b)   (c)   (d)   (e)   (f)

Fig. 4. Segmentation results for ablation experiments on the training set: (a, d) $GASGAN_{noL_{shape}}$; (b, e) $GASGAN_{noL_{cls}}$; (c, f) $GASGAN_{noSamp}$. (a)–(c) results correspond to top row of Fig. 3 and (d)–(f) correspond to its bottom row.

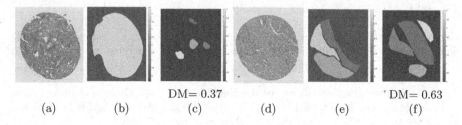

DM= 0.37   DM= 0.63
(a)   (b)   (c)   (d)   (e)   (f)

Fig. 5. Examples of failure cases: (a, d) original image; (b, e) manual segmentation map; (c, f) segmentation map obtained using GASGAN.

segmented. Either case is undesirable for disease diagnosis and quantification. Thus, different components of our cost functions are integral to the method's performance. Excluding one or more of classification loss, geometric loss, and sampling loss adversely affects segmentation performance.

Failure Cases: Figure 5 shows examples of some failure cases where our proposed algorithm's segmentation had very low dice score on the Gleason dataset. The underlying reason is due to various factors such as artifacts or similar-looking structures nearby. In future work on improving the algorithm, we aim to incorporate ways to overcome the challenges posed by dissimilar tissues being too close together.

4   Conclusion

We have proposed a novel approach to generate high-quality synthetic histopathology images for segmenting Gleason graded PCa and other pathological conditions. Our method exploits the inherent geometric relationship between different segmentation labels and uses it to guide the shape generation process. We propose a self-supervised learning approach where a shape restoration network (ShaRe-Net) learns to predict the original segmentation mask from its distorted version. The pre-trained ShaRe-Net is used to extract feature maps and integrate them in the training stage. Comparative results with other synthetic image generation methods show that the augmented dataset from our proposed

GASGAN method outperforms standard data augmentation and other competing methods when applied to the segmentation of PCa (from the Gleason grading challenge). The synergy between shape, classification, and sampling terms leads to improved and robust segmentation and each term is important in generating realistic images.

References

1. Antoniou, A., Storkey, A., Edwards, H.: Data augmentation generative adversarial networks. arXiv preprint arXiv:1711.04340 (2017)
2. Baumgartner, C.F., et al.: PHiSeg: capturing uncertainty in medical image segmentation. In: Shen, D., et al. (eds.) MICCAI 2019. LNCS, vol. 11765, pp. 119–127. Springer, Cham (2019). https://doi.org/10.1007/978-3-030-32245-8_14
3. Bozorgtabar, B., Mahapatra, D., Vray, G., Thiran, J.-P.: SALAD: self-supervised aggregation learning for anomaly detection on X-rays. In: Martel, A.L., et al. (eds.) MICCAI 2020. LNCS, vol. 12261, pp. 468–478. Springer, Cham (2020). https://doi.org/10.1007/978-3-030-59710-8_46
4. Bozorgtabar, B., et al.: Informative sample generation using class aware generative adversarial networks for classification of chest Xrays. Comput. Vis. Image Underst. **184**, 57–65 (2019)
5. Campanella, G., Silva, V., Fuchs, T.: Terabyte-scale deep multiple instance learning for classification and localization in pathology. arXiv preprint arXiv:1805.06983 (2018)
6. Chen, L., Bentley, P., Mori, K., Misawa, K., Fujiwara, M., Rueckert, D.: Self-supervised learning for medical image analysis using image context restoration. Medical Imag. Anal. **58**, 1–12 (2019)
7. Goodfellow, I., et al.: Generative adversarial nets. In: Advances in Neural Information Processing Systems, pp. 2672–2680 (2014)
8. Gupta, L., Klinkhammer, B.M., Boor, P., Merhof, D., Gadermayr, M.: GAN-based image enrichment in digital pathology boosts segmentation accuracy. In: Shen, D., et al. (eds.) MICCAI 2019. LNCS, vol. 11764, pp. 631–639. Springer, Cham (2019). https://doi.org/10.1007/978-3-030-32239-7_70
9. Han, C., et al.: GAN-based synthetic brain MR image generation. In: 2018 IEEE 15th International Symposium on Biomedical Imaging (ISBI 2018), pp. 734–738. IEEE (2018)
10. Ing, N., et al.: Semantic segmentation for prostate cancer grading by convolutional neural networks. In: Medical Imaging 2018: Digital Pathology, pp. 343–355 (2018)
11. Jaderberg, M., Simonyan, K., Zisserman, A., Kavukcuoglu, K.: Spatial transformer networks. In: NIPS (2015)
12. Jamaludin, A., Kadir, T., Zisserman, A.: Self-supervised learning for spinal MRIs. In: Cardoso, M.J., et al. (eds.) DLMIA/ML-CDS 2017. LNCS, vol. 10553, pp. 294–302. Springer, Cham (2017). https://doi.org/10.1007/978-3-319-67558-9_34
13. Karimi, D., Nir, G., Fazli, L., Black, P., Goldenberg, L., Salcudean, S.: Deep learning-based Gleason grading of prostate cancer from histopathology images-role of multiscale decision aggregation and data augmentation. IEEE J. Biomed. Health Inform. **24**(5), 1413–1426 (2020)
14. Kingma, D.P., Ba, J.: Adam: a method for stochastic optimization. arXiv preprint arXiv:1412.6980 (2014)

15. Koohbanani, N.A., Unnikrishnan, B., Khurram, S.A., Krishnaswamy, P., Rajpoot, N.: Self-path: self-supervision for classification of pathology images with limited annotations. arXiv:2008.05571 (2020)
16. Leo, P., et al.: Stable and discriminating features are predictive of cancer presence and Gleason grade in radical prostatectomy specimens: a multi-site study. Sci. Rep. **8**, 1–13 (2018)
17. Li, X., Chen, H., Qi, X., Dou, Q., Fu, C.W., Heng, P.A.: H-DenseUNet: hybrid densely connected UNet for liver and tumor segmentation from CT volumes. IEEE Trans. Med. Imag. **37**(12), 2663–2674 (2018)
18. Lu, M.Y., Chen, R.J., Wang, J., Dillon, D., Mahmood, F.: Semi-supervised histology classification using deep multiple instance learning and contrastive predictive coding. arXiv:1910.10825 (2019)
19. Mahapatra, D., Antony, B., Sedai, S., Garnavi, R.: Deformable medical image registration using generative adversarial networks. In: Proceedings of the IEEE ISBI, pp. 1449–1453 (2018)
20. Mahapatra, D., Bozorgtabar, B., Shao, L.: Pathological retinal region segmentation from oct images using geometric relation based augmentation. In: Proceedings of the IEEE CVPR, pp. 9611–9620 (2020)
21. Mahapatra, D., Bozorgtabar, B., Thiran, J.-P., Shao, L.: Structure preserving stain normalization of histopathology images using self supervised semantic guidance. In: Martel, A.L., et al. (eds.) MICCAI 2020. LNCS, vol. 12265, pp. 309–319. Springer, Cham (2020). https://doi.org/10.1007/978-3-030-59722-1_30
22. Mahapatra, D., Ge, Z.: Training data independent image registration with GANs using transfer learning and segmentation information. In: Proceedings of the IEEE ISBI, pp. 709–713 (2019)
23. Mahapatra, D., Bozorgtabar, B., Thiran, J.-P., Reyes, M.: Efficient active learning for image classification and segmentation using a sample selection and conditional generative adversarial network. In: Frangi, A.F., Schnabel, J.A., Davatzikos, C., Alberola-López, C., Fichtinger, G. (eds.) MICCAI 2018. LNCS, vol. 11071, pp. 580–588. Springer, Cham (2018). https://doi.org/10.1007/978-3-030-00934-2_65
24. Mahapatra, D., Ge, Z.: Training data independent image registration using generative adversarial networks and domain adaptation. Pattern Recogn. **100**, 1–14 (2020, in press)
25. Mahapatra, D., Poellinger, A., Shao, L., Reyes, M.: Interpretability-driven sample selection using self supervised learning for disease classification and segmentation. IEEE TMI, pp. 1–15 (2021)
26. Milletari, F., Navab, N., Ahmadi, S.A.: V-Net: fully convolutional neural networks for volumetric medical image segmentation. In: Proceedings of the International Conference on 3D vision, pp. 565–571 (2016)
27. Muhammad, H., et al.: Towards unsupervised cancer subtyping: predicting prognosis using a histologic visual dictionary. arXiv:1903.05257 (2019)
28. Nielsen, C., Okoniewski, M.: GAN data augmentation through active learning inspired sample acquisition. In: Proceedings of the IEEE Conference on Computer Vision and Pattern Recognition Workshops, pp. 109–112 (2019)
29. Persson, J., et al.: Interobserver variability in the pathological assessment of radical prostatectomy specimens: findings of the laparoscopic prostatectomy robot open (LAPPRO) study. Scand. J. Urol. **48**(2), 160–167 (2014)
30. Tong, J., Mahapatra, D., Bonnington, P., Drummond, T., Ge, Z.: Registration of histopathology images using self supervised fine grained feature maps. In: Albarqouni, S., et al. (eds.) DART/DCL 2020. LNCS, vol. 12444, pp. 41–51. Springer, Cham (2020). https://doi.org/10.1007/978-3-030-60548-3_5

31. Xue, Y., et al.: Synthetic augmentation and feature-based filtering for improved cervical histopathology image classification. In: Shen, D., et al. (eds.) MICCAI 2019. LNCS, vol. 11764, pp. 387–396. Springer, Cham (2019). https://doi.org/10.1007/978-3-030-32239-7_43
32. Zhou, Z., Siddiquee, M.M.R., Tajbakhsh, N., Liang, J.: UNet++: redesigning skip connections to exploit multiscale features in image segmentation. IEEE Trans. Med. Imag. **39**, 1–10 (2019)

Stop Throwing Away Discriminators! Re-using Adversaries for Test-Time Training

Gabriele Valvano[1,2(✉)], Andrea Leo[1], and Sotirios A. Tsaftaris[2]

[1] IMT School for Advanced Studies Lucca, 55100 Lucca, LU, Italy
gabriele.valvano@imtlucca.it
[2] School of Engineering, University of Edinburgh, Edinburgh EH9 3FB, UK

Abstract. Thanks to their ability to learn data distributions without requiring paired data, Generative Adversarial Networks (GANs) have become an integral part of many computer vision methods, including those developed for medical image segmentation. These methods jointly train a segmentor and an adversarial mask discriminator, which provides a data-driven shape prior. At inference, the discriminator is discarded, and only the segmentor is used to predict label maps on test images. But should we discard the discriminator? Here, we argue that the life cycle of adversarial discriminators should not end after training. On the contrary, training stable GANs produces powerful shape priors that we can use to *correct* segmentor mistakes at inference. To achieve this, we develop stable mask discriminators that do not overfit or catastrophically forget. At test time, we fine-tune the segmentor on each individual test instance until it satisfies the learned shape prior. Our method is simple to implement and increases model performance. Moreover, it opens new directions for re-using mask discriminators at inference. We release the code used for the experiments at https://vios-s.github.io/adversarial-test-time-training.

Keywords: GAN · Segmentation · Test-time training · Shape prior

1 Introduction

Semi- and weakly-supervised learning are emerging paradigms for image segmentation [3,34], often involving adversarial training [9] when annotations are sparse or missing. Adversarial training involves two simultaneously trained networks: one focusing on an image generation task, and the other learning to tell apart generated images from real ones. In semantic segmentation, it is standard practice to condition the generator, also termed segmentor, on an input image and optimise it to output realistic and accurate segmentation masks. After training, the discriminator is discarded and the segmentor used for inference.

Electronic supplementary material The online version of this chapter (https://doi.org/10.1007/978-3-030-87722-4_7) contains supplementary material, which is available to authorized users.

Unfortunately, segmentors may under-perform and make errors whenever the test data fall outside the training data distribution (e.g., because acquired with a different scanner or belonging to a different population study). Here we propose a simple mechanism to detect and correct such errors in an end-to-end fashion, re-using components already developed during training.

We embrace an emerging paradigm [11,13,33,36] where a model is fine-tuned on individual test instances without requiring access to other data nor labels. We propose strategies that permit *recycling* an adversarial mask discriminator during inference, thus introducing a data-driven shape prior to correct predictions. Motivated by recent findings of Asano et al. [1], reporting that we can effectively train the early layers' weights of a CNN with just one image, we propose to tune them on a per-testing instance to minimise an adversarial loss. Lastly, contrary to standard post-processing operations, our method can potentially learn from a continuous stream of data [33]. Our **contributions** are: **1)** to the best of our knowledge, this is the first attempt to use adversarial mask discriminators to *detect* and *correct* segmentation mistakes during inference; **2)** we define specific assumptions (and show how to satisfy them) to make the discriminators useful after training; **3)** we report performance increase on several medical datasets.

2   Related Work

Learning from Test Samples. In our work, we use a discriminator to tune a segmentor on the individual *test* images until it predicts realistic masks. The idea of fine-tuning a model on the test samples has recently been introduced by Sun et al. [33] with the name of Test-time Training (TTT). TTT optimises a model by jointly minimising a supervised and an auxiliary self-supervised loss on a training set, such as detecting the rotation angle of an input image. At inference, TTT fine-tunes the model to minimise the auxiliary loss on the individual test instances, thus adapting to potential distribution shifts. Although the model was successful for classification, the authors admit that designing a well-suited auxiliary task is non-trivial. For example, predicting a rotation angle may be less effective for medical image segmentation, where images have different acquisition geometries. Moreover, Sun et al. only test their model "simulating" domain shifts with hand-crafted image corruptions (e.g., noise and blurring) without investigating if TTT can improve segmentation performance.

Following this seminal work, Wang et al. [36] suggested tuning an adaptor network to minimise the test prediction entropy. Unfortunately, CNNs usually make low-entropy overly-confident predictions [10], and entropy minimisation could be sub-optimal for segmentation. More crucially, Wang et al. rely on having access to the *entire* test-set to do the fine-tuning.

Karani et al. [13] recently proposed Test-time Adaptable Neural Networks to extend TTT for image segmentation using a pre-trained mask denoising autoencoder (DAE). At inference, they compute a reconstruction error between the mask generated by a segmentor and its auto-encoded version predicted by the DAE. This error constitutes a test-time loss used to fine-tune a small adaptor

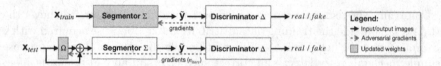

Fig. 1. We re-use GAN discriminators to correct segmentation predictions at inference. The key to our success is training stable and re-usable discriminators, as we detail in Sect. 3.1. At inference, we tune a small convolutional block Ω on each test image $\mathbf{x}$, independently, until the predicted mask $\tilde{\mathbf{y}}$ satisfies the adversarial shape prior. We only need a single test sample to do the fine-tuning.

CNN in front of the segmentor. Once tuned, the adaptor maps the individual test images onto a normalised space which overcomes domain shifts problems for the segmentor. A limitation of this approach is the need to train the mask DAE separately. On the contrary, GANs learn the shape prior and optimise the segmentor in an *end-to-end* fashion. Moreover, tuning the model with a convolutional encoder (the discriminator) rather than an autoencoder has advantages in terms of occupied memory and is faster at inference. Herein, we show that improving performance using a discriminator is also possible and, at the same time, we open a new research direction toward learning re-usable discriminators.

Shape Priors in Deep Learning for Medical Image Segmentation. Incorporating prior knowledge about organ shapes is not uncommon in medical imaging [25]. Several methods introduced shape priors to regularise the training of a segmentor using penalties [5,14], autoencoders [6,26], atlases [7], and adversarial learning [35,37]. Others included shape priors for post-processing, fixing prediction mistakes [18,27]. GANs have become a popular way of introducing shape priors for image segmentation [37], with the advantage of: i) learning the prior directly from data; ii) having a simple model that works well for semi- and weakly-supervised learning; and iii) learning the prior while also training the segmentor, instead of in two separate steps (as happens for autoencoders).

Re-using Adversarial Discriminators. Re-using pre-trained discriminators was proposed to obtain features extractors for transfer learning [8,19,28], or anomaly detectors [24,38]. To the best of our knowledge, their (re-)use to detect segmentor mistakes during inference remains unexplored. We are also not aware of previous use of discriminators for test-time tuning of a segmentor.

3   Method

As summarised in Fig. 1, we consider two stages: i) standard adversarial training; and ii) at inference, image-specific tuning of a small adaptor CNN (Ω) in front of the trained segmentor. In the first stage, we optimise a segmentor Σ to minimise a supervised cost on the annotated data and an adversarial cost on a set of unpaired images. Meanwhile, we train the discriminator Δ to distinguish real from predicted masks. At inference, for each test-image, we tune the adaptor Ω

using the (unsupervised) adversarial loss, and improve performance. For Σ and Ω we use architectures that proved to be effective in segmentation tasks [13,29], while we leave exploring alternative architectures as future work.

Obtaining discriminators re-usable at inference is not trivial and requires specific solutions to overcome crucial challenges. These solutions, with our optimisation strategy and model design, are one major contribution of this work.

In the following, we will use italic lowercase letters to denote scalars s, and bold lowercase for 2D images $\mathbf{x} \in \mathbb{R}^{n \times m}$, where $n, m \in \mathbb{N}$ are the height and width of the image, respectively. Lastly, we adopt capital Greek letters for functions Φ.

3.1   Re-usable Discriminators: Challenges and Proposed Solutions

Challenge 1. To obtain a re-usable discriminator Δ, we must prevent it from *overfitting* and *catastrophically forget*, or its predictions on the masks generated during inference will not be reliable. Generally speaking, this is a challenging task because: GANs can easily memorise data if trained for too long [23].[1] Moreover, the discriminator may forget how unrealistic segmentation masks look like after the segmentor training has converged [30]. Although Δ may work well at training in these cases, it would not generalise to the test data, as we explain below.

If properly trained, a segmentor Σ predicts *realistic* segmentation masks in the latest stages of training. Thus, in standard GANs, we stop training while optimising Δ to tell apart *real* from more and more *real-looking* masks. At convergence, this becomes similar to training the discriminator using only *real* images and labelling them as *real* half the times, as *fake* the other half. At this point, gradients become uninformative, and the discriminator collapses to one of the following cases: **i)** it always predicts its equilibrium point (which in vanilla GANs is the number 0.5, equidistant from the labels *real*: 1, *fake*: 0) but it can still detect unrealistic images; **ii)** it predicts the equilibrium point independently of the input image, forgetting what *fake* samples look like [15,30]; or **iii)** it memorises the real masks (which, differently from the generated ones, appear unchanged since the beginning of training) and it always classifies them as *real*, while classifying *any other input* as *fake*. It is crucial to prevent the behaviours **ii)** and **iii)** to have a re-usable discriminator. For this reason, we use:

– *Fake anchors*: we ensure to expose the discriminator to unrealistic masks (labelled as *fake*) until the end of training. In particular, we train Δ using real masks $\mathbf{y}$, predicted masks $\tilde{\mathbf{y}}$, and corrupted masks $\mathbf{y}_{corr}$. We obtain $\mathbf{y}_{corr}$ by randomly swapping squared patches within the image[2] and adding binary noise to the real masks, as this proved to be a fast and effective strategy to learn robust shape priors in autoencoders [13]. While, towards the end of the training, the discriminator may not distinguish $\mathbf{y}$ from the real-looking $\tilde{\mathbf{y}}$,

---

[1] Memorisation can also happen just in the discriminator. In fact, contrarily to the segmentors, we do not use any additional supervised cost to regularise the discriminator training. We show how to detect memorisation from the losses in the Supplemental.

[2] We use patches having size equal to 10% of the image size.

the exposure to $\mathbf{y}_{corr}$ will prevent forgetting how unrealistic masks look like, providing informative gradients until we stop training.

Challenge 2. An additional challenge is to train *stable* discriminators, which do not change much in the latest training epochs. In other words, we want small oscillations in the discriminator loss. This is necessary because we typically stop training using early stopping criteria on the segmentor loss. Therefore, we want to promote the optimisation of Lipschitz smooth discriminators, avoiding suddenly big gradient updates (thus leaving Δ mostly unchanged between the last few training epochs). To this end, we suggest using:

- *Smoothness Constraints*: we increase discriminator smoothness [4] through Spectral normalisation [21] and *tanh* activations.
- *Discriminator data augmentation*: consisting of random roto-translations, and Instance Noise [22,31], to map similar inputs to the same prediction label. We translate images up to 10% of image pixels on both vertical and horizontal axes, and we rotate them between $0 \div \pi/2$. We generate noise using a Normal distribution with zero mean and 0.1 standard deviation.

3.2 Architectures and Training Objectives for Σ and Δ

We use a UNet [29] segmentor with batch normalisation [12]. Given an input image $\mathbf{x}$, the segmentor Σ predicts a multi-channel label map $\tilde{\mathbf{y}} = \Sigma(\mathbf{x})$. For the annotated images, we minimise the supervised weighted cross-entropy loss:

$$\mathcal{L}(\Sigma) = -\sum_{i=1}^{c} w_i \cdot \mathbf{y}_i \log(\tilde{\mathbf{y}}_i) \tag{1}$$

where i is a class index, c the number of classes, and w_i a class scaling factor used to address the class imbalance problem. The value $w_i = 1 - n_i/n_{tot}$ considers both the total number of pixels n_{tot} and the number of pixels n_i with label i.

As discriminator Δ, we use a convolutional encoder, processing the predicted masks with a series of 5 convolutional layers. Layers use a number of 4×4 filters following the series: 32, 64, 128, 256, 512. After the first two layers, we down-sample the features maps using a stride of 2. We increase smoothness according to Sect. 3.1. Finally, a fully-connected layer integrates the extracted features and predicts a scalar linear output, used to compute the adversarial losses [20]:

$$\min_{\Delta} \left\{ \mathcal{V}_{LS}(\Delta) = \frac{1}{2} E_{\mathbf{y} \sim \mathcal{y}}[(\Delta(\mathbf{y}) - 1)^2] + \frac{1}{2} E_{\mathbf{x} \sim \mathcal{x}}[(\Delta(\Sigma(\mathbf{x})) + 1)^2] \right\}$$
$$\min_{\Sigma} \left\{ \mathcal{V}_{LS}(\Sigma) = \frac{1}{2} E_{\mathbf{x} \sim \mathcal{x}}[(\Delta(\Sigma(\mathbf{x})))^2] \right\} \tag{2}$$

where -1 and $+1$ are the labels for *fake* and *real* images, respectively. As in standard adversarial semi-supervised training, we alternately minimise Eq. 1 on a batch of annotated images and Eq. 2 on a batch of unpaired images and unpaired masks. We use Adam optimiser [16], learning rate: 10^{-4}, and batch size: 12. Training proceeds until the segmentation loss stops decreasing on a validation set.

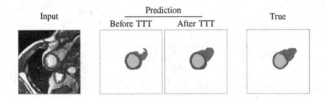

Fig. 2. Effect of Test-time Training (TTT). Re-using a discriminator at inference, we optimise a small input adaptor Ω until the predicted mask becomes realistic. We report additional examples in the Supplemental.

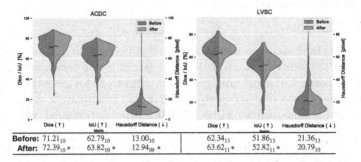

	Dice ($\uparrow$)	IoU ($\uparrow$)	Hausdorff Distance ($\downarrow$)	Dice ($\uparrow$)	IoU ($\uparrow$)	Hausdorff Distance ($\downarrow$)
Before:	71.21_{10}	62.79_{10}	13.00_{10}	62.34_{13}	51.86_{13}	21.36_{13}
After:	72.39_{10} *	63.82_{10} *	12.94_{09} *	63.62_{11} *	52.82_{11} *	20.79_{10}

Fig. 3. Dice ($\uparrow$), IoU ($\uparrow$) and Hausdorff distance ($\downarrow$) obtained **before** and **after** tuning the segmentor on the individual test instances. Arrows show metric improvement directions. Under each violin plot, we also report the average performance (standard deviation as subscript). We always improve the metrics, also in the worst-case scenarios (bottom of the distribution tails for Dice and IoU, upper tails for Hausdorff distance). Asterisks show statistical significance.

3.3 Adversarial Test-Time Training: Adapting Ω

At inference, we do not fine-tune the whole segmentor but only adapt a few convolutional layers at its input. These layers are, according to [1], the most suited for one-shot learning. By keeping the deeper layers of Σ unchanged, we also limit the segmentor flexibility and let it adapt only to changes at lower abstraction levels, ultimately preventing trivial solutions. Thus, we include a shallow convolutional residual block (adaptor Ω) in front of the segmentor, that we tune on the individual test images by minimising $\mathcal{V}_{LS}(\Sigma \equiv \Omega)$ for n_{iter} iterations. The adaptor is the same as in [13] and has 3 convolutional layers with 16 3×3 kernels and activation e^{-T^2/s^2}, being T an input tensor and s a trainable scaling parameter, initialised as 0 and optimised at inference. After tuning Ω, the input to the segmentor is an augmented version of **x** which can be more easily classified. We show qualitative examples in Fig. 2 and in the Supplemental.

4 Experiments

Data. We consider two cardiac MRI datasets, described below.

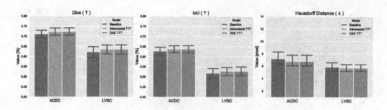

Fig. 4. Test-time Training using an adversarial shape prior vs a prior learned by a pre-trained DAE. In the plots, "Baseline" refers to standard inference of a GAN (i.e. without TTT). Bar plots show average and 95% confidence interval. Both methods lead to similar improvements ($p = 0.01$).

ACDC [2] has multi-scanner images from 100 patients with manual annotations for right and left ventricle, and for left myocardium. We resample data to the average resolution: 1.51 mm^2, and crop/pad them to 224 × 224 pixels. We standardise data using the patient-specific median and interquartile range.

LVSC [32] contains cardiac MRIs of 100 subjects, obtained with different scanners and imaging parameters. There are manual annotations for the left myocardium. We resample images to the average resolution of 1.45 mm^2, and then crop or pad them to 224 × 224 pixel size. We normalise images as in ACDC.

Setup and Evaluation. We divide datasets by patients, using groups of 40% for training, 20% for validation, and 40% for the test set, respectively. Out of the 40% training patients, we consider annotations for one fourth of the training subjects in ACDC and LVSC (10 patients). We treat the remaining data as unpaired and use them for adversarial training (Eq. 2). Notice that the small training sets cannot fully represent the entire data distribution, leading to segmentation errors at inference. We will investigate dealing with larger distribution shifts (e.g. different scanners, etc.) in the future. We analyse performance increases obtained from adversarial Test-time Training. Inspired by [13], we also compare to using a DAE to drive the adaptation (DAEs learn the shape prior separately, we do it while training Σ). We do 3-fold cross-validation and measure performance comparing the predicted segmentation masks and the ground truth labels contained in the test set. We use Dice and IoU scores, and the Hausdorff distance. We assess statistical significance with the non-parametric Wilcoxon test ($p = 0.01$).

4.1   Results and Discussion

We show a qualitative example of test-time adaptation in Fig. 2. In Fig. 3, we represent segmentation performance with violin plots before and after Test-time Training. These plots show the whole distribution of performance values for images in the test set. We observe performance improvements on all metrics and datasets. Importantly, worst-case scenarios (bottom tails of violin plots, for Dice and IoU; top tails for Hausdorff distance) considerably improve, reflecting the desired tendency to correct unrealistic segmentation masks that do not satisfy

Table 1. Ablation Study. We compare the performance of a UNet; a standard GAN; the GAN after adding: smoothness constraints (#1), the proposed regularisation technique: *fake anchors* (#2), and Test-Time Training (#3). Results are average (standard deviation as subscript) Dice scores on the ACDC test set.

UNet	GAN	GAN + #1	GAN + #1 + #2	GAN + #1 + #2 + #3
70.1_{13}	70.0_{12}	70.9_{11}	71.2_{10}	72.4_{10}

the learned adversarial shape prior. Qualitatively, we observed that the model removes scattered false positives and closes holes in the segmentation masks (see Fig. 1 in the Supplementary material).

In Fig. 4, we compare the performance of our method vs using a shape prior separately learned by a DAE, inspired by [13]. Our method achieves similar performance gains to a DAE (no statistically significant differences found), but it has the advantage of *not* requiring a separate pre-training step.

Lastly, we perform an ablation study to analyse the effect regularising the model with *smoothness constraints* and *fake anchors*. As illustrated in Table 1, the techniques improves training and makes the adversarial shape prior stronger. As a result: i) the adversarial training leads to a better segmentor; and ii) the re-usable discriminator further increases model performance.

Computational Aspects. The memory required to store the weights of the model is 90 MB. At inference, our method needs n_{iter} forward and backward passes to correct a segmentation. This is slower than standard inference, where each image requires one forward pass. We find that $n_{iter} > 0$ improves segmentation, but high values (e.g. > 100) overfit the segmentor leading to worse performance. As a compromise, we use $n_{iter} = 50$ (with small temporal overhead: ~10 s/patient on a TITAN Xp GPU). This fixed iteration strategy is also used by previous work [11,13,33], but using image-specific optimal n_{iter} would be useful and potentially increase performance. We leave automated strategies to set n_{iter} as future work.

Limitations. We find that Δ does not penalise wrong predictions that appear realistic but do not correspond to the input image. In fact, the discriminator only evaluates the predicted mask without considering the segmentor input. We highlight that this is also a limitation of [13] and of all methods learning the shape prior only using unpaired masks. We expect that including also the image-related information would improve Test-time Training.

5   Conclusion

We demonstrated that by satisfying simple assumptions, it is possible to re-use adversarial discriminators during inference. In particular, we re-used a mask discriminator to detect and then correct segmentation mistakes made by a segmentor. The proposed method is simple and can be potentially applied to any GAN, increasing its test-time performance on the most challenging images.

More broadly, the possibility of re-using adversarial discriminators to correct generator errors may open opportunities even outside image segmentation. Given their flexibility and the ability to learn data-driven losses, GANs have been widely adopted in medical imaging, from domain adaptation to image synthesis tasks [37]. With improved architectures and regularisation techniques [4,17], we believe adversarial networks will be even more popular in the future. In this context, training stable and re-usable discriminators opens opportunities for an all-round use of the GAN components.

Acknowledgments. This work was partially supported by the Alan Turing Institute (EPSRC grant EP/N510129/1). S.A. Tsaftaris acknowledges the support of Canon Medical and the Royal Academy of Engineering and the Research Chairs and Senior Research Fellowships scheme (grant RCSRF1819\8\25).

References

1. Asano, Y.M., Rupprecht, C., Vedaldi, A.: A critical analysis of self-supervision, or what we can learn from a single image. In: ICLR (2020)
2. Bernard, O., et al.: Deep learning techniques for automatic MRI cardiac multi-structures segmentation and diagnosis: is the problem solved? IEEE TMI **37**, 2514–2525 (2018)
3. Cheplygina, V., de Bruijne, M., Pluim, J.P.: Not-so-supervised: a survey of semi-supervised, multi-instance, and transfer learning in medical image analysis. MIA **54**, 280–296 (2019)
4. Chu, C., Minami, K., Fukumizu, K.: Smoothness and stability in GANs. In: ICLR (2020)
5. Clough, J., Byrne, N., Oksuz, I., Zimmer, V.A., Schnabel, J.A., King, A.: A topological loss function for deep-learning based image segmentation using persistent homology. IEEE Trans. Pattern Anal. Mach. Intell. (2020). https://doi.org/10.1109/TPAMI.2020.3013679
6. Dalca, A.V., Guttag, J.V., Sabuncu, M.R.: Anatomical priors in convolutional networks for unsupervised biomedical segmentation. In: CVPR (2018)
7. Dalca, A.V., Yu, E., Golland, P., Fischl, B., Sabuncu, M.R., Iglesias, J.E.: Unsupervised deep learning for Bayesian brain MRI segmentation. In: MICCAI (2019)
8. Donahue, J., Krähenbühl, P., Darrell, T.: Adversarial feature learning. In: ICLR (2017)
9. Goodfellow, I.J., et al.: Generative adversarial nets. In: NeurIPS (2014)
10. Guo, C., Pleiss, G., Sun, Y., Weinberger, K.Q.: On calibration of modern neural networks. In: ICML (2017)
11. He, Y., Carass, A., Zuo, L., Dewey, B.E., Prince, J.L.: Autoencoder based self-supervised test-time adaptation for medical image analysis. Med. Image Anal. **72**, 102136 (2021). https://doi.org/10.1016/j.media.2021.102136. ISSN 1361-8415
12. Ioffe, S., Szegedy, C.: Batch normalization: accelerating deep network training by reducing internal covariate shift. In: ICML (2015)
13. Karani, N., Erdil, E., Chaitanya, K., Konukoglu, E.: Test-time adaptable neural networks for robust medical image segmentation. Med. Image Anal. **68**, 101907 (2021). https://doi.org/10.1016/j.media.2020.101907. ISSN 1361-8415
14. Kervadec, H., Dolz, J., Tang, M., Granger, E., Boykov, Y., Ayed, I.B.: Constrained-CNN losses for weakly supervised segmentation. MIA **54**, 88–99 (2019)

15. Kim, Y., Kim, M., Kim, G.: Memorization precedes generation: learning unsupervised GANs with memory networks. In: ICML (2018)
16. Kingma, D.P., Ba, J.: Adam: a method for stochastic optimization. In: ICLR (2015)
17. Kurach, K., Lučić, M., Zhai, X., Michalski, M., Gelly, S.: A large-scale study on regularization and normalization in GANs. In: ICML (2019)
18. Larrazabal, A.J., Martínez, C., Glocker, B., Ferrante, E.: Post-DAE: anatomically plausible segmentation via post-processing with denoising autoencoders. IEEE TMI **39**, 3813–3820 (2020)
19. Mao, X., Su, Z., Tan, P.S., Chow, J.K., Wang, Y.H.: Is discriminator a good feature extractor? arXiv preprint arXiv:1912.00789 (2019)
20. Mao, X., Li, Q., Xie, H., Lau, R., Wang, Z., Smolley, S.P.: On the effectiveness of least squares generative adversarial networks. IEEE Trans. Pattern Anal. Mach. Intell. **41**(12), 2947–2960 (2019). https://doi.org/10.1109/TPAMI.2018.2872043
21. Miyato, T., Kataoka, T., Koyama, M., Yoshida, Y.: Spectral normalization for generative adversarial networks. In: ICLR (2018)
22. Müller, R., Kornblith, S., Hinton, G.E.: When does label smoothing help? In: NeurIPS (2019)
23. Nagarajan, V., Raffel, C., Goodfellow, I.: Theoretical insights into memorization in GANs. In: NeurIPS Workshop (2018)
24. Ngo, P.C., Winarto, A.A., Kou, C.K.L., Park, S., Akram, F., Lee, H.K.: Fence GAN: towards better anomaly detection. In: ICTAI (2019)
25. Nosrati, M.S., Hamarneh, G.: Incorporating prior knowledge in medical image segmentation: a survey. arXiv:1607.01092 (2016)
26. Oktay, O., et al.: Anatomically constrained neural networks (ACNNs), application to cardiac image enhancement and segmentation. IEEE TMI **37**, 384–395 (2017)
27. Painchaud, N., Skandarani, Y., Judge, T., Bernard, O., Lalande, A., Jodoin, P.M.: Cardiac MRI segmentation with strong anatomical guarantees. In: MICCAI (2019)
28. Radford, A., Metz, L., Chintala, S.: Unsupervised representation learning with deep convolutional generative adversarial networks (2016)
29. Ronneberger, O., Fischer, P., Brox, T.: U-Net: convolutional networks for biomedical image segmentation. In: Navab, N., Hornegger, J., Wells, W.M., Frangi, A.F. (eds.) MICCAI 2015. LNCS, vol. 9351, pp. 234–241. Springer, Cham (2015). https://doi.org/10.1007/978-3-319-24574-4_28
30. Shrivastava, A., Pfister, T., Tuzel, O., Susskind, J., Wang, W., Webb, R.: Learning from simulated and unsupervised images through adversarial training. In: CVPR, pp. 2107–2116 (2017)
31. Sønderby, C.K., Caballero, J., Theis, L., Shi, W., Huszár, F.: Amortised MAP inference for image super-resolution. In: ICLR (2017)
32. Suinesiaputra, A., et al.: A collaborative resource to build consensus for automated left ventricular segmentation of cardiac MR images. MIA **18**, 50–62 (2014)
33. Sun, Y., Wang, X., Liu, Z., Miller, J., Efros, A.A., Hardt, M.: Test-time training with self-supervision for generalization under distribution shifts. In: ICML (2020)
34. Tajbakhsh, N., Jeyaseelan, L., Li, Q., Chiang, J.N., Wu, Z., Ding, X.: Embracing imperfect datasets: a review of deep learning solutions for medical image segmentation. MIA **63**, 101693 (2020)
35. Valvano, G., Leo, A., Tsaftaris, S.A.: Learning to segment from scribbles using multi-scale adversarial attention gates. IEEE Trans. Med. Imaging **40**(8), 1990–2001 (2021). https://doi.org/10.1109/TMI.2021.3069634
36. Wang, D., Shelhamer, E., Liu, S., Olshausen, B., Darrell, T.: Tent: fully test-time adaptation by entropy minimization. In: ICLR (2021)

37. Yi, X., Walia, E., Babyn, P.: Generative adversarial network in medical imaging: a review. Med. Image Anal. **58**, 101552 (2019). https://doi.org/10.1016/j.media.2019.101552
38. Zenati, H., Romain, M., Foo, C.S., Lecouat, B., Chandrasekhar, V.: Adversarially learned anomaly detection. In: ICDM (2018)

Transductive Image Segmentation: Self-training and Effect of Uncertainty Estimation

Konstantinos Kamnitsas[1]([✉]), Stefan Winzeck[1], Evgenios N. Kornaropoulos[2,5],
Daniel Whitehouse[2], Cameron Englman[2], Poe Phyu[3], Norman Pao[4],
David K. Menon[2], Daniel Rueckert[1,6], Tilak Das[3], Virginia F. J. Newcombe[2],
and Ben Glocker[1]

[1] Department of Computing, Imperial College London, London, UK
`konstantinos.kamnitsas12@imperial.ac.uk`
[2] Division of Anaesthesia, Department of Medicine, University of Cambridge,
Cambridge, UK
[3] Department of Radiology, Cambridge University Hospitals NHS Foundation Trust,
Cambridge, UK
[4] Department of Emergency Medicine, Cambridge University Hospitals NHS
Foundation Trust, Cambridge, UK
[5] Clinical Sciences, Diagnostic Radiology, Lund University, Lund, Sweden
[6] Klinikum Rechts der Isar, Technical University of Munich, Munich, Germany

Abstract. Semi-supervised learning (SSL) uses unlabeled data during
training to learn better models. Previous studies on SSL for medical
image segmentation focused mostly on improving model generalization
to unseen data. In some applications, however, our primary interest is not
generalization but to obtain optimal predictions on a specific unlabeled
database that is fully available during model development. Examples
include population studies for extracting imaging phenotypes. This work
investigates an often overlooked aspect of SSL, transduction. It focuses
on the quality of predictions made on the unlabeled data of interest when
they are included for optimization during training, rather than improving generalization. We focus on the self-training framework and explore
its potential for transduction. We analyze it through the lens of Information Gain and reveal that learning benefits from the use of calibrated
or under-confident models. Our extensive experiments on a large MRI
database for multi-class segmentation of traumatic brain lesions shows
promising results when comparing transductive with inductive predictions. We believe this study will inspire further research on transductive
learning, a well-suited paradigm for medical image analysis.

1 Introduction

The predominant paradigm for developing ML models is supervised training
using labeled data. Because labels are scarce in many applications, these models
often suffer from generalization issues on real-world heterogeneous data. To alleviate this, semi-supervised learning (SSL) [5], aims to extract additional useful

S. Albarqouni et al. (Eds.): DART 2021/FAIR 2021, LNCS 12968, pp. 79–89, 2021.
https://doi.org/10.1007/978-3-030-87722-4_8

information from unlabeled data provided during training. Most commonly, the trained model is afterwards applied to *new, previously unseen* unlabeled data at test time, predicts by *induction*, and its generalization is measured by the accuracy on this new data. Model predictions made for the unlabeled data during training, as part of the optimization, are commonly discarded. For certain applications, however, the objective is to obtain predictions on a particular unlabeled database which may be *fully available at time of model development*, rather than generalization to other data afterwards. Applications of this type include population analysis on retrospective data to extract image phenotypes (e.g. size or location of pathology). In such a setting, it is desirable to *directly* optimize predictions for the specific unlabeled data, which is the concept of *transduction*. In this context, this study takes a fresh look on SSL and evaluates the potential of transduction for image segmentation, which has been previously under-explored.

Transductive learning is a form of SSL. Formally, SSL approaches assume two databases are available during model development: a *labeled* database $D_L = \{X_L \cup Y_L\} = \{(x_{L,i}, y_{L,i})\}_{i=1}^N$ and an *unlabeled* database $D_U = \{X_U\} = \{x_{U,i}\}_{i=1}^M$, with $y_{L,i} \in \mathcal{Y} = \{1, ..., C\}$ and C the number of classes. Common SSL approaches construct model $f_\theta(x) = p(y|x; \theta) \in \mathbb{R}^C$ and learn parameters θ such that it will approximate distribution $q(y|x)$ that generates the labels. This is commonly done by jointly minimizing a supervised cost $J_s(X_L, Y_L, f_\theta)$ (e.g. cross entropy) and an unsupervised cost, $J_u(X_U, \theta)$. For SSL focusing on *induction*, the goal is to obtain optimized model f_θ' which *generalizes* to new, unseen and unlabeled *target* (or *test*) data D_T. In contrast, in SSL focusing on *transduction*, it is assumed that the unlabeled *target* data is available during training (i.e. $D_U = D_T$). The goal is then to obtain the best possible predictions $Y_T' = \{y_{T,i}'\}_{i=1}^M$ that approximate the unknown true labels $Y_T = \{y_{T,i}\}_i^M$. Some methods, such as those employed in this study, can be used both for induction and transduction. However, in transduction the predictions on D_T are directly optimized during training, and may thus be better than predictions via induction, as the latter is subject to degradation due to distributional shift and the generalization gap [5,22].

Research on transductive SSL spans decades, with prime example the transductive SVM [22]. In medical imaging, transduction has been mostly used for learning over graphs [7,23] and few-shot learning [2]. Related work also explored transductive learning of a meta-learner for combining predictions from multiple models [25]. We are instead interested in the common setting where a relation graph between samples does not exist, significant amounts of labels are available (not few-shot) and we investigate whether and how can transduction reliably benefit standard segmentation models. In these settings, most prior work on SSL developed inductive methods and evaluated generalization to unseen data (e.g. [1,6,12,15,24]). Benefits by SSL were primarily shown when labeled data are very limited (few tens of images) and diminish with more labels [1,6,15,24]. In practical settings where more labeled data are available, the predominant paradigm is still supervised learning (e.g. large scale studies). Therefore, SSL methods that reliably offer improvements in such practical settings are still desirable.

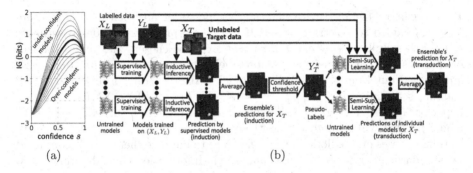

Fig. 1. a) Expected IG when a sample predicted with confidence s is assigned a 1-hot pseudo-label can be computed as $IG = \delta s H - (1 - \delta s)H$, where H the entropy in the predicted class-posteriors. Lines represent IG for different values of δ. $\delta = 1$ models perfect calibrated s, $0 < \delta < 1$ over-confident and $\delta > 1$ under-confident. For illustration, here H was calculated for 7-class problem, with s probability of most probable class and assuming $(1 - s)/6$ for remaining classes. b) Self-training with ensembles for improved confidence calibration and learning. Predictions for target data are obtained via transduction in the end of 2nd training phase.

This study makes the following contributions: (a) We present a theoretical analysis of the SSL framework of self-training via the lenses of Information Gain and reveal that learning benefits from well-calibrated or under-confident predictive uncertainty. This motivates us to introduce model ensembling within the framework, which is currently the most reliable method for obtaining calibrated uncertainty. (b) We assess quality of predictions obtained via transductive SSL and show consistent improvements over induction with our proposed framework. (c) We perform extensive evaluation on a large multi-center database for the challenging task of multi-class segmentation of traumatic brain injuries (TBI), including a blinded assessment via manual segmentation refinement, and demonstrate that transductive SSL with our framework can provide consistent improvements over induction, even when hundreds of labeled data are used. We believe this study will motivate further research into transductive learning, which is a suitable, but currently overlooked, paradigm for medical image analysis.

2   Methodology

We analyse and improve *self-training* [21], a popular framework for inductive SSL with neural networks [1,17]. Hence its transductive potential is of interest.

2.1   Transductive Learning via Self-training

In this framework, model f_θ (e.g. a neural net) is first trained via supervised cost $J_s(X_L, Y_L, \theta)$ to obtain θ'. Then, $f_{\theta'}$ is applied to unlabeled data, which in our transductive setting are *target* data X_T. The predicted labels

$Y_T' = \arg\max_c(f_{\theta'}(X_T))$ with highest class-posterior are then used as *pseudo-labels* $Y_T^\star$ added to the training set. If the model provides confidence estimate s_i for i-th sample, a *confidence threshold* t can be chosen, and only confident predictions ($s_i > t$) are used as pseudo-labels (in segmentation, uncertain pixels are masked out from the loss). The model is then re-trained on the extended training set minimizing $J_{total,ps} = J_s(X_L, Y_L, \theta) + \beta J_s(X_T, Y_T^\star, \theta)$, with β a hyper-parameter. This gives new parameters θ''. Inductive SSL strives to learn model $f_{\theta''}$ that will *generalize* better than $f_{\theta'}$ on *new* data. Instead, transductive learning focuses on predictions $Y_T'' = f_{\theta''}(X_T)$ to improve them over Y_T' obtained via standard supervision. We now analyse via Information Gain the framework, which reveals how uncertainty calibration of $f_{\theta'}$ influences learning, guiding us to improve it.

2.2   Analysing Information Gain and Improving Self-training

Various works argue that SSL cannot gain information about the label generating process $p(y|x)$ solely from unlabeled data without priors or assumptions [4,5,20]. Then, *how* does this framework benefit learning? A rigorous theory does not exist. We conjecture that information useful for learning is given to the system via a *subtle assumption* when pseudo-labels $Y_T^\star$ are created from class-posteriors. We analyse this hypothesis through the lens of Information Gain (IG). IG is defined as the change of entropy H of a system from a prior to a new state due to a new condition: $IG = H_{old} - H_{new}$. Predicted posteriors for sample x_i have entropy $H_i = -\sum_{c=1}^{C} p(c|x_i; \theta') \log p(c|x_i; \theta')$. A 1-hot posterior has 0 entropy. When a pseudo-label is created via $y_{T,i}^\star = \arg\max_c p(c|x_i; \theta')$, an *assumption* is made that the class with highest posterior is correct and its posterior *should* be 1. This generates $IG_i = H_i - 0 = H_i$ bits of info. If the prediction is correct, the pseudo-label matches the true unknown label, $y_{T,i}^\star = y_{T,i}$, and the obtained H_i bits of information will facilitate learning. If the prediction is wrong, $y_{T,i}^\star \neq y_{T,i}$, the system is given H_i bits of wrong information that hinder learning. We show next how confidence calibration relates to IG and how gains can be maximized.

A model's confidence score $s \in [0, 1]$ is considered *well calibrated* if $s \approx \frac{\sum_i \mathbb{1}[y_i' = y_i]\mathbb{1}[s_i' = s]}{\sum_i \mathbb{1}[s_i' = s]} \forall s$ (for discretized s). That is, confidence s is equal to the ratio of correct ($y_i' = y_i$) over all predictions made with confidence $s = s_i'$. As confidence score we use the model's maximum class posterior, $s_i = \max_c p(y = c|x_i; \theta)$ [11]. In this case, if we *assume* model confidence is well-calibrated, we can estimate the *expected* IG given to the system *on average* when a sample predicted with confidence s is assigned a 1-hot pseudo-label: $IG_{i,s} = sH_i - (1-s)H_i$. We plot this function in Fig. 1a. Standard neural nets, commonly used with self-training, are over-confident in practice [9,10,16]. Therefore we also plot the function $IG_{i,s} = \delta s H_i - (1 - \delta s)H_i$, where δ models the level of over or under-confidence. Lower δ represents lower percentage of correct predictions than estimated by s (over-confidence). Figure 1a shows that expected IG_i is positive for samples above a certain confidence and negative below. This supports the common practice of using confidence threshold t when creating pseudo-labels. Importantly, we find that for over-confident models, not only this threshold increases, but also the expected positive IG_i decreases! This means that in self-training, using well-calibrated or under-confident models a) leads to higher gains and better learning than with over-confident models; b) allows a wider range of values for threshold t without injecting negative IG to the system. The latter is practically important as choosing hyper-parameters is difficult for methods using unlabeled data [19].

Guided by the above findings, we improve the framework by introducing the currently most reliable method for obtaining calibrated uncertainty with neural networks, model ensembling [13,16]. Figure 1b shows the framework with network ensembles. Individual models are trained with different weight initialization and data sampling, found sufficient for better uncertainty estimates (Fig. 2c). Ensembling is also known to improve predictive accuracy, benefiting quality of pseudo-labels. We will demonstrate in Sect. 3 empirical evidence aligned with our above two theoretical observations: Self-training via ensembling a) provides performance gains for different values of t the magnitude of which follows trend aligned with what our theoretical analysis suggests (Fig. 1a), b) provides high performance for a wide range of t making configuration easier and more reliable.

3   Experimental Evaluation

3.1   Data and Model Configuration

We evaluate transduction on multi-class TBI segmentation in the following setup.

DB1: This database was acquired in one clinic using 3 scanners. Consists of 180 subjects who underwent MRI including T1w, T2w, FLAIR, and SWI or GRE (SWI & GE used interchangeably here). It includes ~85% moderate/severe and ~15% mild cases. DB1 labels are used for training in all following experiments.

DB2: This includes 101 subjects, acquired in 9 clinics with 4 scanner models, all different from DB1. T1w, T2w, FLAIR and SWI scans are available for all patients. DB2 consists of ~35% moderate/severe and ~65% mild cases. These distribution shifts between DB1 and DB2 make for a challenging benchmark. DB2 is used as *target* data D_T in following experiments, except for those in Sect. 3.4, where its labels are used for training and evaluation is done on DB3.

DB3: This database includes scans of 265 subjects from 14 sites (9 overlap with DB2, 5 new). Sequences and severity are similar to DB2. This database was unlabeled, kept unseen during model development. We used DB3 for blinded comparison of induction versus transduction after model development (Sect. 3.4).

Manual Annotations: Experts annotated any abnormality visible in DB1 and DB2. Here, we consider 6 classes: core, oedema, petechial hemorrhages, intraventricular hemorrhages, non-TBI lesions, monitoring probe.

Pre-processing: We registered each image to the corresponding T1w, resampled to 1 mm isotropic resolution, and performed z-score intensity normalization.

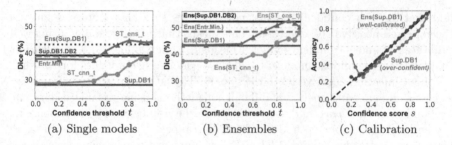

(a) Single models   (b) Ensembles   (c) Calibration

Fig. 2. (a, b) Average Dice (%) over all 6 TBI classes achieved by different methods. Shown as function of confidence threshold t for self-training methods. (c) Reliability diagram [9] of a single CNN *Sup.DB1* and an ensemble *Ens(Sup.DB1)*.

Main Model: We use a 3D Convolutional Network (CNN), DeepMedic, previously used for TBI segmentation [14,18]. We use the 'wide' variant with default hyper-parameters (from: https://github.com/deepmedic/deepmedic, v0.8.4).

Compared Methods and Configuration: Besides self-training, we explore transduction with Entropy Minimization (EM), originally developed for inductive SSL [8]. Moreover, training a CNN on pseudo-labels by ensemble with no confidence threshold t (ST_ens_0 below) is equal to ensemble-distillation on unlabelled data [3]. Striving for reliability, we chose the studied methods for their simplicity, to avoid hyper-parameter tuning that is impractical in SSL [19]. Except t that we study below, the only hyper-parameter is the weight of unsupervised vs supervised cost, which we set to 1 for self-training and EM.

3.2   Comparing Supervised and Transductive Learning

Each experiment below is repeated for 10 seeds. Figure 2 summarizes results. Shown is mean Dice over all classes and seeds, as space is limited for per-class analysis.

Sup.DB1: The supervised baseline are 10 models trained with full supervision on DB1, evaluated on DB2. Models achieve 28.1% Dice on average over all classes. The low score is due to the tiny size of mild TBI lesions where few false positives greatly impact Dice (Fig. 3b), and domain shift between DB1–DB2.

Sup.DB1.DB2: We also evaluate how well a supervised model would perform if labels for DB2 were available for training. We train 10 models using 100% DB1 and 80% DB2 labels, and evaluate on remaining 20% of DB2. We repeat for 5 folds of DB2. SSL without labels on DB2 is not expected to outperform this, but hopefully approach it.

Entr.Min.: We perform transductive learning by training 10 models with supervision over DB1 and entropy minimization [8] over unlabelled DB2 images. We evaluate predictions made on DB2 at end of training. Transduction with EM improves over baseline *Sup.DB1* (Fig. 2a).

ST_cnn_t: We evaluate transduction via self-training with pseudo-labels made by a single CNN. We create pseudo-labels using confidence threshold t on posteriors of a supervised CNN (*Sup.DB1*), and re-train a CNN on the extended training set. We repeat for 10 models of *Sup.DB1* and 7 values of t to investigate its effect. Using $t = 0$ equals to using segmentations of *Sup.DB1* as pseudo-labels because no prediction was less confident. Figure 2 shows improvements over *Sup.DB1*. Performance reaches EM only for best choice of t. High t values are required, aligned with our theoretical findings about over-confident CNNs (Fig. 1a).

Pseudo-Labels from Ensemble: We assess performance of a CNN when trained with pseudo-labels derived from confidence estimates of an ensemble. We make ensemble **Ens(Sup.DB1)** combining 10 baseline CNNs (*Sup.DB1*). Figure 2c shows the ensemble is indeed well-calibrated. We obtain pseudo-labels using threshold t on ensemble's posteriors for DB2. We re-train 10 CNNs using DB1 labels and DB2 pseudo-labels. Transductive predictions are obtained. We show average Dice of 10 seeds as **ST_ens_t**. We repeat for varying t. CNN retrained using ensemble segmentations as pseudo-labels (*ST_ens_0*, $t = 0$, ala ensemble distillation [3]) outperform *ST_cnn_t*, thanks to the better segmentation of ensemble, which is not surprising. Most importantly, treating confidence via threshold t offers performance gains that follow trend similar to what we derived for IG with well-calibrated models (Fig. 1a), supporting our theoretical results. Close-to-optimal performance is achieved with wide range of t, in contrast to *ST_cnn_t*, explained by the wider range of t with positive IG of the former according to our findings (Fig. 1a). This facilitates the choice of a reliable t value. Finally, without labels for DB2, *ST_ens_t* outperforms inductive *Sup.DB1.DB2*, likely thanks to direct optimization of predictions on DB2 by transductive learning.

Results by Ensembles: If ensemble is used to make pseudo-labels, then effective self-training should improve an ensemble. To assess this, we average transductive predictions from 10 models *ST_ens_t* per t, obtaining transductive predictions of self-trained ensemble **Ens(ST_ens_t)** (Fig. 1b). We compare with ensembles of CNNs re-trained with pseudo-labels from single CNNs (**Ens(ST_cnn_t)**), entropy minimization (**Ens(Entr.Min)**) and supervision on DB1∪DB2 (**Ens(Sup.DB1.DB2)**). Figure 2b shows that self-training improves the ensemble, with its transductive results (*Ens(ST_ens_t)*) exhibit lower and less stable performance. Importantly, this holds for all values $t > 0.5$, aligned with our findings for well-calibrated models (Fig. 1a). In contrast, ensembles retrained on pseudo-labels from individual CNNs (*Ens(ST_cnn_t)*) reach the baseline ensemble only for optimal t. Transductive ensemble trained via entropy minimization also surpasses the inductive ensemble. The best transductive ensemble approaches inductive *Ens(Sup.DB1.DB2)* that requires labels for DB2.

Method	Dice % Induct.	Transd.
Sup.DB1	28.1	-
Entr.Min.	36.4	38.1
ST_ens_0.7	39.7	43.3
ST_ens_0.9	41.9	44.8
Ens(Sup.DB1)	43.3	-
Ens(Entr.Min.)	47.4	48.4
Ens(ST_ens_0.7)	48.8	50.8
Ens(ST_ens_0.9)	50.0	53.3

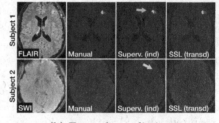

(a) Inductive vs transductive SSL   (b) Example predictions

Fig. 3. a) Comparison of transduction and induction. We show average Dice% over all TBI classes and 10 seeds. Differences (Ind/Transd) are significant ($p < 0.05$). b) Example predictions by induction ($Ens(Sup.DB1)$) and transduction ($Ens(ST\_ens\_0.7)$). Yellow arrow shows edema mistaken as related to monitoring probe. Cyan arrow shows false-positive TBI core. Both corrected by transductive SSL. Mild TBI lesions are often tiny, hence Dice degrades even with small errors. (Color figure online)

Table 1. Dice% between predicted and manually refined lesions on DB3. Transduction overlaps more than induction ($p < 0.05$ for average over classes, Avg).

Method	Core	Oed.	Non-TBI	Probe	Petech.	Intrav.	Avg.
Induction, fully supervised	75.7	77.9	84.4	85.7	53.4	83.9	76.8
Transduction, self-training	79.7	79.8	91.9	82.4	58.8	83.9	79.4

3.3 Comparing Inductive and Transductive Semi-supervised Learning

We compare transduction and induction with the same model after self-training. To do this, we divide DB2 in 2 equal folds. We start with pseudo-labels for DB2 from $Ens(Sup.DB1)$ (Sect. 3.2). For 10 seeds, we perform self-training using 100% DB1 labels and 50% pseudo-labeled DB2. After training, we obtain transductive predictions for the DB2 fold used in self-training and inductive predictions for the other fold. We switch DB2 folds and repeat, to obtain these predictions for whole DB2 (ST_ens_t). Finally, we ensemble inductive and transductive predictions (separately) from 10 seeds, to assess ensemble performance ($Ens(ST\_ens\_t)$). We repeat for $t = 0.7$ and 0.9 to assess along the wider effective range. We repeat similar procedure for the entropy minimization method. Results are shown in Fig. 3a. Transductive SSL outperforms inductive SSL in all cases.

3.4 Blinded Comparison via Manual Refinement of Segmentations

SSL is often found beneficial when labels are limited but its reliability is still questioned in practical settings, where supervised learning is preferred, e.g. large studies with more labels (Sect. 1). To address this, we compare transductive SSL with supervised training when using all 281 labeled cases of DB1∪DB2, largest segmented cohort reported for MRI TBI, and predict 265 cases of multi-center

DB3, which were unlabeled and held-out during model development for objective evaluation. We perform blinded comparison via manual segmentation refinement.

Because manual refinement is time consuming, we defined a process that requires only one refinement per image. We obtain *inductive* predictions on DB3 using *Ens(Sup.DB1.DB2)*, which is supervised on $DB1 \cup DB2$ (Sect. 3.2). From these predictions we make pseudo-labels with $t = 0.7$ (choice based on Fig. 1a). Using $DB1 \cup DB2$ labels and $DB3$ pseudo-labels we train an ensemble, similar to *Ens(ST_ens_0.7)* (Sect. 3.2), and obtain *transductive* predictions. We then create an 'in-between' segmentation per image by averaging inductive and transductive posteriors (ala ensembling). These latter segmentations were manually refined by clinicians with TBI expertise to meet the highest standards for follow-up studies on TBI phenotyping. Experts were unaware how the predictions were created.

Finally, we evaluated overlap of manually-refined segmentations with inductive and transductive predictions. Table 1 shows the results. Transduction clearly out-performs inductive supervised learning. These results show that transductive SSL can facilitate large-scale studies by reducing effort for manual refinement.

4 Conclusion

We explored the potential of transduction for segmentation. We showed with extensive experiments that if test data are available in advance, transduction outperforms induction by supervision or SSL. We also presented theoretical and empirical evidence that using well-calibrated or under-confident models facilitates self-training. Future research should evaluate transduction on other data and perform analysis with multiple metrics, beyond Dice, to capture the method's effect more comprehensively. Having set the first stone with this study, future works should explore transductive segmentation with other SSL methods. We believe these results will inspire further research in transductive SSL, which is well-suitable for retrospective medical image analysis studies.

Acknowledgements. This work received funding from the UKRI London Medical Imaging & Artificial Intelligence Centre for Value Based Healthcare. VFJN is funded by an Academy of Medical Sciences/The Health Foundation Clinical Scientist Fellowship. DKM is supported by the National Institute for Health Research (NIHR, UK) through the Cambridge NIHR Biomedical Research Centre, and by a European Union Framework Program 7 grant (CENTER-TBI; Grant agreement 602150).

References

1. Bai, W., et al.: Semi-supervised learning for network-based cardiac MR image segmentation. In: Descoteaux, M., Maier-Hein, L., Franz, A., Jannin, P., Collins, D.L., Duchesne, S. (eds.) MICCAI 2017. LNCS, vol. 10434, pp. 253–260. Springer, Cham (2017). https://doi.org/10.1007/978-3-319-66185-8_29
2. Boudiaf, M., Masud, Z.I., Rony, J., Dolz, J., Piantanida, P., Ayed, I.B.: Transductive information maximization for few-shot learning. In: Advances in Neural Information Processing Systems (2020)

3. Bucilua, C., Caruana, R., Niculescu-Mizil, A.: Model compression. In: Proceedings of the 12th ACM SIGKDD International Conference on Knowledge Discovery and Data Mining, pp. 535–541 (2006)
4. Castro, D.C., Walker, I., Glocker, B.: Causality matters in medical imaging. Nat. Commun. **11**(1), 1–10 (2020)
5. Chapelle, O., Scholkopf, B., Zien, A.: Semi-Supervised Learning. MIT Press, Cambridge (2006)
6. Cui, W., et al.: Semi-supervised brain lesion segmentation with an adapted mean teacher model. In: Chung, A.C.S., Gee, J.C., Yushkevich, P.A., Bao, S. (eds.) IPMI 2019. LNCS, vol. 11492, pp. 554–565. Springer, Cham (2019). https://doi.org/10.1007/978-3-030-20351-1_43
7. De, J., Li, H., Cheng, L.: Tracing retinal vessel trees by transductive inference. BMC Bioinform. **15**(1), 1–20 (2014)
8. Grandvalet, Y., Bengio, Y., et al.: Semi-supervised learning by entropy minimization. In: CAP, pp. 281–296 (2005)
9. Guo, C., Pleiss, G., Sun, Y., Weinberger, K.Q.: On calibration of modern neural networks. In: International Conference on Machine Learning, pp. 1321–1330. PMLR (2017)
10. Hein, M., Andriushchenko, M., Bitterwolf, J.: Why ReLU networks yield high-confidence predictions far away from the training data and how to mitigate the problem. In: Proceedings of the IEEE/CVF Conference on Computer Vision and Pattern Recognition, pp. 41–50 (2019)
11. Hendrycks, D., Gimpel, K.: A baseline for detecting misclassified and out-of-distribution examples in neural networks. In: International Conference on Learning Representations (2017)
12. Huang, R., Noble, J.A., Namburete, A.I.L.: Omni-supervised learning: scaling up to large unlabelled medical datasets. In: Frangi, A.F., Schnabel, J.A., Davatzikos, C., Alberola-López, C., Fichtinger, G. (eds.) MICCAI 2018. LNCS, vol. 11070, pp. 572–580. Springer, Cham (2018). https://doi.org/10.1007/978-3-030-00928-1_65
13. Jungo, A., Balsiger, F., Reyes, M.: Analyzing the quality and challenges of uncertainty estimations for brain tumor segmentation. Front. Neurosci. **14**, 282 (2020)
14. Kamnitsas, K., et al.: Efficient multi-scale 3D CNN with fully connected CRF for accurate brain lesion segmentation. Med. Image Anal. **36**, 61–78 (2017)
15. Kervadec, H., Dolz, J., Granger, É., Ben Ayed, I.: Curriculum semi-supervised segmentation. In: Shen, D., et al. (eds.) MICCAI 2019. LNCS, vol. 11765, pp. 568–576. Springer, Cham (2019). https://doi.org/10.1007/978-3-030-32245-8_63
16. Lakshminarayanan, B., Pritzel, A., Blundell, C.: Simple and scalable predictive uncertainty estimation using deep ensembles. In: Advances in Neural Information Processing Systems (2017)
17. Lee, D.H., et al.: Pseudo-label: the simple and efficient semi-supervised learning method for deep neural networks. In: Workshop on challenges in representation learning. In: ICML, vol. 3 (2013)
18. Monteiro, M., et al.: Multiclass semantic segmentation and quantification of traumatic brain injury lesions on head CT using deep learning: an algorithm development and multicentre validation study. Lancet Digital Health **2**(6), e314–e322 (2020)
19. Oliver, A., Odena, A., Raffel, C.A., Cubuk, E.D., Goodfellow, I.: Realistic evaluation of deep semi-supervised learning algorithms. In: Advances in Neural Information Processing Systems, pp. 3235–3246 (2018)
20. Schölkopf, B., Luo, Z., Vovk, V.: Empirical inference: Festschrift in honor of vladimir n. vapnik (2013)

21. Scudder, H.: Probability of error of some adaptive pattern-recognition machines. IEEE Trans. Inf. Theory **11**(3), 363–371 (1965)
22. Vapnik, V.: Statistical Learning Theory. A Wiley-Interscience publication, Wiley (1998). https://books.google.co.uk/books?id=GowoAQAAMAAJ
23. Wang, Z., et al.: Progressive graph-based transductive learning for multi-modal classification of brain disorder disease. In: Ourselin, S., Joskowicz, L., Sabuncu, M.R., Unal, G., Wells, W. (eds.) MICCAI 2016. LNCS, vol. 9900, pp. 291–299. Springer, Cham (2016). https://doi.org/10.1007/978-3-319-46720-7_34
24. Yu, L., Wang, S., Li, X., Fu, C.-W., Heng, P.-A.: Uncertainty-aware self-ensembling model for semi-supervised 3D left atrium segmentation. In: Shen, D., et al. (eds.) MICCAI 2019. LNCS, vol. 11765, pp. 605–613. Springer, Cham (2019). https://doi.org/10.1007/978-3-030-32245-8_67
25. Zheng, H., et al.: A new ensemble learning framework for 3D biomedical image segmentation. In: Proceedings of the AAAI Conference on Artificial Intelligence, vol. 33, pp. 5909–5916 (2019)

Unsupervised Domain Adaptation with Semantic Consistency Across Heterogeneous Modalities for MRI Prostate Lesion Segmentation

Eleni Chiou[1,2(✉)], Francesco Giganti[3,4], Shonit Punwani[5], Iasonas Kokkinos[2], and Eleftheria Panagiotaki[1,2]

[1] Centre for Medical Image Computing, UCL, London, UK
eleni.chiou.17@ucl.ac.uk
[2] Department of Computer Science, UCL, London, UK
[3] Department of Radiology, UCLH NHS Foundation Trust, London, UK
[4] Division of Surgery and Interventional Science, UCL, London, UK
[5] Centre for Medical Imaging, Division of Medicine, UCL, London, UK

Abstract. Any novel medical imaging modality that differs from previous protocols e.g. in the number of imaging channels, introduces a new domain that is heterogeneous from previous ones. This common medical imaging scenario is rarely considered in the domain adaptation literature, which handles shifts across domains of the same dimensionality. In our work we rely on stochastic generative modeling to translate across two heterogeneous domains at pixel space and introduce two new loss functions that promote semantic consistency. Firstly, we introduce a semantic cycle-consistency loss in the source domain to ensure that the translation preserves the semantics. Secondly, we introduce a pseudo-labelling loss, where we translate target data to source, label them by a source-domain network, and use the generated pseudo-labels to supervise the target-domain network. Our results show that this allows us to extract systematically better representations for the target domain. In particular, we address the challenge of enhancing performance on VERDICT-MRI, an advanced diffusion-weighted imaging technique, by exploiting labeled mp-MRI data. When compared to several unsupervised domain adaptation approaches, our approach yields substantial improvements, that consistently carry over to the semi-supervised and supervised learning settings.

Keywords: Unsupervised domain adaptation · Pseudo-labeling · Entropy minimization · Lesion segmentation · VERDICT-MRI · mp-MRI

1 Introduction

Domain adaptation transfers knowledge from a label-rich 'source' domain to a label-scarce or unlabeled 'target' domain. This allows us to train robust

© Springer Nature Switzerland AG 2021
S. Albarqouni et al. (Eds.): DART 2021/FAIR 2021, LNCS 12968, pp. 90–100, 2021.
https://doi.org/10.1007/978-3-030-87722-4_9

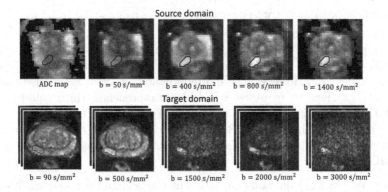

Fig. 1. Example of heterogeneous domains: standard DW-MRI (source domain) consists of 5 input channels (4 b-values and the ADC map) while VERDICT-MRI consists of 15 input channels (5 b-values in 3 orthogonal directions).

models targeted towards novel medical imaging modalities or acquisition protocols with scarce supervision, if any. Recent unsupervised domain adaptation methods typically achieve this either by minimizing the discrepancy of the feature and/or output space for the two domains or by learning a mapping between the two domains at the raw pixel space. Either way, these approaches usually consider moderate domain shifts where the dimensionality of the input feature-space between the source and the target domain is identical.

In this work we address the challenge of adapting across two heterogeneous domains where both the distribution and the dimensionality of the input features are different (Fig. 1). Inspired by [9], we rely on stochastic translation [12] to align the two domains at pixel-level; [9] shows that stochastic translation yields clear improvements in heterogeneous domain adaptation tasks compared to deterministic, CycleGAN-based [32] translation approaches. However, these improvements have been obtained with semi-supervised learning, where a few labeled target-domain images are available, whereas our goal is unsupervised domain adaptation. To this end we introduce a *semantic cycle-consistency* loss on the cycle-reconstructed source images; if a source image is translated to the target domain and then back to the source domain, we require that critical structures are preserved. We also introduce a *pseudo-labeling* loss that allows us to use the unlabeled target data to supervise the target-domain segmentation network. In particular we translate the target data to the source domain, predict their labels according to a pre-trained source-domain segmentation network and use the generated pseudo-labels to supervise the target-domain segmentation network. This allows us to use exclusively target-domain statistics and train highly discriminative models.

We demonstrate the effectiveness of our approach in prostate lesion segmentation and an advanced diffusion-weighted MRI (DW-MRI) method called VERDICT-MRI (Vascular, Extracellular and Restricted Diffusion for Cytometry in Tumours) [20,21]. Compared to the naive DW-MRI from multiparametric (mp)-MRI acquisitions, VERDICT-MRI has a richer acquisition protocol to probe the underlying microstructure and reveal changes in tissue features similar to histology. VERDICT-MRI has shown promising results in clinical settings, discriminating normal from malignant tissue [7,20,26] and identifying specific Gleason grades [15]. However, the limited amount of available labeled training data does not allow us to utilize data-driven approaches which could directly exploit the information in the raw VERDICT-MRI [6,8]. On the other hand, labeled, large scale clinical mp-MRI datasets exist [1,18]. In this work we exploit labeled mp-MRI data to train a segmentation network that performs well on VERDICT-MRI.

Related Work: Most domain adaptation methods align the two domains either by extracting domain-invariant features or by aligning the two domains at the raw pixel space. Ren et al. [23] and Kamnitsas et al. [16], rely on adversarial training to align the feature distributions between the source and the target domain for medical image classification and segmentation respectively. Pixel-level approaches [3,9,13,30,31], use GAN-based methods [12,32] to align the source and the target domains at pixel level. Chen et al. [4] align simultaneously the two domains at pixel- and feature-level by utilizing adversarial training. Ouyang et al. [19] combine a variational autoencoder (VAE)-based feature prior matching and pixel-level adversarial training to learn a domain-invariant latent space which is exploited during segmentation. Similarly, [28] perform pixel-level adversarial training to extract content-only images and use them to train a segmentation model that operates well in both domains. Other studies exploit unlabeled target domain data during the discriminative training. Bateson et al. [2] and Guodong et al. [29] use entropy minimization on the prediction of target domain as an extra regularization while [17] propose a teacher-student framework to train a model using labeled and unlabeled target data as well as labeled source data.

2   Method

2.1   Problem Formulation

We consider the problem of domain adaptation in prostate lesion segmentation. Let $\mathcal{X}_S \subset \mathbb{R}^{H \times W \times C_S}$ be a set of N_S source images and $\mathcal{Y}_S \subset \{0,1\}^{H,W}$ their segmentation masks. The sample $X_S \in \mathcal{X}_S$ is a $H \times W \times C_S$ image and the entry $Y_S^{(h,w)}$ of the mask Y_S provides the label of voxel (h,w) as a one-hot vector. Let also $\mathcal{X}_T \subset \mathbb{R}^{H \times W \times C_T}$ be a set of N_T unlabeled target images. Sample $X_T \in \mathcal{X}_T$ is an $H \times W \times C_T$ image.

Stochastic Translation with Semantic Cycle-Consistency Regularization

We rely on stochastic translation [9,12] to learn the mapping between the two domains and introduce a semantic cycle-consistency loss to enforce the cross-domain mapping to preserve critical structures.

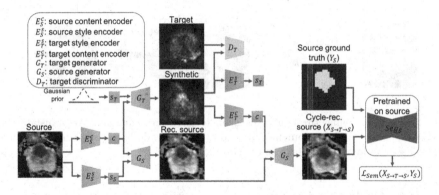

Fig. 2. We force a network for stochastic translation across domains to preserve semantics through a semantic segmentation-based loss. The image-to-image translation network translates source-domain images to the style of the target domain by combining a domain-invariant content code c with a random code s_T. We introduce a semantic cycle-consistency loss, $\mathcal{L}_{Sem}$, on the cycle-reconstructed images that ensures that the prostate lesions are successfully preserved.

The image-to-image translation network (Fig. 2) consists of content encoders E_S^c, E_T^c, style encoders E_S^s, E_T^s, generators G_S, G_T and domain discriminators D_S, D_T for both domains. The content encoders E_S^c, E_T^c extract a domain-invariant content code $c \in \mathcal{C}$ ($E_S^c : \mathcal{X}_S \to \mathcal{C}$, $E_T^c : \mathcal{X}_T \to \mathcal{C}$) while the style encoders E_S^s, E_T^s extract domain-specific style codes $s_S \in \mathcal{S}_S$ ($E_S^s : \mathcal{X}_S \to \mathcal{S}_S$) and $s_T \in \mathcal{S}_T$ ($E_T^s : \mathcal{X}_T \to \mathcal{S}_T$). Image-to-image translation is performed by combining the content code ($c = E_S^c(X_S)$) extracted from a given input ($X_S \in \mathcal{X}_S$) and a random style code s_T sampled from the target-style space. We note that the random style-code sampled from a Gaussian distribution represents structures that cannot be accounted by a deterministic mapping and results in one-to-many translation. We train the networks with a loss function consisting of domain adversarial, self-reconstruction, latent reconstruction and semantic cycle-consistency losses.

Domain Adversarial Loss

$$\mathcal{L}_{GAN}^T = \mathbb{E}_{c_S \sim \mathcal{C}, s_T \sim \mathcal{S}_T}[\log(1 - D_T(G_T(c_S, s_T)))] + \mathbb{E}_{X_T \sim \mathcal{X}_T}[\log(D_T(X_T))].$$

Self-reconstruction Loss

$$\mathcal{L}_{recon}^S = \mathbb{E}_{X_S \sim \mathcal{X}_S}[\|G_S(E_S^c(X_S), E_S^s(X_S)) - X_S\|_1].$$

Latent Reconstruction Loss

$$\mathcal{L}_{recon}^{c_S} = \mathbb{E}_{X_S \sim \mathcal{X}_S, s_T \sim \mathcal{S}_T}[\|E_T^c(G_T(E_S^c(X_s), s_T)) - E_S^c(X_s)\|_1].$$
$$\mathcal{L}_{recon}^{S_T} = \mathbb{E}_{X_S \sim \mathcal{X}_S, s_T \sim \mathcal{S}_T}[\|E_T^s(G_T(E_S^c(X_s), s_T)) - s_T\|_1].$$

Cycle-Consistency Loss

$$\mathcal{L}_{cyc}^S = \mathbb{E}_{X_S \sim \mathcal{X}_S, s_T \sim \mathcal{S}_T}[\|G_S(E_T^c(G_T(E_S^c(X_S), s_T)), E_S^s(X_S)) - X_S\|_1].$$
$\mathcal{L}_{GAN}^S$, $\mathcal{L}_{recon}^T$, $\mathcal{L}_{recon}^{c_T}$, $\mathcal{L}_{recon}^{s_S}$, $\mathcal{L}_{cyc}^T$ are defined in a similar way.

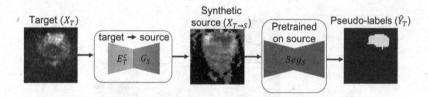

Fig. 3. Pseudo-labeling through translation to the source domain: We translate the target data to the source domain and predict their pseudo-labels according to a pre-trained source-domain segmentation network Seg_S.

Semantic Cycle-Consistency Loss. Recent studies [3,9,13,31] enforce semantic consistency between the real source and the synthetic target images by exploiting a target-domain segmentation network trained on a few available labeled target-domain images. However, in the unsupervised scenario, where there is no supervision available for the target domain, such approach is not feasible. To this end we introduce a semantic cycle-consistency loss or lesion segmentation loss on the cycle-reconstructed source images $X_{S \to T \to S}$; if a source image is translated to the target domain and then back to the source domain, we require that critical structures, corresponding to lesions, are preserved. The naive cycle-consistency loss, introduced in [32], penalizes inconsistencies in the entire image and may fail to preserve small structures corresponding to lesions. In contrast our semantic cycle-consistency loss penalizes inconsistencies in the label space enforcing the translation network to preserve the lesions. The semantic cycle-consistency loss is a soft generalization of the dice score given by

$$\mathcal{L}_{Sem} = -\frac{2\sum_{h,w} P^{(h,w,1)} Y_S^{(h,w,1)}}{\sum_{h,w}(P^{(h,w,1)} + Y_S^{(h,w,1)})}, \tag{1}$$

where $P^{(h,w,1)}$ is the predictive probability of class 1 for voxel (h, w) provided by the pre-trained source network Seg_S. The full objective is given by

$$\min_{E_S^c, E_S^s, E_T^c, E_T^s, G_S, G_T} \max_{D_S, D_T} \lambda_{GAN}(\mathcal{L}_{GAN}^S + \mathcal{L}_{GAN}^T) + \lambda_x(\mathcal{L}_{recon}^S + \mathcal{L}_{recon}^T)$$
$$+ \lambda_c(\mathcal{L}_{recon}^{cS} + \mathcal{L}_{recon}^{cT}) + \lambda_s(\mathcal{L}_{recon}^{sS} + \mathcal{L}_{recon}^{sT}) + \lambda_{cyc}(\mathcal{L}_{cyc}^S + \mathcal{L}_{cyc}^T) \tag{2}$$
$$+ \lambda_{sem}\mathcal{L}_{sem},$$

where $\lambda_{GAN}, \lambda_x, \lambda_c, \lambda_s, \lambda_{cyc}, \lambda_{sem}$ control the importance of each term.

Pseudo-Labeling Through Translation to the Source
We generate pseudo-labels for the target images by translating them to the source domain and predicting their labels according to the pre-trained source-domain segmentation network Seg_S, trained on the labeled source data (Fig. 3).

Given a synthetic source image $X_{T \to S}$ and the segmentation network Seg_S we obtain a soft-segmentation map $P_{X_{T \to S}} = Seg_S(X_{T \to S})$, where each vector $P_{X_{T \to S}}^{(h,w)}$ corresponds to a probability distribution over classes. Assuming that

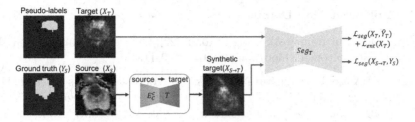

Fig. 4. We use data that have exclusively target-domain statistics to train the target segmentation network (Seg_T). We translate the source data to the target domain and supervise Seg_T using the ground-truth segmentation masks. We also use target pseudo-labels to supervise Seg_T.

high-scoring pixel-wise predictions on synthetic source samples are correct, we obtain a segmentation mask $\hat{Y}_T$ by selecting high-scoring pixels with a fixed threshold. Each entry $\hat{Y}_T^{(h,w)}$ can be either a discrete one-hot vector for high-scoring pixels or a zero-vector for low-scoring pixels. The pseudo-labeling configuration is defined as follows

$$\hat{Y}_T^{(h,w,c)} = \begin{cases} 1, & \text{if } c = \arg\max_c P_{X_{T\to S}}^{(h,w)} \text{ and } P_{X_{T\to S}}^{(h,w,c)} > \text{threshold} \\ 0, & \text{otherwise.} \end{cases} \tag{3}$$

Segmentation Network

The target-domain segmentation network, Seg_T, is an encoder-decoder network [5,24]. We supervise Seg_T using both the synthetic target images and the corresponding source labels and the real target images and their pseudo-labels (Fig. 4). Given an image X and its segmentation mask Y, the segmentation loss is defined as

$$\mathcal{L}_{Seg}(X,Y) = -\frac{2\sum_{h,w} P^{(h,w,1)} Y^{(h,w,1)}}{\sum_{h,w}(P^{(h,w,1)} + Y^{(h,w,1)})}, \tag{4}$$

where $P^{(h,w,1)}$ is the predictive probability of class 1 for voxel (h,w).

As in recent studies [2,27], to further regularize the network on the target-domain data for which we have not obtained pseudo-labels, we apply entropy-based regularization. The loss $\mathcal{L}_{ent}$ is defined as follows

$$\mathcal{L}_{Ent}(X_T) = \sum_{h,w} \frac{-1}{\log C} \sum_{c=1}^{C} P_{X_T}^{(h,w,c)} \log P_{X_T}^{(h,w,c)} \tag{5}$$

The full objective is given by $\min_{Seg_T} L_{Seg} + \mathcal{L}_{Ent}$.

2.2   Implementation Details

We implement our framework using Pytorch [22]. Our code is publicly available at https://github.com/elchiou/SemanticConsistUDA.

Image-to-Image Translation Network: The content encoders consist of convolutional layers and residual blocks followed by instance normalization [25]. The style encoders consist of convolutional layers followed by fully connected layers. The decoders include residual blocks followed by upsampling and convolutional layers. The residual blocks are followed by adaptive instance normalization (AdaIN) [11] layers to adjust the style of the output image. The discriminators consist of convolutional layers. For training we use Adam optimizer, a batch size of 32, a learning rate of 0.0001 and set losses weights to $\lambda_{GAN} = 1$, $\lambda_x = 10$, $\lambda_c = 1$, $\lambda_s = 1$, $\lambda_{sem} = 10$. We train the translation network for 50000 iterations.

Segmentation Network: The encoder of the segmentation network is a standard ResNet [10] consisting of convolutional layers while the decoder consists of upsampling and convolutional layers. For training we use stochastic gradient decent and a batch size of 32. We split the training set into 80% training and 20% validation to select the learning rate, the number of iterations and the threshold to perform pseudo-labeling.

3   Datasets

VERDICT-MRI: We use VERDICT-MRI data from 60 men. The DW-MRI was acquired with pulsed-gradient spin-echo sequence (PGSE) and an optimized imaging protocol for VERDICT prostate characterization with 5 b-values (90, 500, 1500, 2000, 3000 s/mm^2) in 3 orthogonal directions [14]. The DW-MRI sequence was acquired with a voxel size of $1.25 \times 1.25 \times 5$ mm^3, 5 mm slice thickness, 14 slices, and field of view of 220×220 mm^2 and the images were reconstructed to a 176×176 matrix size. A radiologist contoured the lesions on VERDICT-MRI using mp-MRI for guidance.

DW-MRI from mp-MRI Acquisition: We use DW-MRI data from 80 patients from the ProstateX challenge dataset [18]. Three b-values were acquired $(50, 400, 800 \text{ s/mm}^2)$, and the ADC map and a b-value image at b $= 1400$ s/mm^2 were calculated by the scanner. The DW-MRI data were acquired with a single-shot echo planar imaging sequence with a voxel size of $2 \times 2 \times 3.6$ mm^3, 3.6 mm slice thickness. Since the ProstateX dataset provides only the position of the lesion, a radiologist manually annotated the lesions on the ADC map using as reference the provided position.

4   Results

We evaluate the performance based on the average recall, precision, dice similarity coefficient (DSC), and average precision (AP) across 5-folds.

Table 1. Average recall, precision, dice similarity coefficient (DSC), and average precision (AP) across 5 folds. The results are given in mean (± std) format.

Model	Recall	Precision	DSC	AP
VERDICT-MRI (Oracle)	66.2 (8.1)	70.5 (9.9)	68.9 (9.2)	72.1 (10.4)
VERDICT-MRI + Synth (MUNIT + $\mathcal{L}_{Sem}$)	71.1 (8.9)	72.5 (10.4)	72.1 (8.7)	76.7 (9.6)
mp-MRI + EntMin (ADVENT [27])	50.8 (12.3)	48.0 (11.4)	49.8 (13.0)	51.4 (13.9)
Synth (MUNIT)	51.5 (13.3)	60.6 (11.9)	53.6 (12.7)	60.2 (13.0)
Synth (MUNIT + $\mathcal{L}_{Sem}$)	55.1 (13.9)	62.4 (12.8)	55.3 (10.9)	62.0 (13.4)
Synth (MUNIT + $\mathcal{L}_{Sem}$) + EntMin (ours)	54.7 (11.5)	**69.2** (10.3)	57.1 (10.8)	63.4 (12.8)
Synth (MUNIT + $\mathcal{L}_{Sem}$) + PsLab (ours)	59.8 (10.1)	64.8 (11.1)	61.5 (10.3)	64.9 (10.1)
Synth (MUNIT + $\mathcal{L}_{Sem}$) + EntMin + PsLab (ours)	**61.4** (9.9)	66.9 (10.7)	**62.1** (9.8)	**65.6** (10.9)

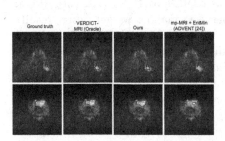

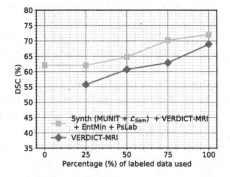

Fig. 5. Lesion segmentation results for two patients - the proposed approach performs well on the target domain despite the fact that it does not utilize any manual target annotations during training.

Fig. 6. Performance as we vary the percentage of labeled target data used for training. We observe that our method improves with more supervision and the improvements introduced by our method over the baseline of target-only training carry over all the way to the fully-supervised regime.

We compare our approach to several baselines. i) VERDICT-MRI: train using VERDICT-MRI only. ii) VERDICT-MRI + Synth (MUNIT + $\mathcal{L}_{Sem}$) : train using real VERDICT-MRI and the synthetic VERDICT-MRI obtained from MUNIT with semantic cycle-consistency loss. iii) mp-MRI + EntMin (ADVENT [27]): train the model by minimizing the segmentation loss, $\mathcal{L}_{Seg}(X_S, Y_S)$, on the raw mp-MRI and the entropy loss, $\mathcal{L}_{Ent}(X_T)$, on VERDICT-MRI, an approach proposed in [2,27,29]. iv) Synth (MUNIT): use the naive MUNIT to map from source to target and train only on the synthetic data. v) Synth (MUNIT + $\mathcal{L}_{Sem}$): use MUNIT with semantic cycle-consistency loss to translate from source to target. vi) Synth (MUNIT + $\mathcal{L}_{Sem}$) + EntMin: use (v) and entropy-based regularization on VERDICT-MRI data. vii) Synth (MUNIT + $\mathcal{L}_{Sem}$) + PsLab: use (v) and pseudo-labels to train the segmentation network on real VERDICT-MRI. viii) Synth (MUNIT + $\mathcal{L}_{Sem}$) + EntMin

+ PsLab: use (vi) and pseudo-labels to train the segmentation network on real VERDICT-MRI.

We report the results in Table 1. We observe that the performance is poor when the segmentation network is trained on the mp-MRI and VERDICT-MRI data (mp-MRI + EntMin (ADVENT [27])). However, we observe that when we train the network with synthetic VERDICT-MRI and real VERDICT-MRI (Synth (MUNIT + $\mathcal{L}_{Sem}$) + EntMin) the performance improves. This justifies our assumption that pixel-level alignment is beneficial in cases where there is a large distribution shift. The performance further improves when we use pseudo-labels obtained from confident predictions (Synth (MUNIT + $\mathcal{L}_{Sem}$) + EntMin + PsLab). We also observe that compared to the naive MUNIT without the semantic cycle-consistency loss (Synth (MUNIT)) our approach (Synth (MUNIT + $\mathcal{L}_{Sem}$)) performs better since it successfully preserves the lesions. When combining real and synthetic data (VERDICT-MRI + Synth (MUNIT + $\mathcal{L}_{Sem}$)) to train the network in a fully-supervised manner we get better results compared to the oracle, where we use only the real VERDICT-MRI. In Fig. 5 we present lesion segmentation results produced by the different models for two patients. The results indicate that the proposed approach performs well despite the fact that it does not use any manual annotations during training.

So far we have considered only the unsupervised case. However, our approach can also be used in a semi-supervised setting. To evaluate the performance of our method when labeled target data is available, we perform additional experiments varying the percentage of labeled data; we use the pseudo-labels (PsLab) and entropy minimization (EntMin) for the unlabeled data. Figure 6 shows that the performance of our method improves as the percentage of real data increases and always outperforms the baseline that is trained only on the target domain.

5   Conclusion

In this work we propose a domain adaptation approach for lesion segmentation. Our approach relies on appearance alignment along with pseudo-labeling to train a target domain classifier using exclusively target domain statistics. We demonstrate the effectiveness of our approach for lesion segmentation on VERDICT-MRI which is an advanced imaging technique for cancer characterization. However, the proposed work is a general method that can be extended to other applications where there is a large distribution gap between the source and the target domain.

References

1. Ahmed, H.U., et al.: Diagnostic accuracy of multi-parametric MRI and TRUS biopsy in prostate cancer (PROMIS): a paired validating confirmatory study. Lancet **389**, 815–822 (2017)

2. Bateson, M., Kervadec, H., Dolz, J., Lombaert, H., Ben Ayed, I.: Source-relaxed domain adaptation for image segmentation. In: Martel, A.L., et al. (eds.) MICCAI 2020. LNCS, vol. 12261, pp. 490–499. Springer, Cham (2020). https://doi.org/10.1007/978-3-030-59710-8_48

3. Cai, J., Zhang, Z., Cui, L., Zheng, Y., Yang, L.: Towards cross-modal organ translation and segmentation: a cycle and shape consistent generative adversarial network. MedIA **52**, 174–184 (2019)

4. Chen, C., Dou, Q., Chen, H., Qin, J., Heng, P.A.: Synergistic image and feature adaptation: towards cross-modality domain adaptation for medical image segmentation. In: AAAI (2019)

5. Chen, L.-C., Zhu, Y., Papandreou, G., Schroff, F., Adam, H.: Encoder-decoder with atrous separable convolution for semantic image segmentation. In: Ferrari, V., Hebert, M., Sminchisescu, C., Weiss, Y. (eds.) ECCV 2018. LNCS, vol. 11211, pp. 833–851. Springer, Cham (2018). https://doi.org/10.1007/978-3-030-01234-2_49

6. Chiou, E., Giganti, F., Bonet-Carne, E., Punwani, S., Kokkinos, I., Panagiotaki, E.: Prostate cancer classification on VERDICT DW-MRI using convolutional neural networks. In: MLMI (2018)

7. Chiou, E., Giganti, F., Punwani, S., Kokkinos, I., Panagiotaki, E.: Automatic classification of benign and malignant prostate lesions: a comparison using VERDICT DW-MRI and ADC maps. In: ISMRM (2019)

8. Chiou, E., Giganti, F., Punwani, S., Kokkinos, I., Panagiotaki, E.: Domain adaptation for prostate lesion segmentation on VERDICT-MRI. In: ISMRM (2020)

9. Chiou, E., Giganti, F., Punwani, S., Kokkinos, I., Panagiotaki, E.: Harnessing uncertainty in domain adaptation for MRI prostate lesion segmentation. In: Martel, A.L., et al. (eds.) MICCAI 2020. LNCS, vol. 12261, pp. 510–520. Springer, Cham (2020). https://doi.org/10.1007/978-3-030-59710-8_50

10. He, K., Zhang, X., Ren, S., Sun, J.: Deep residual learning for image recognition. In: CVPR (2016)

11. Huang, X., Belongie, S.: Arbitrary style transfer in real-time with adaptive instance normalization. In: ICCV (2017)

12. Huang, X., Liu, M.-Y., Belongie, S., Kautz, J.: Multimodal unsupervised image-to-image translation. In: Ferrari, V., Hebert, M., Sminchisescu, C., Weiss, Y. (eds.) ECCV 2018. LNCS, vol. 11207, pp. 179–196. Springer, Cham (2018). https://doi.org/10.1007/978-3-030-01219-9_11

13. Jiang, J., et al.: Tumor-aware, adversarial domain adaptation from CT to MRI for lung cancer segmentation. In: Frangi, A.F., Schnabel, J.A., Davatzikos, C., Alberola-López, C., Fichtinger, G. (eds.) MICCAI 2018. LNCS, vol. 11071, pp. 777–785. Springer, Cham (2018). https://doi.org/10.1007/978-3-030-00934-2_86

14. Johnston, E., Chan, R.W., Stevens, N., Atkinson, D., Punwani, S., Hawkes, D.J., Alexander, D.C.: Optimised VERDICT MRI protocol for prostate cancer characterisation. In: ISMRM (2015)

15. Johnston, E.W., et al.: VERDICT-MRI for prostate cancer: intracellular volume fraction versus apparent diffusion coefficient. Radiology **291**, 391–397 (2019)

16. Kamnitsas, K., et al.: Unsupervised domain adaptation in brain lesion segmentation with adversarial networks. In: IPMI (2017)

17. Li, K., Wang, S., Yu, L., Heng, P.-A.: Dual-teacher: integrating intra-domain and inter-domain teachers for annotation-efficient cardiac segmentation. In: Martel, A.L., et al. (eds.) MICCAI 2020. LNCS, vol. 12261, pp. 418–427. Springer, Cham (2020). https://doi.org/10.1007/978-3-030-59710-8_41

18. Litjens, G., Debats, O., Barentsz, J., Karssemeijer, N., Huisman, H.: Computer-aided detection of prostate cancer in MRI. TMI **33**, 1083–1092 (2014)

19. Ouyang, C., Kamnitsas, K., Biffi, C., Duan, J., Rueckert, D.: Data efficient unsupervised domain adaptation for cross-modality image segmentation. In: Shen, D., et al. (eds.) MICCAI 2019. LNCS, vol. 11765, pp. 669–677. Springer, Cham (2019). https://doi.org/10.1007/978-3-030-32245-8_74

20. Panagiotaki, E., et al.: Microstructural characterization of normal and malignant human prostate tissue with vascular, extracellular, and restricted diffusion for cytometry in tumours magnetic resonance imaging. Investigate Radiol. 50, 218–227 (2015)

21. Panagiotaki, E., et al.: Noninvasive quantification of solid tumor microstructure using VERDICT MRI. Cancer Res. 74, 1902–1912 (2014)

22. Paszke, A., et al.: Automatic differentiation in PyTorch. In: Autodiff Workshop, NIPS (2017)

23. Ren, J., Hacihaliloglu, I., Singer, E.A., Foran, D.J., Qi, X.: Adversarial domain adaptation for classification of prostate histopathology whole-slide images. In: Frangi, A.F., Schnabel, J.A., Davatzikos, C., Alberola-López, C., Fichtinger, G. (eds.) MICCAI 2018. LNCS, vol. 11071, pp. 201–209. Springer, Cham (2018). https://doi.org/10.1007/978-3-030-00934-2_23

24. Ronneberger, O., Fischer, P., Brox, T.: U-Net: convolutional networks for biomedical image segmentation. In: Navab, N., Hornegger, J., Wells, W.M., Frangi, A.F. (eds.) MICCAI 2015. LNCS, vol. 9351, pp. 234–241. Springer, Cham (2015). https://doi.org/10.1007/978-3-319-24574-4_28

25. Ulyanov, D., Vedaldi, A., Lempitsky, V.S.: Improved texture networks: maximizing quality and diversity in feed-forward stylization and texture synthesis. In: CVPR (2017)

26. Valindria, V., Palombo, M., Chiou, E., Singh, S., Punwani, S., Panagiotaki, E.: Synthetic q-space learning with deep regression networks for prostate cancer characterisation with verdict. In: ISBI (2021)

27. Vu, T.H., Jain, H., Bucher, M., Cord, M., Perez, P.: ADVENT: adversarial entropy minimization for domain adaptation in semantic segmentation. In: CVPR (2019)

28. Yang, J., Dvornek, N.C., Zhang, F., Chapiro, J., Lin, M.D., Duncan, J.S.: Unsupervised domain adaptation via disentangled representations: application to cross-modality liver segmentation. In: Shen, D., et al. (eds.) MICCAI 2019. LNCS, vol. 11765, pp. 255–263. Springer, Cham (2019). https://doi.org/10.1007/978-3-030-32245-8_29

29. Zeng, G., et al.: Entropy guided unsupervised domain adaptation for cross-center hip cartilage segmentation from MRI. In: Martel, A.L., et al. (eds.) Medical Image Computing and Computer Assisted Intervention. MICCAI 2020. LNCS, vol. 12261, pp. 447–456. Springer, Cham (2020). https://doi.org/10.1007/978-3-030-59710-8_44

30. Zhang, Y., Miao, S., Mansi, T., Liao, R.: Task driven generative modeling for unsupervised domain adaptation: application to X-ray image segmentation. In: Frangi, A.F., Schnabel, J.A., Davatzikos, C., Alberola-López, C., Fichtinger, G. (eds.) MICCAI 2018. LNCS, vol. 11071, pp. 599–607. Springer, Cham (2018). https://doi.org/10.1007/978-3-030-00934-2_67

31. Zhang, Z., Yang, L., Zheng, Y.: Translating and segmenting multimodal medical volumes with cycle and shape consistency generative adversarial network. In: CVPR (2018)

32. Zhu, J., Park, T., Isola, P., Efros, A.A.: Unpaired image-to-image translation using cycle-consistent adversarial networks. In: ICCV (2017)

Cohort Bias Adaptation in Aggregated Datasets for Lesion Segmentation

Brennan Nichyporuk[1,2(✉)], Jillian Cardinell[1,2], Justin Szeto[1,2],
Raghav Mehta[1,2], Sotirios Tsaftaris[3,4], Douglas L. Arnold[5,6], and Tal Arbel[1,2]

[1] Centre for Intelligent Machines, McGill University, Montreal, Canada
brennann@cim.mcgill.ca
[2] MILA (Quebec Artificial Intelligence Institute), Montreal, Canada
[3] Institute for Digital Communications, School of Engineering,
University of Edinburgh, Edinburgh, UK
[4] The Alan Turing Institute, London, UK
[5] Department of Neurology and Neurosurgery, McGill University, Montreal, Canada
[6] NeuroRx Research, Montreal, Canada

Abstract. Many automatic machine learning models developed for focal pathology (e.g. lesions, tumours) detection and segmentation perform well, but do not generalize as well to new patient cohorts, impeding their widespread adoption into real clinical contexts. One strategy to create a more diverse, generalizable training set is to naively pool datasets from different cohorts. Surprisingly, training on this *big data* does not necessarily increase, and may even reduce, overall performance and model generalizability, due to the existence of cohort biases that affect label distributions. In this paper, we propose a generalized affine conditioning framework to learn and account for cohort biases across multi-source datasets, which we call Source-Conditioned Instance Normalization (SCIN). Through extensive experimentation on three different, large scale, multi-scanner, multi-centre Multiple Sclerosis (MS) clinical trial MRI datasets, we show that our cohort bias adaptation method (1) improves performance of the network on pooled datasets relative to naively pooling datasets and (2) can quickly adapt to a new cohort by fine-tuning the instance normalization parameters, thus learning the new cohort bias with only 10 labelled samples.

Keywords: Deep learning · Multiple sclerosis · Instance norm conditioning · Trial merging · Segmentation · Label style · Cohort bias

1 Introduction

A large number of deep learning (DL) models have been developed and successfully applied to the contexts of healthy structure and pathology segmentation (e.g. brain tumours), with good performance on a number of public datasets [11,16]. However, DL methods are generally not able to overcome large mismatches between the training and testing distributions. Several recent papers

© Springer Nature Switzerland AG 2021
S. Albarqouni et al. (Eds.): DART 2021/FAIR 2021, LNCS 12968, pp. 101–111, 2021.
https://doi.org/10.1007/978-3-030-87722-4_10

have focused on addressing poor performance and generalizability resulting from differences in imaging acquisition parameters, resolutions, and scanners differences [1,12,20], missing modalities and sequences [5,17], or overcoming labelling style differences between individual raters [2,10,18,21].

In the context of focal pathology (e.g. lesions, tumours) detection and segmentation, the inability to generalize methods to new patient cohorts seriously impedes their widespread adoption into real clinical contexts. However, the problem of generalization is much more subtle in this context, particularly given the lack of real "ground truth" labels. Even in cohorts with similar image acquisition parameters and labelling protocols, significant biases will still exist due to differences between patient populations. Indeed, even if the same set of raters label both cohorts, the probability of labeling any given voxel can still depend on rater knowledge about the overall distribution of each patient population. One obvious strategy to create a more diverse, generalizable training set is to naively pool datasets from different cohorts. However, this will not necessarily increase, and may even decrease, overall performance and model generalizability due to biases that ultimately affect the label distribution of each cohort.

In this work, we propose a generalized conditioning framework to learn and account for cohort biases, and associated annotation *style*, across multi-source pooled datasets. The approach entails the use of conditional instance normalization layers [4] to condition the network on auxiliary cohort information in an approach we call Source-Conditioned Instance Normalization (SCIN). This allows the network to leverage a pooled dataset while providing cohort-specific segmentations. A key advantage of the proposed approach is that it can be adapted to a new clinical dataset if provided with a small subset of labelled samples. Experiments are performed on three different large-scale, multi-scanner, multi-center multiple sclerosis (MS) clinical trial datasets. In particular, we show that our cohort bias adaptation method can (1) learn an annotation protocol specific to each cohort that makes up aggregated dataset without loss in performance, (2) improves performance of the network on pooled datasets relative to naively pooling datasets, and (3) can quickly adapt an already conditioned DL model on a new clinical trial dataset with only 10 labeled samples, by fine-tuning the conditional instance normalization parameters. Finally, to illustrate that our model is accounting for complex cohort biases, we artificially create a sub-trial cohort without small lesions labelled, and show that our model is able to learn this cohort bias and account for it.

2   Related Work

Several papers have focused on accounting for the differences in image acquisition parameters or missing modalities [1,12,20]. However, in clinical trials, these variables are usually kept constant or are vigorously normalized. On the other hand, several papers have examined label bias, focusing on inter-rater bias or modelling an individual rater's label style [6,9,13,22]. In these cases, each rater is provided the same cohort to label, and inter-rater biases can be attributed to

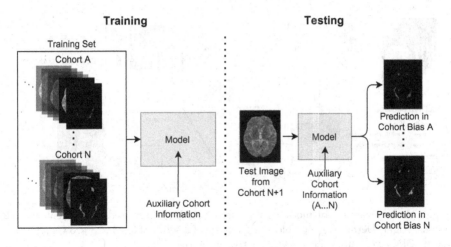

Fig. 1. System overview showing training on the left and testing on the right. The left shows how we train with multiple cohorts and use auxiliary cohort information to learn the associated bias. On the right is how we use cohort information during testing to generate multiple labels for an image in a desired style.

differences in experience level or lesion boundary uncertainty [10]. Furthermore, in many cases a simplifying assumption, that each rater is a noisy reflection of ground truth, is made [8,23]. Despite these shortcomings, inter-rater studies have stressed the importance of considering biases that can affect labels, with Schwartzman *et al.* [18] showing that rater bias in training samples is actually amplified by neural networks. Other researchers have also found that multiple labels with varying biases can provide important insights into uncertainty estimation [2,21].

3   Methods

We propose a framework that is able to learn cohort-specific biases which are ultimately manifested in the 'ground truth' labels. To do this, we employ the conditional instance normalization mechanism proposed by Vincent *et al.* [4] to model cohort biases with source-specific parameters. In this paper, we refer to the whole approach as Source-conditioned Instance Normalization (SCIN), and the conditioning mechanism proposed by Vincent et al. as Conditional Instance Normalization (CIN) [4]. In practice, this involves scaling/shifting the normalized activations at each layer by using the set of affine parameters corresponding to a given sample's source cohort during the forward pass. As a result, source-specific affine parameters are learned only from samples that make up the cohort itself. Mathematically, the mechanism is represented by the following equation:

$$\mathrm{CIN}(z) = \gamma_s \left(\frac{z - \mu(z)}{\sigma(z)} \right) + \beta_s$$

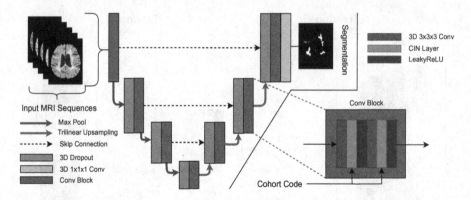

Fig. 2. Left: Overview of modified nnUNet [7] architecture used to segment MS T2 lesions. Right: Detail of a conv block. It consists of a series of 3D $3 \times 3 \times 3$ Convolution Layer, CIN layer, and a LeakyReLU activation layer.

where γ_s and β_s are the affine parameters specific to cohort source s, and where $\mu(z)$ and $\sigma(z)$ represent the per-channel mean and standard deviation, respectively. Other than cohort-specific parameters, all other network parameters are learned from all samples regardless of cohort identity. This allows the approach to pool multiple datasets, leveraging the aggregated dataset to learn common features while still taking into account cohort-specific biases. A full system overview can be found in Fig. 1.

4   Implementation Details

4.1   Network Architecture and Training Parameters

We utilize a modified version of nnUNet [7] for MS T2-weighted lesion segmentation. Specifically, we add dropout and CIN as depicted in Fig. 2. All models in this paper are trained using Binary Cross Entropy (BCE).

4.2   Data Set

We make use of a large proprietary MS dataset from three different clinical trials, where each trial contains multi-modal MR sequences of patients with different disease subtypes, and/or at different stages of disease. Specifically, Trial-A is a late-stage Secondary-Progressive (SPMS) dataset with 1000 patients, Trial-B is a Relapsing Remitting (RRMS) dataset with 1000 patients, and Trial-C is an early-stage SPMS dataset with 500 patients. Each patient sample consists of 5 MR sequences (T1-weighted, T1-weighted with gadolinium contrast agent, T2-weighted, Fluid Attenuated Inverse Recovery, and Proton Density). All MR sequences were acquired at 1 mm × 1 mm × 3 mm resolution. T2 lesion labels were generated at the end of the clinical trial, and were produced through an

external process where trained expert annotators manually corrected a proprietary automated segmentation method. All data was processed using similar
processing pipeline(s), but at different points in time (i.e. the clinical trials were
not all held at the same time) and with different versions of the baseline segmentation method. Although different expert raters corrected the labels, there was
overlap between raters across trials and all raters were trained to follow a similar labelling protocol. All three trial datasets are divided into non-overlapping
training (60%), validation (20%), and testing (20%) sets.

4.3   Evaluation Metrics

Automatic T2 lesion segmentation methods are evaluated with voxel-level segmentation, and for the small lesion removal experiments in Sect. 5.3, lesion-level
detection metrics. Specifically, we use DICE score [15] for voxel-level segmentation and F1-score for lesion-level detection analysis [3].[1] All methods utilized in
this work output a voxel-level segmentation mask. A connected component analysis is performed on the voxel-based segmentation mask to group lesion voxels
in an 18-connected neighbourhood [14]. Given that MS lesions vary significantly
in size, we report results stratified by lesion size.

5   Experiments and Results

We perform three different sets of experiments to demonstrate the usefulness of
the proposed SCIN approach. In the first experiment, we show that SCIN is able
to strategically pool diverse datasets with differing cohort biases. The second
experiment demonstrates the clinical utility of SCIN to adapt to new cohort
biases with limited available labeled data. Finally, we show that SCIN is able
to model complex cohort biases by simulating a type of cohort bias where small
lesions (10 voxels or less) were not labeled.

5.1   Trial Conditioning

Experiments in this section aim to show how the proposed SCIN approach allows
for pooling of data from multiple cohorts while taking into account cohort specific
biases. We use two different clinical trial datasets (Trial-A and Trial-B) for these
experiments. These two trials were collected several years apart with patients
of different disease subtypes. Given that each trial is processed independently
and at different points in time, minor differences in site/scanner distribution
and annotation style will exist. Together, the patient population, site/scanner
distribution, and annotation style create a distinct cohort bias, which we aim to
account for with the proposed method.

We train four different models on these datasets: (i) a model trained on only
Trial-A, with Instance Norm (IN) [19], (ii) a model trained on only Trial-B (with

---

[1] Operating point (threshold) is chosen based on the best voxel-level DICE.

Table 1. Dice scores shown on Trial-A and Trial-B test sets for models trained with different combinations of Trial-A and Trial-B training sets. Trial-A and Trial-B training sets each contain 600 patients.

	Model	Train Set		Conditioned On		Test Performance	
		Trial-A	Trial-B	Trial-A	Trial-B	Trial-A	Trial-B
1	Single-Trial	✓		-		0.793	0.689
2	Single-Trial		✓	-		0.715	0.803
3	Naive-Pooling	✓	✓	-		0.789	0.748
4	SCIN-Pooling	✓	✓	✓		0.794	0.700
5		✓	✓		✓	0.725	0.797

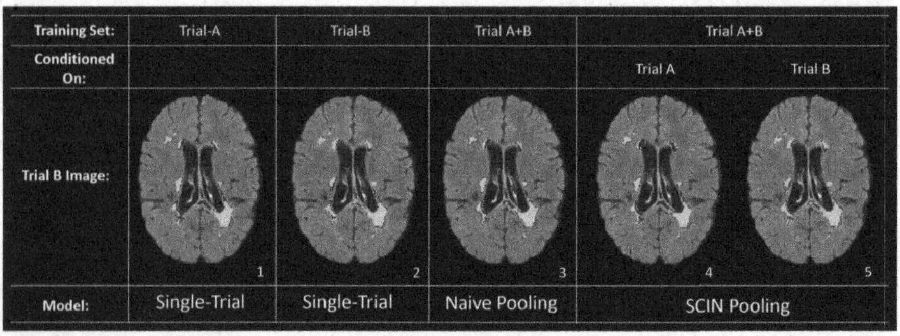

Fig. 3. Qualitative lesion segmentation labels (red is false positives, blue is false negatives, green is true positives) superimposed onto a FLAIR test image from Trial B. The results are based on the models from Rows 1–5 (left to right) of Table 1. (Color figure online)

IN), (iii) a model trained on a naively pooled dataset consisting of Trial-A and Trial-B (with IN), and (iv) a model trained on both Trial-A and Trial-B using the SCIN approach. All four models were tested on the same held-out test set from both trials.

Table 1 depicts the performance of the aforementioned models on the hold-out test sets. Results indicate that models trained on only one trial (Row-1 and Row-2) generalize poorly when tested on the other trial. A model trained on the naively pooled dataset, consisting of data from both trials (Row-3), shows better generalization across trials, but still falls short of the performance achieved by each trial-specific model, especially in the case of the Trial-B dataset. At first glance, this might appear surprising given the expectation that a model trained on a larger, pooled dataset should generally perform better relative to a model that has access to less data. However, given the knowledge that biases can exist between cohorts, it is no mystery why the naively pooled model would be unable to generate a labeling consistent with the cohort of the sample, given that the cohort/labeling bias cannot be identified from the image alone. On the

Table 2. Dice scores shown on the Trial-C test set from the Naive-pooling and SCIN-pooling models trained on Trial A and B. Dice scores are also shown for fine-tuned versions of those models, where the IN parameters were tuned using 10 Trial-C samples.

	Model	Fine-Tuned on Trial-C	Conditioned on			Test Performance
			Trial-A	Trial-B	Trial-C	
1	Naive-Pooling			-		0.774
2		✓		-		0.819
3	SCIN-Pooling		✓			0.763
4				✓		0.806
5		✓			✓	0.834

other hand, a single SCIN-pooling model conditioned on each trial (Row-4 and Row-5), is able to learn trial-specific parameters to model the bias specific to the trial in question, improving performance relative to the naively-pooled model (Row-3). Note that using the incorrect set of trial-specific parameters with the SCIN-Pooling model (Row-4 and Row-5) results in a performance decline similar to that observed when testing the Single-Trial models (Row-1 and Row-2) on the corresponding unseen trial. This simply serves as a sanity check, and confirms that the proposed method effectively models the cohort bias of each dataset.

Qualitative results for labels produced by different models on a single Trial-B test case are shown in Fig. 3. From this, we can see that generating labels on a Trial-B test case using a Single-Trial model trained on the Trial-A dataset (Image-1) leads to an increased number of false positive and false negative voxels. This is also true when testing a naively pooled model (Image-3). On the other hand, the proposed SCIN-pooled model conditioned on Trial-B (Image 5) does not suffer from a significant degradation in segmentation quality, showing that the SCIN approach enables leveraging multiple datasets with different cohort biases without significant performance decrements. Visually, note that the labeling style of the SCIN-pooled model is similar to that of the corresponding Single-Trial model, showing that SCIN-pooled method is able to learn a cohort-specific bias for each trial.

5.2   Fine-Tuning to New Cohort Bias

The second set of experiments aims to mimic a clinical situation where large datasets (Trial-A and Trial-B) have a known cohort bias. A new small dataset (Trial-C) is provided with an unidentified cohort bias. We take two pre-trained models (Naively-Pooled and SCIN-pooled) from the previous set of experiments (see Sect. 5.1), and fine-tune the affine parameters of the IN/CIN layers with only 10 labeled samples from the Trial-C dataset. Segmentations are performed on a hold-out test set from Trial-C. Similar to Experiment 1, time of collection and disease subtype were the primary differences between the three trials (along with minor differences in scanner/site distribution and labeling protocol).

Table 2 depicts the results for this set of experiments. We can see that the performance of the naively-pooled model improves when the IN parameters of the

Table 3. Voxel based Dice scores and small lesion detection F1 scores shown on Trial-C (Trial-Orig) held-out test set using models trained on different combinations of the original dataset (Trial-Orig, 150 training patients) and the dataset with missing small lesions (Trial-MSL, 150 training patients)

	Model	Train Set		Conditioned On		Test Performance	
		Trial-Orig	Trial-MSL	Trial-Orig	Trial-MSL	Sm Lesion F1	Voxel Dice
1	Single-Trial	✓		-		0.795	0.844
2	Single-Trial		✓	-		0.419	0.837
3	Naive-Pooling	✓	✓	-		0.790	0.797
4	SCIN-Pooling	✓	✓	✓		0.784	0.854
5		✓	✓		✓	0.496	0.850

model are fine-tuned (Row-2) compared to no fine-tuning (Row-1). Furthermore, a SCIN-pooled model shows good Trial-C performance when conditioned on Trial-B (Row-4), indicating that Trial-C has similar cohort biases. By fine-tuning the trial-specific CIN layer parameters of this model on 10 samples of Trial-C, we are able to then condition the model on Trial-C during test time. This leads to the highest performance improvement (Row-5) over all models, including fine-tuning the naively pooled model (Compare Row-2 and Row-5). This shows that with SCIN, we can more effectively learn features common to both Trail-A and Trial-B, resulting in better performance after fine-tuning on a hold-out trial.

5.3   Accounting for Complex Cohort Biases - Missing Small Lesions

The final set of experiments examine whether the SCIN approach is able to learn complex non-linear cohort biases. Accordingly, we isolate biases that arise solely from different labelling protocols while keep all other factors, such as disease stage and time of collection, constant. To that end, we utilize a held-out clinical trial dataset (Trial-C) and artificially modify half of the dataset by removing small lesions (10 voxels or less) from the provided labels. This can be thought of as being equivalent to a labeling protocol that misses or ignores small lesions (Trial-MSL). The labels of the remaining half of the dataset are not modified in any way (Trial-Orig).

Table 3 shows the results for this set of experiments on a non-modified Trial-C test set (Trial-Orig). We report detailed results specific to the detection of small lesions in order to examine whether the proposed strategy learns to account for a labeling style that ignores small lesions. The results in Row-1 show that when both train set and test set have the same labeling protocol (Trial-Orig), the Single-Trial model performs well according to both lesion-level detection and voxel-level segmentation metrics. On the other hand, when there is a significant shift in the labeling protocol between the train and test set, the Single-Trial model trained on the Trial-MSL dataset (Row-2) exhibits poor small lesion detection performance. The degradation in small lesion detection performance is expected as Trial-MSL has small lesions labeled as background, while the test set (Trial-Orig) has small lesions marked as lesions. The Naively-Pooled model (Row-3), which is trained on both Trial-Orig and Trial-MSL, learns to completely ignore the bias of the Trial-MSL dataset as measured by lesion-level detection performance. However, voxel-level segmentation performance suffers significantly. On the other hand, a model trained using the SCIN approach is able to adapt to the difference in labeling styles and exhibit good lesion-level detection and voxel-level segmentation performance when conditioned on the appropriate cohort (Row-4). Looking at SCIN-Pooling conditioned on Trial-MSL (Row-5), we see that SCIN is able to learn the Trial-MSL label bias quite effectively, and is able to ignore small lesions while maintaining voxel-level segmentation performance. This shows that SCIN is able to model complex non-linear labeling biases – its not just a matter of over or under segmentation.

6   Conclusions

In this paper, we proposed SCIN, an approach that learns source-specific IN parameters, effectively modeling the bias of each cohort present in an aggregated dataset. We show that the IN parameters of a pre-trained SCIN model can be fine-tuned, allowing the model to learn the bias of an independent cohort with very little data. We demonstrate that the biases learned can be non-linear, resulting in complex differences in the segmentation outputs corresponding to each cohort. Most importantly, we show that proposed method makes it possible to train a high performance model on an aggregate dataset, avoiding the performance penalty observed with naive pooling. Overall, SCIN is simple to implement, and can potentially benefit any application that wishes to leverage a large aggregated dataset in the context of segmentation.

Acknowledgement. This work was supported by an award from the International Progressive MS Alliance (PA-1603-08175) and by funding from the Canada Institute for Advanced Research (CIFAR) Artificial Intelligence Chairs program. S.A. Tsaftaris acknowledges the support of Canon Medical and the Royal Academy of Engineering and the Research Chairs and Senior Research Fellowships scheme (grant RCSRF1819 \8\25).

References

1. Biberacher, V., et al.: Intra-and interscanner variability of magnetic resonance imaging based volumetry in multiple sclerosis. Neuroimage **142**, 188–197 (2016)
2. Chotzoglou, E., Kainz, B.: Exploring the relationship between segmentation uncertainty, segmentation performance and inter-observer variability with probabilistic networks. In: Zhou, L., et al. (eds.) LABELS/HAL-MICCAI/CuRIOUS -2019. LNCS, vol. 11851, pp. 51–60. Springer, Cham (2019). https://doi.org/10.1007/978-3-030-33642-4_6
3. Commowick, O., et al.: Objective evaluation of multiple sclerosis lesion segmentation using a data management and processing infrastructure. Sci. Rep. **8**(1), 1–17 (2018)

4. Vincent, D., Jonathon, S., Manjunath, K.: A learned representation for artistic style. In: 5th International Conference on Learning Representations, ICLR 2017, Conference Track Proceedings, Toulon, France, 24–26 April 2017 (2017). OpenReview.net

5. Havaei, M., Guizard, N., Chapados, N., Bengio, Y.: HeMIS: hetero-modal image segmentation. In: Ourselin, S., Joskowicz, L., Sabuncu, M.R., Unal, G., Wells, W. (eds.) MICCAI 2016. LNCS, vol. 9901, pp. 469–477. Springer, Cham (2016). https://doi.org/10.1007/978-3-319-46723-8_54

6. Heller, N., Dean, J., Papanikolopoulos, N.: Imperfect segmentation labels: how much do they matter? In: Stoyanov, D., et al. (eds.) LABELS/CVII/STENT - 2018. LNCS, vol. 11043, pp. 112–120. Springer, Cham (2018). https://doi.org/10.1007/978-3-030-01364-6_13

7. Isensee, F., Kickingereder, P., Wick, W., Bendszus, M., Maier-Hein, K.H.: No new-net. In: Crimi, A., Bakas, S., Kuijf, H., Keyvan, F., Reyes, M., van Walsum, T. (eds.) BrainLes 2018. LNCS, vol. 11384, pp. 234–244. Springer, Cham (2019). https://doi.org/10.1007/978-3-030-11726-9_21

8. Ji, W., et al.: Learning calibrated medical image segmentation via multi-rater agreement modeling. In: Proceedings of the IEEE/CVF Conference on Computer Vision and Pattern Recognition, pp. 12341–12351 (2021)

9. Joskowicz, L., Cohen, D., Caplan, N., Sosna, J.: Inter-observer variability of manual contour delineation of structures in CT. Eur. Radiol. **29**(3), 1391–1399 (2019)

10. Jungo, A., et al.: On the effect of inter-observer variability for a reliable estimation of uncertainty of medical image segmentation. In: Frangi, A.F., Schnabel, J.A., Davatzikos, C., Alberola-López, C., Fichtinger, G. (eds.) MICCAI 2018. LNCS, vol. 11070, pp. 682–690. Springer, Cham (2018). https://doi.org/10.1007/978-3-030-00928-1_77

11. Kamnitsas, K., et al.: Efficient multi-scale 3D CNN with fully connected CRF for accurate brain lesion segmentation. Med. Image Anal. **36**, 61–78 (2017)

12. Karani, N., Chaitanya, K., Baumgartner, C., Konukoglu, E.: A lifelong learning approach to brain MR segmentation across scanners and protocols. In: Frangi, A.F., Schnabel, J.A., Davatzikos, C., Alberola-López, C., Fichtinger, G. (eds.) MICCAI 2018. LNCS, vol. 11070, pp. 476–484. Springer, Cham (2018). https://doi.org/10.1007/978-3-030-00928-1_54

13. Kohl, S.A.A., et al.: A hierarchical probabilistic U-Net for modeling multi-scale ambiguities. arXiv preprint arXiv:1905.13077 (2019)

14. Nair, T., Precup, D., Arnold, D.L., Arbel, T.: Exploring uncertainty measures in deep networks for multiple sclerosis lesion detection and segmentation. Med. Image Anal. **59**, 101557 (2020)

15. Powers, D.: Evaluation: from precision, recall and f-factor to ROC, informedness, markedness & correlation. Mach. Learn. Technol. **2**, 01 (2008)

16. Ronneberger, O., Fischer, P., Brox, T.: U-Net: convolutional networks for biomedical image segmentation. In: Navab, N., Hornegger, J., Wells, W.M., Frangi, A.F. (eds.) MICCAI 2015. LNCS, vol. 9351, pp. 234–241. Springer, Cham (2015). https://doi.org/10.1007/978-3-319-24574-4_28

17. Shen, Y., Gao, M.: Brain tumor segmentation on MRI with missing modalities. In: Chung, A.C.S., Gee, J.C., Yushkevich, P.A., Bao, S. (eds.) IPMI 2019. LNCS, vol. 11492, pp. 417–428. Springer, Cham (2019). https://doi.org/10.1007/978-3-030-20351-1_32

18. Shwartzman, O., Gazit, H., Shelef, I., Riklin-Raviv, T.: The worrisome impact of an inter-rater bias on neural network training. arXiv preprint arXiv:1906.11872 (2019)

19. Ulyanov, D., Vedaldi, A., Lempitsky, V.: Instance normalization: the missing ingredient for fast stylization, 07 2016
20. Van Opbroek, A., Ikram, M.A., Vernooij, M.W., De Bruijne, M.: Transfer learning improves supervised image segmentation across imaging protocols. IEEE Trans. Med. Imaging **34**(5), 1018–1030 (2014)
21. Vincent, O., Gros, C., Cohen-Adad, J.: Impact of individual rater style on deep learning uncertainty in medical imaging segmentation. arXiv preprint arXiv:2105.02197 (2021)
22. Warfield, S.K., Zou, K.H., Wells, W.M.: Simultaneous truth and performance level estimation (staple): an algorithm for the validation of image segmentation. IEEE Trans. Med. Imaging **23**(7), 903–921 (2004)
23. Zhang, L., et al.: Disentangling human error from the ground truth in segmentation of medical images. arXiv preprint arXiv:2007.15963 (2020)

Exploring Deep Registration Latent Spaces

Théo Estienne[1,2](✉), Maria Vakalopoulou[1], Stergios Christodoulidis[1],
Enzo Battistella[1,2], Théophraste Henry[2], Marvin Lerousseau[1,2],
Amaury Leroy[2,4], Guillaume Chassagnon[3], Marie-Pierre Revel[3],
Nikos Paragios[4], and Eric Deutsch[2]

[1] Université Paris-Saclay, CentraleSupélec, Mathématiques et Informatique pour la Complexité et les Systèmes, Inria Saclay, 91190 Gif-sur-Yvette, France
theo.estienne@centralesupelec.fr
[2] Université Paris-Saclay, Institut Gustave Roussy, Inserm, Radiothérapie Moléculaire et Innovation Thérapeutique, 94800 Villejuif, France
[3] Departement de Radiology, Hôpital Cochin, AP-HP Centre, Université de Paris, 27 Rue du Faubourg Saint-Jacques, 75014 Paris, France
[4] TheraPanacea, Pépiniere Santé Cochin, Paris, France

Abstract. Explainability of deep neural networks is one of the most challenging and interesting problems in the field. In this study, we investigate the topic focusing on the interpretability of deep learning-based registration methods. In particular, with the appropriate model architecture and using a simple linear projection, we decompose the encoding space, generating a new basis, and we empirically show that this basis captures various decomposed anatomically aware geometrical transformations. We perform experiments using two different datasets focusing on lungs and hippocampus MRI. We show that such an approach can decompose the highly convoluted latent spaces of registration pipelines in an orthogonal space with several interesting properties. We hope that this work could shed some light on a better understanding of deep learning-based registration methods.

Keywords: Deep learning-based medical image registration ·
Deformable registration · Explainability

1 Introduction

Deep learning methods provide the state of the art performance for various applications currently. This is due to their inherent property to generate highly abstract representations hierarchically. These representations are building on top of each other, making it possible to encode highly non-linear manifolds.

Electronic supplementary material The online version of this chapter (https://doi.org/10.1007/978-3-030-87722-4_11) contains supplementary material, which is available to authorized users.

© Springer Nature Switzerland AG 2021
S. Albarqouni et al. (Eds.): DART 2021/FAIR 2021, LNCS 12968, pp. 112–122, 2021.
https://doi.org/10.1007/978-3-030-87722-4_11

Even though such hierarchies can outperform traditional methods, they lack explainability, making their translation difficult to solve real-life problems. This drawback is of great significance in the medical field and especially for the algorithms that are intended to be adapted to clinical practice, addressing problems of precision medicine [2,12]. For these reasons, it is essential to identify ways to understand better the high throughput operations that are applied.

Recently, with the introduction of the differentiable spatial transformer [13], trainable deep learning registration methods are becoming more and more popular, reducing computational times while reporting similar to traditional methods performance [1,17,25]. Meanwhile, the deformation field, which is one of the products of deformable registration methods, has been shown to encode not only the spatial correspondences but also clinical relevant information that could add valuable aspects to a variety of problems related to survival assessment or anomaly detection [18]. Indeed, encoding information between subjects can be very informative for various medical tasks such as medical image segmentation [9]. However, according to our knowledge, there are not many efforts focusing on understanding and analysing this encoding information which could initiate the explainability of deep learning-based registration methods.

In this study, we propose a framework for interpreting the encoded representation of deep learning-based registration methods. In particular, with the appropriate model architecture and by using a simple linear projection, we decompose the encoding space, generating a new basis that captures various geometrical operations. This decomposed encoding space is then driving the generation of the deformation field. The contributions of this work are twofold: *(i)* to the best of our knowledge, this study is one of the first to explore the explainability of deep learning-based registration methods through their encodings using linear projections, *(ii)* we show empirically, using two different datasets, one focusing on lungs and the other on the brain hippocampus that our projections are associated with different types of deformations and in particular rigid transformations. We hope that this work can highlight the very challenging topic of explainability of deep neural networks.

2   Related Work

Explaining how deep neural networks function is a matter of extensive research the recent years. GradCam [21] is one of the most popular methods that can provide some insights on deep neural networks for many applications, including medical imaging. GradCam highlights the region of the original input that contributes the most to the final prediction, producing coarse heatmaps based on the gradients. Similar to GradCam, there are many additional methods based on the gradient [3,24,27] that are commonly used for the explainability of the models. Moreover, in [10] the authors proposed a general framework of explanations as meta-predictors while they also reinterpret the network's saliency providing a natural generalisation of the gradient-based saliency techniques. Even though such approaches can provide information on where the models attend, they can be mostly utilised in classification or detection schemes.

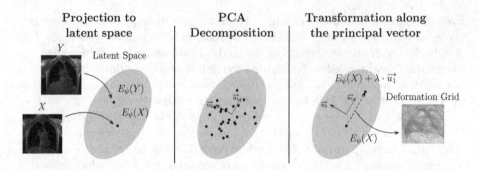

Fig. 1. Overall overview of our proposed framework. The different subjects (X,Y) are projected on the latent representation by the encoder E_ψ and then a linear decomposition of this latent space is calculated to identify a new vector space $(\boldsymbol{u}_i)$.

Representation disentangling methodologies is a concurrent field of research also investigating explainability topics. Such approaches are mainly focusing on generating interpretable latent representations by enforcing several constraints. This can be achieved either using architecture tricks [20,22] or with appropriate loss functions [4,14]. In medical image computing, several studies focus on approaches for generating disentangled or decomposed representations. In [19] for example, the authors proposed a multimodal image registration method by decomposing the volumes into a common latent shape space and separate latent appearance spaces via image-to-image translation approach and generative models. Our method shares many common points with the approaches mentioned above, yet it focuses on exploring the registration latent space decomposition.

3   Methodology

Deep learning-based registration methods have received much attention in the last few years [1,25]. Formally, let us consider two volumes, the moving M and the fixed F. The goal of deep learning-based registration methods is to obtain the best parameters θ^* for the network g_θ that will map most accurately M to F using the predicted deformation grid Φ. The network g_θ usually is composed of an encoding E_ψ and a decoding D_ω part.

There are multiple ways to fuse the input volumes in deep learning-based registration approaches. Most of the methods use an early fusion strategy on which the two volumes are concatenated before they pass through the g_θ. However, some methods investigate late fusion strategies [7,11] where the two volumes pass independently through the encoder, and their merging operation is achieved in the encoding representation using various operations such as concatenation or subtraction. Thanks to this formulation, each volume has a unique encoding representation. In this study, we adopt the second strategy using the subtraction operation to encode each volume independently and calculate its latent space's linear decomposition. In Fig. 1, the overall scheme is presented.

3.1 Deep Learning-Based Registration Scheme

To perform our experiments and obtain our embeddings, we defined a network based on a 3D UNet architecture [5]. The encoder and the decoder are composed of a fixed number of blocks with 3D convolution layers (stride 3, padding 1), instance normalisation layer and leaky ReLU activation function. The down and up-sampling operations are performed with a 3D convolution layer with stride and padding of 2. One of the main differences in our architecture was the absence of skip connections. Indeed, we want to enforce that all information passes through the last encoding layer without any leak due to the skip connections. This modification led us to reduce the downsampling operations from four to three for the lung dataset, to maintain the spatial resolution of the bottleneck.

Different formulations have been proposed to generate the deformation from deep learning schemes, such as displacement field formulation [1], diffeomorphic formulations [6,15] and formulations based on the spatial gradients [25]. In this work, we focused on the last one, with our network regressing the spatial gradients $\nabla_x \Phi_x$, $\nabla_y \Phi_y$ and $\nabla_z \Phi_z$, while the final deformation field is obtained through a cumulative sum operation. We also followed the symmetric formulation proposed in [8], predicting both the forward and backward deformations: $\nabla \Phi_{M \to F} = D_\omega(E_\psi(M) - E_\psi(F))$ and $\nabla \Phi_{F \to M} = D_\omega(E_\psi(M) - E_\psi(F))$.

The network was trained with a combination of four losses, one focusing on the intensity similarity using normalised cross-correlation ($\mathcal{L}_{sim}$), one focusing on anatomical structures using dice loss ($\mathcal{L}_{seg}$) and two losses for regularisation of the displacements. The first one was the Jacobian loss which is exploited on different works such as [16,17,26] ($\mathcal{L}_{jac}$) and the second one enforcing smooth gradients similar to [8] ($\mathcal{L}_{smooth}$). As such our final loss is: $\mathcal{L} = (\mathcal{L}_{sim} + \mathcal{L}_{seg} + \alpha \mathcal{L}_{smooth} + \beta \mathcal{L}_{jac})_{M \to F} + (\mathcal{L}_{sim} + \mathcal{L}_{seg} + \alpha \mathcal{L}_{smooth} + \beta \mathcal{L}_{jac})_{F \to M}$ with α and β being the weights of the regularisation losses.

3.2 Decomposition of Latent Space

Let $A_{train} = \{X_i | i \in [0, n]\}$ be the set of our n training samples. The proposed formulation apply the encoder independently to each volume, and thus we can obtain the set of latent vectors: $E_\psi(A_{train}) = \{E_\psi(X_i) | i \in [0, n]\}$. Then, we decompose this space using principal components analysis (PCA). That way, we obtain a set of principal vectors $\mathcal{U}_K = (\overrightarrow{u_1}, \cdots, \overrightarrow{u_K})$ with K being a hyperparameter fixing the number of principal components. It worth noting that each vector $\boldsymbol{u}_i$ has the same size as the activation map of the encoder's last layer. This size depends on the number of channels, the size of the input images and the number of downsampling operations. We flatten each encoding representation from its four dimensions representation (channel dimension and the three spatial dimensions) to a one-dimensional array to perform the PCA. Thus, the PCA is not calculated channel-wise, but all the channels are considered together. Each principal vector $\boldsymbol{u}_i$ can be converted to a deformation grid ϕ_i using the corresponding decoder D_ω: $\phi_i = D_\omega(\boldsymbol{u}_i)$. Therefore, we obtained a set of elementary transformations $\{\phi_i\}_{i=1 \cdots K}$. These elementary transformations generate a basis that can be used to approximate and decompose every new deformation. Using such a decomposition, we can obtain a representation in small dimensions of every training volume X_i. These representations are obtained by the projection of $E_\psi(X_i)$ to each principal vector: $a_i^j = E_\psi(X_i) \cdot \vec{u_j}$. For every volume of our training set we have the approximation: $E_\psi(X_i) \approx \sum_{j=1}^{K} a_i^j \vec{u_j}$. After calculating the vector of the principal components $\mathcal{U}_K$ with the training set, we projected each image of the validation set to obtain its PCA representation.

3.3   Implementation and Training Details

The Adam optimiser was used for our training, with a constant learning rate set to $1e^{-4}$, a batch size equal to 4 and 8 for lung and hippocampus, respectively. Our models were trained for 600 epochs, and it last approximately 4 and 9 hours for the lung and hippocampus dataset. Concerning data augmentation, we applied random flip, rotation, translation and zoom. Moreover, the weights of the different loss components were set to 1 except the loss for smoothness set to $\alpha = 0.1$ for both datasets and the weight for the jacobian loss β that was discarded for the hippocampus dataset. During the training process, we registered random pairs of different patients. Our training has been performed using the framework PyTorch and one GPU card Nvidia Tesla V100 with 32G memory. The PCA decomposition was calculated using the library scikit-learn, and the number of principal components K was set to 32. Using 32 components, our decomposition covered 95% and 93% of the variance ratio for the lung and hippocampus dataset, respectively, while 42% and 62% are covered by the first four components for each dataset, respectively.

4   Experiments and Results

We performed our experiments on two different datasets, one public and one private. Starting with the public dataset, we conduct experiments with the hippocampus[1] [23]. This dataset comprises 394 MRI with the segmentations of two small structures, the head and the body hippocampus. The images have been cropped around the hippocampus into small patches of $64 \times 64 \times 64$ voxels. The second dataset is composed of 41 lung MRI patients (12 healthy and 29 diseased with pulmonary fibrosis) together with their lung segmentations. Each patient had been acquired in two states, the inspiration and the expiration. Each volume has been resampled to 1.39 mm on the x and z-axis and 1.69 on the y-axis and cropped to $128 \times 64 \times 128$ volumes. The same normalisation strategy has been applied for the two datasets: $\mathcal{N}(0,1)$ standardisation, clip to $[-5, 5]$ to remove outliers values and min-max normalisation to $(0, 1)$. Both datasets were split into training and validation, resulting in 200 and 60 patients for hippocampus and 28 and 13 patients for the lung dataset.

---

[1] http://medicaldecathlon.com/.

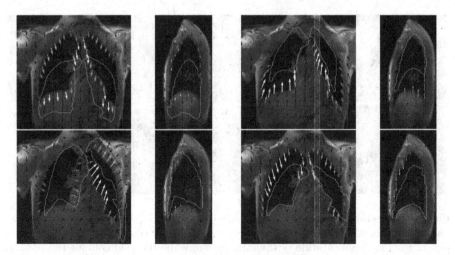

Fig. 2. Visualisation of the displacements following the first four principal components. For each component, we depicted coronal and sagittal views. In red, the contours of the lungs of the M, and in gold, the $\mathcal{W}(D_\omega(\lambda\overrightarrow{u_i}), M)$ lung's contours. The deformation field is represented with arrows. The arrows' norm is represented with a colour map, red being the smallest and white the largest. Other patients and components are displayed on supplementary materials. (Color figure online)

As the first step of our evaluation, we benchmarked the performance of the registration network g_θ, on which our decomposition is based on. More specifically, we obtained a Dice coefficient of 0.90 ± 0.04 for the lungs and 0.76 ± 0.05 for the hippocampus, while the initial unregistered cases reported a Dice of 0.74 ± 0.14 and 0.59 ± 0.15 respectively. Moreover, we calculated the registration for g_θ with the skip connections to measure their impact on the registration. The Dice is then equal to 0.92 ± 0.02 and 0.85 ± 0.03 respectively. Thus, by removing the skip-connections, we decrease the performance of the registration, slightly on the lungs, more importantly, on the hippocampus. However, both strategies register the pair of volumes properly.

4.1 Qualitative Evaluation

To understand and evaluate the calculated components of $\mathcal{U}_K$ per dataset, we perform a qualitative analysis. In particular, for each principal vector $\overrightarrow{u_i}$, we calculated the corresponding deformation $\{\phi_i\}_{i=1\cdots K}$ and we applied to the moving image M together with its corresponding segmentation map M_{seg}. More formally, the deformed contour correspond to $\mathcal{W}(D_\omega(\lambda\overrightarrow{u_i}), M_{seg})$ with $\mathcal{W}$ being the warping operation and λ the parameter to control the strength of the displacements for better visualisation.

In Fig. 2, we show the principal components obtained for one validation subject for the lung dataset. Interestingly, one can observe that each ϕ_i corresponds to a different elementary transformation. More precisely, the 1st component is

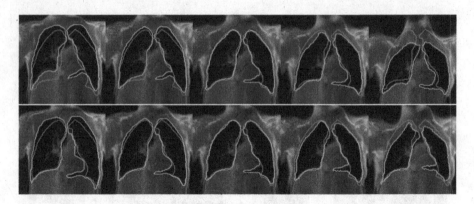

Fig. 3. Representation of the deformed MR together with its lung contours following the first and fourth principal components u_1, u_4. The red contour represents the position of the lung segmentation of the input image while the gold contour the position of the deformed lung. The values of lambda range from: -200, -100, 0, 100 and 200 (left to right). Negative values of lambda correspond to an upward translation, while positive values to a downward translation. (Color figure online)

associated with translation, the 2nd with a deformation focusing on the bottom of the lungs, the 3rd with a deformation on the right lung focusing also on the heart region and lastly the 4th with a deformation focusing on the top region of the lung and shoulders.

In Fig. 3, we show the effect of the values of λ. In the figure, we present the lung contours of the scaled component (in red) and the corresponding component of the warped of the first ad third components. As we have indicated, the $1st$ component is associated with translation, which we can also be observed in this visualisation. In particular, for this experiment we sample λ from the values $\{-200, -100, 0, 100, 200\}$. One can observe that we retrieve a near identity deformation for a value of 0, while for negative and positive values, the lung moves up and down, respectively. On the other hand, the fourth component is responsible for deforming the shoulders and the top of the lungs. Indeed, one can

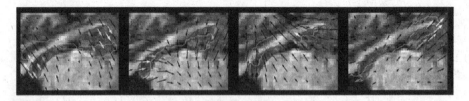

Fig. 4. Visualisation of the displacements following the first four principal components. We depicted a sagittal view of one patient of the validation set of the hippocampus dataset. We represented the ground truth hippocampus contours (red) and the deformed one (gold), following the principal components. (Color figure online)

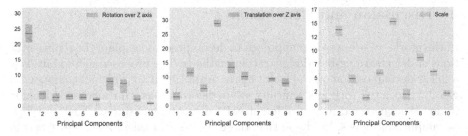

Fig. 5. Visualisation of the differences between the components of a reference image and the same image to which we applied a predefined transformation. The first ten components have been displayed. From right to left: rotation, translation along Z-axis and scaling.

observe that through the different λ values, the top lungs region is the one that reports the most changes. In Fig. 4, similarly, the 4 deformations produced by the first 4 principal components of the hippocampus dataset are presented. In this case, the 1st component seems to capture rotation on the sagittal plane, the 2nd translation and shrinking towards the bottom right, while the 3rd seems to be the same operation towards the top left corner. Finally, the 4th seems to be related to scaling, inflating both the hippocampus's head and tail. We observed that the decomposition of the two datasets created different elementary transformations ϕ_i, with transformations closer to affine for the hippocampus and more complex for the lung.

Finally, to verify the obtained decomposition, we performed a case study for all the validation subjects of the hippocampus dataset. More specifically, we applied some predefined translation using 10 pixels on the z axis, rotation using 20° on the z axis and scaling using a factor of 0.2, transforming each subject X to X'. Then we calculated the difference between the projection of $E_\psi(X)$ and $E_\psi(X')$ on the PCA decomposition. In Fig. 5, a box plot for all the validation subjects of the absolute difference is presented. Specifically, the amount $||a^j_{E_\psi(X)} - a^j_{E_\psi(X')}||$ is shown for each principal component j, with $a^j_{E_\psi(X)}$ being the projection of $E_\psi(X)$ on the principal vectors $\mathcal{U}_K$, for the three different applied deformations. One can observe that for rotation and translation, only one component is significantly different from the rest. In the case of scaling, however, two components seem to be more activated. Moreover, these findings are in accordance with Fig. 4 for the rotation and translation. In supplementary materials, we upgraded the Fig. 5 by comparing the network with and without skip-connections. Contrary to our proposed formulation, many components are activated with the skip-connections, demonstrating the necessity of removing them to have a good decomposition.

5   Discussion and Conclusion

In this work, we proposed an approach to decompose and explain the representations of deep learning-based registration methods. The proposed method utilises a linear decomposition on the latent space projecting it to principal components closely associated with anatomically aware deformations. Our method's dynamics are demonstrated in two different MRI datasets, focusing on lung and hippocampus anatomies. We hope that these results will take some steps towards a better understanding of latent representations learned by the deep learning registration architectures. We also explored a direct application of the PCA on the deformation's grid instead of the latent representation. However, we did not observe any qualitative correlations with types of deformations, which is the case for our proposed formulation. One of the main limitations of our approach is the difficulty of quantitative evaluation. Our future steps include the more extensive evaluation of our method, including new anatomies such as abdominal volumes and its clinical significance. More specifically, we want to apply our approach to multi-temporal follow-up of patients, monitoring diseases' progression.

Funding

This work has been partially funding by the ARC: Grant SIGNIT201801286, the Fondation pour la Recherche Médicale: Grant DIC20161236437, SIRIC-SOCRATE 2.0, ITMO Cancer, Institut National du Cancer (INCa) and Amazon Web Services (AWS).

References

1. Balakrishnan, G., Zhao, A., Sabuncu, M.R., Guttag, J., Dalca, A.V.: Voxelmorph: a learning framework for deformable medical image registration. IEEE Trans. Med. Imag. **38**(8), 1788–1800 (2019)
2. Castro, D.C., Walker, I., Glocker, B.: Causality matters in medical imaging. Nat. Commun. **11**(1), 1–10 (2020)
3. Chattopadhay, A., Sarkar, A., Howlader, P., Balasubramanian, V.N.: Gradcam++: Generalized gradient-based visual explanations for deep convolutional networks. In: 2018 IEEE Winter Conference on Applications of Computer Vision (WACV), pp. 839–847. IEEE (2018)
4. Chen, X., Duan, Y., Houthooft, R., Schulman, J., Sutskever, I., Abbeel, P.: Infogan: interpretable representation learning by information maximizing generative adversarial nets. arXiv preprint arXiv:1606.03657 (2016)
5. Çiçek, Ö., Abdulkadir, A., Lienkamp, S.S., Brox, T., Ronneberger, O.: 3D U-Net: learning dense volumetric segmentation from sparse annotation. In: Ourselin, S., Joskowicz, L., Sabuncu, M.R., Unal, G., Wells, W. (eds.) MICCAI 2016. LNCS, vol. 9901, pp. 424–432. Springer, Cham (2016). https://doi.org/10.1007/978-3-319-46723-8_49

6. Dalca, A.V., Balakrishnan, G., Guttag, J., Sabuncu, M.R.: Unsupervised learning for fast probabilistic diffeomorphic registration. In: Frangi, A.F., Schnabel, J.A., Davatzikos, C., Alberola-López, C., Fichtinger, G. (eds.) MICCAI 2018. LNCS, vol. 11070, pp. 729–738. Springer, Cham (2018). https://doi.org/10.1007/978-3-030-00928-1_82

7. Estienne, T., et al.: Deep learning-based concurrent brain registration and tumor segmentation. Front. Comput. Neurosci. **14**, 17 (2020)

8. Estienne, T., et al.: Deep learning based registration using spatial gradients and noisy segmentation labels. In: Segmentation. Classification, and Registration of Multi-Modality Medical Imaging Data, pp. 87–93. Lecture Notes in Computer Science, Cham (2021). https://doi.org/10.1007/978-3-030-71827-5

9. Estienne, T., et al.: U-ReSNet: ultimate coupling of registration and segmentation with deep nets. In: Shen, D., et al. (eds.) MICCAI 2019. LNCS, vol. 11766, pp. 310–319. Springer, Cham (2019). https://doi.org/10.1007/978-3-030-32248-9_35

10. Fong, R.C., Vedaldi, A.: Interpretable explanations of black boxes by meaningful perturbation. In: Proceedings of the IEEE International Conference on Computer Vision, pp. 3429–3437 (2017)

11. Heinrich, M.P.: Closing the gap between deep and conventional image registration using probabilistic dense displacement networks. In: Shen, D., et al. (eds.) MICCAI 2019. LNCS, vol. 11769, pp. 50–58. Springer, Cham (2019). https://doi.org/10.1007/978-3-030-32226-7_6

12. Holzinger, A., Langs, G., Denk, H., Zatloukal, K., Müller, H.: Causability and explainability of artificial intelligence in medicine. Wiley Interdiscip. Rev. Data Mining Knowl. Discov. **9**(4), e1312 (2019)

13. Jaderberg, M., Simonyan, K., Zisserman, A., Kavukcuoglu, K.: Spatial transformer networks. arXiv:1506.02025 [cs], February 2016

14. Kingma, D.P., Welling, M.: Auto-encoding variational bayes. arXiv preprint arXiv:1312.6114 (2013)

15. Krebs, J., Delingette, H., Mailhé, B., Ayache, N., Mansi, T.: Learning a probabilistic model for diffeomorphic registration. IEEE Trans. Med. Imag. **38**(9), 2165–2176 (2019)

16. Kuang, D., Schmah, T.: FAIM - A ConvNet method for unsupervised 3d medical image registration. arXiv:1811.09243 [cs], June 2019

17. Mok, T.C., Chung, A.: Fast symmetric diffeomorphic image registration with convolutional neural networks. In: Proceedings of the IEEE/CVF Conference on Computer Vison and Pattern Recognition, pp. 4644–4653 (2020)

18. Ou, Y., et al.: Deformable registration for quantifying longitudinal tumor changes during neoadjuvant chemotherapy. Mag. Reson. Med. **73**(6), 2343–2356 (2015)

19. Qin, C., Shi, B., Liao, R., Mansi, T., Rueckert, D., Kamen, A.: Unsupervised deformable registration for multi-modal images via disentangled representations. In: Chung, A.C.S., Gee, J.C., Yushkevich, P.A., Bao, S. (eds.) Information Processing in Medical Imaging, pp. 249–261 (2019)

20. Sahasrabudhe, M., Shu, Z., Bartrum, E., Alp Guler, R., Samaras, D., Kokkinos, I.: Lifting autoencoders: Unsupervised learning of a fully-disentangled 3d morphable model using deep non-rigid structure from motion. In: Proceedings of the IEEE/CVF International Conference on Computer Vision Workshops, (2019)

21. Selvaraju, R.R., Cogswell, M., Das, A., Vedantam, R., Parikh, D., Batra, D.: Gradcam: Visual explanations from deep networks via gradient-based localization. In: Proceedings of the IEEE International Conference on Computer Vision, pp. 618–626 (2017)

22. Shu, Z., Sahasrabudhe, M., Guler, R.A., Samaras, D., Paragios, N., Kokkinos, I.: Deforming autoencoders: Unsupervised disentangling of shape and appearance. In: Proceedings of the European Conference on Computer Vision (ECCV), pp. 650–665 (2018)
23. Simpson, A.L., et al.: A large annotated medical image dataset for the development and evaluation of segmentation algorithms. arXiv preprint arXiv:1902.09063 (2019)
24. Springenberg, J.T., Dosovitskiy, A., Brox, T., Riedmiller, M.: Striving for simplicity: The all convolutional net. arXiv preprint arXiv:1412.6806 (2014)
25. Stergios, C., et al.: Linear and deformable image registration with 3D convolutional neural networks. In: Stoyanov, D., et al. (eds.) RAMBO/BIA/TIA-2018. LNCS, vol. 11040, pp. 13–22. Springer, Cham (2018). https://doi.org/10.1007/978-3-030-00946-5_2
26. Zhang, S., Liu, P.X., Zheng, M., Shi, W.: A diffeomorphic unsupervised method for deformable soft tissue image registration. Comput. Biol. Med. **120**, 103708 (2020). https://doi.org/10.1016/j.compbiomed.2020.103708
27. Zhou, B., Khosla, A., Lapedriza, A., Oliva, A., Torralba, A.: Learning deep features for discriminative localization. In: Proceedings of the IEEE Conference on Computer Vision and Pattern Recognition, pp. 2921–2929 (2016)

Learning from Partially Overlapping Labels: Image Segmentation Under Annotation Shift

Gregory Filbrandt[1], Konstantinos Kamnitsas[1], David Bernstein[2],
Alexandra Taylor[3], and Ben Glocker[1(✉)]

[1] Department of Computing, Imperial College London, London, UK
`b.glocker@imperial.ac.uk`
[2] Joint Department of Physics, The Institute of Cancer Research and The Royal
Marsden NHS Foundation Trust, London, UK
[3] Department of Radiotherapy, The Royal Marsden NHS Foundation Trust,
London, UK

Abstract. Scarcity of high quality annotated images remains a limiting factor for training accurate image segmentation models. While more and more annotated datasets become publicly available, the number of samples in each individual database is often small. Combining different databases to create larger amounts of training data is appealing yet challenging due to the heterogeneity as a result of differences in data acquisition and annotation processes, often yielding incompatible or even conflicting information. In this paper, we investigate and propose several strategies for learning from partially overlapping labels in the context of abdominal organ segmentation. We find that combining a semi-supervised approach with an adaptive cross entropy loss can successfully exploit heterogeneously annotated data and substantially improve segmentation accuracy compared to baseline and alternative approaches.

1 Introduction

Obtaining sufficient amounts of high quality and accurate annotations in the context of image segmentation remains a major bottleneck due to the time-consuming nature of the expert labelling task. In recent years, an increasing amount of publicly available data has become available (e.g., brain MRI [9] or abdominal CT [7]), often through the efforts of organizing computational challenges and benchmarks. However, these public datasets are often either limited in size or specific to a particular anatomy or pathology of interest (e.g., brain [10] or liver tumours [1]). Pooling data from different studies to form larger datasets that are suitable for training automated segmentation methods is appealing yet challenging, due to the inherent heterogeneity of the available annotations. The set of labels from different datasets may be partially overlapping, but more

---

G. Filbrandt and K. Kamnitsas—Equal contribution.

S. Albarqouni et al. (Eds.): DART 2021/FAIR 2021, LNCS 12968, pp. 123–132, 2021.
https://doi.org/10.1007/978-3-030-87722-4_12

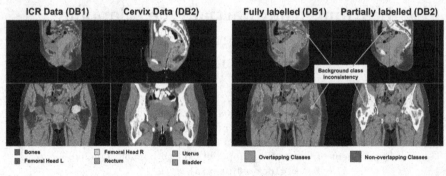

(a) Heterogeneous labels in abdominal CT   (b) Label-contradiction problem

Fig. 1. a) Annotation shift between the two databases in this study. DB1 has 6 structures annotated, whereas only 3 of them are annotated on DB2. b) Label contradiction problem due to a structure being labeled differently in the two databases (e.g. spine is part of background in DB2).

importantly, may yield conflicting information due to differences in the annotation protocols (a problem also known as 'annotation shift' [2]). Learning from such heterogeneous data is an open problem in machine learning for imaging.

Here, we investigate this challenge of learning under annotation shift in the context of automated segmentation of abdominal CT for the application of radiotherapy planning. In this study, we consider two datasets, one internal and one external, with partially overlapping and contradicting labels, as illustrated in Fig. 1. Our goal is to devise learning strategies that can successfully exploit the available information from the different datasets with the aim to improve segmentation accuracy. While focusing on abdominal CT, our results should be of interest for other medical imaging applications and modalities.

Related Work: Learning from heterogeneous data poses a variety of challenges. Previous works investigated learning in the existence of different input distributions across databases, known as the domain (or acquisition) shift. This was approached via domain adaptation [5], augmentation [3], feature-matching [4] or their combination. This study instead focuses on the issue of annotation shift, where we assume that the total set of labels is $\mathcal{Y}$, but each database $\mathcal{D}_k$ has been annotated with a possibly different subset of labels, $\mathcal{Y}_k \subseteq \mathcal{Y}$. This problem has been commonly explored under the assumption that the label sets $\mathcal{Y}_k$ are *disjoint* [3,13], which can be formulated as *multi-task learning* [3,11,12,15]. A common approach is to construct a model that predicts a separate output for each label set $\mathcal{Y}_k$, for example, using a multi-head neural net. If a single output is desired (e.g. a joint segmentation map) then the multiple outputs need to be fused via task-specific choice of aggregation rules (e.g. brain lesion taking precedence over a brain anatomy segmentation map [3]). The multi-task approach has also been tried for problems where label sets $\mathcal{Y}_k$ may partially overlap. In this case, a means for fusing the separate output per label set in a single prediction is required, often done as a post-processing step, such as Non Maximal Suppression [15]. These fusion steps that are external to the model do not facilitate learning. Instead of altering the model to predict a different output per label set, an alternative approach is to predict outputs that can take any value of the total label set, $\mathcal{Y}$, and develop a learning method that can process data with annotations belonging to different subsets $\mathcal{Y}_k$. Some related works explored how to incorporate prior knowledge about structures of interest (e.g. [16]). Priors may not exist for various tasks, such as for the shape of certain pathologies. Therefore we focus on purely learning-based methods. Such a general training objective for learning with two disjoint label sets has been proposed in [13], termed *adaptive cross entropy* (ACE). When making predictions for training image x_i^k with manual annotation y_i^k from label set $\mathcal{Y}_k$, it considers the labels outside its own label set ($\mathcal{Y}_{o,k} = \mathcal{Y} - \mathcal{Y}_k$) as part of the background class, alleviating the label contradiction in the application of the cost function (Fig. 1). This method has been originally proposed under the assumption that the label sets are disjoint, specifically for joint learning of brain structures and lesions from different databases. The original ACE, however, does not facilitate improvements for non-overlapping labels $\mathcal{Y}_{o,k}$ when processing background pixels, which we build upon.

Contributions: This study explores how to train a segmentation model to predict output $y \in \mathcal{Y}$ from the total label-set, using databases $\mathcal{D}_k$ with label-sets $\mathcal{Y}_k$ that may partially overlap, $\mathcal{Y}_k \cap \mathcal{Y}_l \neq \varnothing$. We adapt ACE to this setting and extend it by interpreting voxels of the background class as unlabeled samples, adopting ideas from semi-supervised learning to improve learning from them. We find that ACE facilitates learning of overlapping classes, whereas semi-supervision provides benefits by also regularizing non-overlapping classes. Experiments with two databases of abdominal CT with partially overlapping label sets show that combining these complementary approaches improves segmentation.

2   Learning from Heterogeneously Labeled Data

2.1   Problem Definition and Label-Contradiction Issue

Assume a set of classes of interest described by the *total* label-set $\mathcal{Y}$, with cardinality $|\mathcal{Y}| = C$ number of classes. We define as $\mathcal{D}_k = \{(x_i^k, y_i^k)\}_{i=1}^{N_k}$ different databases, where x_i^k is the i-th sample (image for classification or pixel for segmentation) of $\mathcal{D}_k$ and y_i^k its true label. We do not assume disjoint label-sets as in previous works [3,13], but rather investigate the general case where only a subset $\mathcal{Y}_k \subseteq \mathcal{Y}$ of labels has been annotated for the k-th database, and different label-sets $\mathcal{Y}_k$ may be overlapping arbitrarily. We denote with $\mathcal{Y}_{o,k} = \mathcal{Y} - \mathcal{Y}_k$ the set of labels that have *not* been annotated in database k. Finally, we define as $\hat{y}_{bg}$ the *background* class, that is not an element of $\mathcal{Y}$.

In this general setting, we wish to create a model $f(x_i^k, \theta) \in \mathbb{R}^{C+1}$ that predicts a posterior probability for each of the C classes and the background $\hat{y}_{bg}$, and we denote with $f(x_i^k, \theta)_y$ the posterior of class y. We wish to train such a model with all available databases $\{\mathcal{D}_k\}_{k=1}^K$.

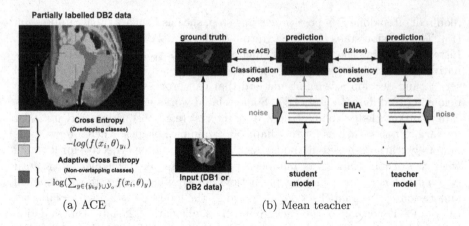

(a) ACE   (b) Mean teacher

Fig. 2. a) Adaptive Cross Entropy (ACE) loss as it varies between overlapping and non-overlapping classes. b) Key components of the Mean Teacher [14], one of the SSL approaches explored in this study.

In this setting, a label-contradiction problem arises due to inconsistent use of the background class for annotation protocols of each database: As is the common annotation practice, when annotating $\mathcal{D}_k$, all samples that in reality belong to one of the non-annotated label-set $\mathcal{Y}_{o,k}$ are assigned the background label $\hat{y}_{bg}$. As a result, the same sample may be given class $y_c \in \mathcal{Y}_k$ according to one annotation protocol, and $\hat{y}_{bg}$ according to other. Figure 1 illustrates this. Consequently, standard learning frameworks, such as training a neural network f with cross entropy (CE), will assign contradicting penalties to predictions about samples of the same content (e.g. same anatomy) when processing samples from different databases. We below describe the methods we studied to alleviate this.

2.2 Adaptive Cross Entropy for Learning from Data with Heterogeneous Annotations

To train a model while avoiding the label-contradiction problem due to the differing definition of the 'background' class, we need a learning framework that treats this class differently per database. Such a learning objective is *adaptive cross entropy* (ACE) [13]. It has been originally formulated for learning with two databases and the assumption that their label-sets $\mathcal{Y}_k$ are disjoint. We observe that ACE can be straightforwardly generalized to the case of learning from any number of databases with potentially overlapping label-sets:

$$J_{ace}(\{\mathcal{D}_k\}_{k=1}^{K}) = \sum_{k=1}^{K}\sum_{i} H_{ace}(x_i^k, y_i^k) \tag{1}$$

$$H_{\text{ace}}(x_i^k, y_i^k) = \begin{cases} -\log(f(x_i^k, \theta)_{y_i^k}) & \text{if } y_i^k \in \mathcal{Y}_k \\ -\log(\sum_{y \in \{\hat{y}_{bg}\} \cup \mathcal{Y}_{o,k}} f(x_i^k, \theta)_y) & \text{otherwise} \end{cases} \tag{2}$$

Here, H_{ace} is entropy per sample and J_{ace} the total cost across all databases. Intuitively, for every database $\mathcal{D}_k$, ACE behaves similar to CE for annotated samples. For samples that the annotation protocol of $\mathcal{D}_k$ leaves non-annotated (as background), it sums up the predicted probabilities for non-annotated classes $\mathcal{Y}_{o,k}$ and class $\hat{y}_{bg}$, forcing them to sum up to 1 (minimize $-log$). Figure 2a illustrates this. How does this facilitate learning? It does not penalize a model for predicting any of the non-annotated classes $\mathcal{Y}_{o,k}$ for samples that have not been annotated and, therefore, not contradicting information learned from other databases where these classes are labeled. As a result, it enables making use of all available supervision signal from any sample annotated across databases.

We identify that for samples not annotated in $\mathcal{D}_k$ (i.e. considered background in $\mathcal{D}_k$), ACE does not explicitly encourage better predictions for one of the non-annotated classes $\mathcal{Y}_{o,k}$. In fact, the lower part of Eq. 2 will be minimized for any combination of posteriors that sum up to 1. We improve this by adopting ideas from semi-supervised learning and introducing them to ACE, as described next.

2.3   Learning from Non-annotated Regions via Mean Teacher

We here interpret samples that are assigned the background class in each database $\mathcal{D}_k$ as unlabeled samples, and investigate the integration of semi-supervised learning (SSL) in a framework for learning from heterogeneously labeled databases.

We study one of the most successful recent methods for SSL, the Mean Teacher (MT) [14]. In a SSL setting, it assumes a labeled $\mathcal{D}_L$ and an unlabeled database $\mathcal{D}_U$. It benefits from unlabeled data by learning model parameters θ such that predictions are consistent regardless perturbations of the input or the model parameters. This has been shown to improve generalization.

This is accomplished in MT via complementing a standard classification cost J_{cl} (e.g. cross entropy H_{ce}) with a consistency cost J_{con}. The original definition of MT's cost function [14] for SSL is given by the following:

$$J_{mt}(x_i, y_i) = J_{cl}(x_i, y_i) + J_{con}(x_i) \tag{3}$$

$$J_{cl}(x_i, y_i) = H_{ce}(x_i, y_i) = \begin{cases} -\log(f_{stu}(x_i, \theta)_{y_i}) & \text{if } x_i \in \mathcal{D}_L \\ 0 & \text{if } x_i \in \mathcal{D}_U \end{cases} \tag{4}$$

$$J_{con}(x_i) = (f_{stu}(x_i, \theta) - f_{tea}(x_i, \theta))^2 \qquad \forall x_i \tag{5}$$

The consistency cost J_{con} is defined via two perturbations of the sample's x_i embedding: the embeddings by the *student* f_{stu} and the *teacher* f_{tea}. The perturbed embeddings are the result of two components. First, the student uses the current state of model parameters θ, whereas the teacher uses an exponential moving average (EMA) of their values, θ_{ema}. The assumption is that EMA over parameters improves predictions similar to an implicit ensemble, and hence it will enforce the student to predict better. Secondly, student and teacher embeddings are computed via different perturbations of the signal. In our settings, as commonly done, this is computed for different values of dropout masks between f_{stu} and f_{tea}. Our ablation study (Sect. 3) will investigate the influence of both.

The above formulation cannot be straightforwardly applied for the general case of partially annotated databases, because CE would suffer from the label-contradiction problem for the background class (Sect. 2.1). We extend the framework to this setting by combining it with ACE. This can be done by using ACE (H_{ace}, Eq. 2) as the classification loss in Eq. 4, instead of CE (H_{ce}). This combines benefits of learning from all samples x_i^k that are annotated for each database $\mathcal{D}_k$, with the use of consistency loss $J_{con}(x_i^k)$ for all samples $x_i^k \in \{D_k\}_{k=1}^K$, which includes non-annotated (background) samples. We hypothesize that the latter will offer orthogonal benefits to those from ACE, improving predictions of non-annotated samples in each database. The following empirical investigations investigates this hypothesis.

3   Experiments

3.1   Data and Model Configuration

DB1: This is an internal database consisting of 40 3D CT scans of the abdominal region of patients with cervical cancer. The scans consist of between 183 and 331 axial slices with 512×512 pixel resolution. They were acquired with a full-bladder drinking protocol with patients in supine position. 20 samples were randomly chosen for training, and the remaining 20 used for testing. DB1 is considered fully annotated in our experiments, defining $\mathcal{Y} = \mathcal{Y}_1$ with 6 labeled classes, which consist of: Bladder, Rectum, Uterus, Bones, left and right Femoral heads.

DB2: Partially annotated, public database consisting of 30 3D CT scans of the abdominal region of patients with cervical cancer from the Synapse benchmark [7]. The scans consist of between 125 and 237 axial slices of 512×512 pixels. They were acquired with a full-bladder drinking protocol with most patients in prone and some in supine position. This database was used only for training. This is considered the partially annotated database with 3 of the 6 classes labeled. Therefore, $\mathcal{Y}_2$ here consists of: Bladder, Rectum and Uterus. The remaining Bones, left and right Femoral head classes are non-overlapping ($\mathcal{Y}_{o,2}$).

Pre-processing: All images were resampled to 2 mm isotropic resolution followed by intensity capping (-200 to $+200$) and normalisation ($\mu = 0$ and $\sigma = 1$). Scans were reoriented to simulate supine patient position where necessary.

Main Model: We use a 3D CNN, DeepMedic, previously used for a variety of segmentation tasks with promising performance [6]. We employ the 'wide' model variant publicly available https://github.com/deepmedic/deepmedic, v0.8.4) and otherwise use the default hyper-parameters and model architecture.

Configuration of Methods: Hyper-parameters of the explored methods were set based on original works. Additional settings include the maximum weight of the consistency cost (set to 1.0) and its "warm-up" period (10 training epochs starting from zero in the third epoch linearly increasing to the maximum weight). This was found to improve training convergence in preliminary experiments.

Table 1. Dice % from studied methods. (* significant difference vs SL.1, $p < 0.05$)

Class	SL1	SL12	ACE	PL	MT	ACE/PL	ACE/MT
Bladder	89.6	90.7	90.7	90.6	89.2	91.0	91.5
Rectum	73.7	76.1	75.2	77.3	78.3	76.7	78.2
Uterus	60.1	67.4	67.9	67.9	66.8	68.4	68.8
Bones	87.9	81.1	88.3	88.7	89.2	88.6	88.2
Fem.Head L	87.5	85.1	88.2	88.3	90.3	88.9	89.8
Fem.Head R	87.6	84.9	88.1	88.2	88.9	88.5	88.8
Overlapping	74.5	78.1*	77.9*	78.6*	78.1*	78.7	79.5*
Non-Overlap.	87.7	83.7	88.2*	88.4*	89.5*	88.7*	88.9*
Total mean	81.1	80.9	83.1*	83.5*	83.8*	83.7*	84.2*

3.2 Results

All the below experiments were repeated for 3 seeds. We report average performance on DB1 test data (Dice%) for all methods in Table 1.

Baselines: We first evaluate a DeepMedic model trained only with supervised learning on fully labeled DB1 data. This **SL1** method performed well for segmenting abdominal tumours and organs, marking a suitable point for baseline comparison. The **SL12** method naively uses both databases for training a model with CE. Results for SL12 show clear improvements for overlapping classes over SL1. Performance for non-overlapping classes, however, is negatively affected. We hypothesise this is due to label contradiction across databases.

Adaptive Cross Entropy: We assess how well **ACE** [13] mitigates the effect of label contradiction. We train DeepMedic with ACE using both DB1 and DB2. Accuracy for overlapping classes is maintained as with SL12, without losing accuracy for non-overlapping classes compared to SL1, confirming its effectiveness.

Pseudo-labelling: As additional comparison, we apply the pseudo-labelling SSL approach [8]. Here, predictions for DB2 from supervised SL1 are combined with partial annotations of DB2 to generate *pseudo-labels* for DB2. This is done by over-writing the background class in the manual annotations for pixels where the model predicted a non-annotated class ($Y_{o,2}$). Then, a new model is trained using DB1 labels and DB2 pseudo-labels. This **PL** approach shows small improvements over ACE on average. We note that, contrary to ACE, it cannot be easily generalised to K databases as it requires K initial models and fusion

Class	ACE	ACE+	MT_t	MT_s	MT_s+
Bladder	90.7	90.6	89.6	89.6	89.2
Rectum	75.2	76.2	77.0	76.8	78.3
Uterus	67.9	67.5	63.7	63.7	66.8
Bones	88.3	88.3	88.3	88.3	89.2
Fem.Head L	88.2	88.2	88.8	88.9	90.3
Fem.Head R	88.1	87.3	89.1	89.0	88.9
Overlapping	77.9	78.1	76.8	76.7	78.1
Non-Overlap.	88.2	87.9	88.7	88.7	89.5
Total Mean	83.1	83.0	82.8	82.7	83.8

(a) Results (Dice%) of ablation study   (b) ACE/MT predictions

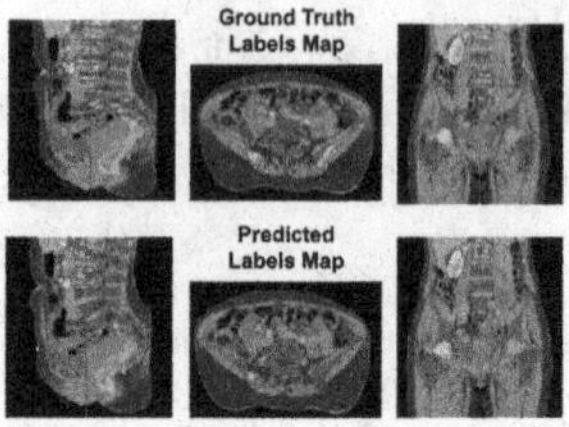

Fig. 3. (a) Ablation study on mean teacher. We find that predictions via (EMA) teacher (MT_t) and student (MT_s) parameters are similar. Training with more dropout (MT_s+) improves the framework. The baseline ACE does not improve by increased dropout (ACE+). Therefore we conclude it is the interplay of consistency loss and perturbations that MT benefits from. (b) Example of results obtained by best method, combination of ACE and Mean Teacher.

of their predictions to form a single pseudo-label, which is not trivial. We also combine PL with ACE, simply by using predictions from the ACE method to create pseudo-labels. This **ACE/PL** approach improves over PL and ACE.

Mean Teacher: We first evaluate the **MT** approach in a purely semi-supervised fashion. We train MT via Eq. 4, using CE on DB1 as labeled $\mathcal{D}_L$, and DB2 as *completely unlabeled* $\mathcal{D}_U$ via the consistency loss only. In addition to EMA as a signal perturbation, 50% dropout is used in all layers except the first 2. MT shows clear improvements over SL1, and modest improvements over ACE and PL, even though it does not use *any* labels from DB2, contrary to ACE and PL.

Combined ACE and Mean Teacher: Finally, we evaluate the proposed combination of ACE with MT, **ACE/MT**, taking advantage of their complementary nature. We use the DB2 partial labels directly within the MT framework through the ACE loss, instead of CE. The results of ACE/MT show best overall performance across all studied methods. Overall, ACE/MT improves over SL1 baseline by 3% Dice, and over the most recent approach for this problem, ACE, by 1% Dice, supporting that SSL provides complementary benefits.

Ablation Study: We perform an ablation study on MT to test whether benefits are provided due to EMA or perturbation via dropout. Results (Dice%) are summarized in Fig. 3a. Specifically, we first train a model with MT using dropout 50% only on the 2 last hidden layers. We report performance of predictions made using the student parameters $(\mathbf{MT}_s)$ and the (EMA) teacher parameters $(\mathbf{MT}_t)$. We also trained MT with more perturbation, using dropout 50% on all layers except first 2 $(\mathbf{MT}+)$. We find that the EMA parameters make no difference. In contrast, additional dropout offers substantial improvements. We test whether the additional dropout benefits ACE, and find no improvements (**ACE** vs **ACE+**). Therefore, we conclude it is the interplay of MT's consistency loss with perturbation that leads to the method's high performance.

4   Conclusion

This study investigated several strategies for learning from databases that were annotated via different annotation protocols, resulting in partial overlapping sets of labels. In the process, we identified that a semi-supervised learning approach, Mean Teacher [14], offers complementary benefits with a recently proposed approach for the task, adaptive cross entropy [13]. We demonstrated that their combination is elegant and effective, outperforming its individual components. Experiments on the task of segmenting anatomical structures in abdominal CT for cervical cancer radiotherapy planning demonstrated that this proposed combined approach can successfully leverage an internal and a public database with partial overlap of labels. It achieved a +3% Dice score improvement over a supervised model trained using only the internal database, which was specifically made for radiotherapy planning. Our results demonstrate the potential of these methods, which enable leveraging public, heterogeneously annotated datasets in order to overcome the scarcity of high quality annotated data.

Acknowledgements. This work received funding from the UKRI London Medical Imaging & Artificial Intelligence Centre for Value Based Healthcare and the European Research Council (ERC) under the European Union's Horizon 2020 research and innovation programme (grant agreement No 757173, project MIRA, ERC-2017-STG). AT receives a grant from Lady Garden Foundation.

References

1. Bilic, P., et al.: The liver tumor segmentation benchmark (LiTS). arXiv preprint arXiv:1901.04056 (2019)
2. Castro, D.C., Walker, I., Glocker, B.: Causality matters in medical imaging. Nature Commun. **11**(1), 1–10 (2020)
3. Dorent, R., et al.: Learning joint segmentation of tissues and brain lesions from task-specific hetero-modal domain-shifted datasets. Med. Image Anal. **67**, 101862 (2021)
4. Dou, Q., Liu, Q., Heng, P.A., Glocker, B.: Unpaired multi-modal segmentation via knowledge distillation. IEEE Trans. Med. Imaging **39**(7), 2415–2425 (2020)
5. Kamnitsas, K., et al.: Unsupervised domain adaptation in brain lesion segmentation with adversarial networks. In: Niethammer, M., et al. (eds.) IPMI 2017. LNCS, vol. 10265, pp. 597–609. Springer, Cham (2017). https://doi.org/10.1007/978-3-319-59050-9_47
6. Kamnitsas, K., et al.: Efficient multi-scale 3d CNN with fully connected CRF for accurate brain lesion segmentation. Med. Image Anal. **36**, 61–78 (2017)
7. Landman, B., Xu, Z., Igelsias, J.E., Styner, M., Langerak, T.R., Klein, A.: 2015 MICCAI multi-atlas labeling beyond the cranial vault - workshop and challenge (2015). https://www.synapse.org/#!Synapse:syn3193805/wiki/217790
8. Lee, D.H., et al.: Pseudo-label: the simple and efficient semi-supervised learning method for deep neural networks. In: Workshop on Challenges in Representation Learning, ICML, vol. 3 (2013)

9. Marcus, D.S., Wang, T.H., Parker, J., Csernansky, J.G., Morris, J.C., Buckner, R.L.: Open access series of imaging studies (OASIS): cross-sectional MRI data in young, middle aged, nondemented, and demented older adults. J. Cogn. Neurosci. **19**(9), 1498–1507 (2007)

10. Menze, B.H., et al.: The multimodal brain tumor image segmentation benchmark (BRATS). IEEE Trans. Med. Imag. **34**(10), 1993–2024 (2014)

11. Moeskops, P., et al.: Deep learning for multi-task medical image segmentation in multiple modalities. In: Ourselin, S., Joskowicz, L., Sabuncu, M.R., Unal, G., Wells, W. (eds.) MICCAI 2016. LNCS, vol. 9901, pp. 478–486. Springer, Cham (2016). https://doi.org/10.1007/978-3-319-46723-8_55

12. Rajchl, M., Pawlowski, N., Rueckert, D., Matthews, P.M., Glocker, B.: NeuroNet: fast and robust reproduction of multiple brain image segmentation pipelines. arXiv preprint arXiv:1806.04224 (2018)

13. Roulet, N., Slezak, D.F., Ferrante, E.: Joint learning of brain lesion and anatomy segmentation from heterogeneous datasets. In: International Conference on Medical Imaging with Deep Learning, pp. 401–413. PMLR (2019)

14. Tarvainen, A., Valpola, H.: Mean teachers are better role models: weight-averaged consistency targets improve semi-supervised deep learning results. arXiv preprint arXiv:1703.01780 (2017)

15. Yan, K., et al.: Learning from multiple datasets with heterogeneous and partial labels for universal lesion detection in CT. IEEE Trans. Med. Imag. (2020)

16. Zhou, Y., et al.: Prior-aware neural network for partially-supervised multi-organ segmentation. In: Proceedings of the IEEE/CVF International Conference on Computer Vision, pp. 10672–10681 (2019)

Unsupervised Domain Adaption via Similarity-Based Prototypes for Cross-Modality Segmentation

Ziyu Ye, Chen Ju, Chaofan Ma, and Xiaoyun Zhang$^{(\boxtimes)}$

Cooperative Medianet Innovation Center, Shanghai Jiao Tong University,
Shanghai, China
{ziyu_ye,ju_chen,chaofanma,xiaoyun.zhang}@sjtu.edu.cn

Abstract. Deep learning models have achieved great success on various vision challenges, but a well-trained model would face drastic performance degradation when applied to unseen data. Since the model is sensitive to domain shift, unsupervised domain adaption attempts to reduce the domain gap and avoid costly annotation of unseen domains. This paper proposes a novel framework for cross-modality segmentation via similarity-based prototypes. In specific, we learn class-wise prototypes within an embedding space, then introduce a similarity constraint to make these prototypes representative for each semantic class while separable from different classes. Moreover, we use dictionaries to store prototypes extracted from different images, which prevents the class-missing problem and enables the contrastive learning of prototypes, and further improves performance. Extensive experiments show that our method achieves better results than other state-of-the-art methods.

Keywords: Unsupervised domain adaption · Medical image segmentation · Prototype · Contrastive learning

1 Introduction

Medical image segmentation is a pixel-wise classification task, which is the basis of many clinical applications [2]. Though deep neural networks have made significant progress in medical image analysis [13,24], most supervised works have the assumption that enough annotated data is collected, which is prohibitively difficult in reality. In clinical scenarios, data collection is time-consuming and laborious, and pixel-wise annotations require expert knowledge of doctors. Hence, unsupervised domain adaption (UDA) is introduced as an annotation-efficient method to help cross-modality medical image segmentation [1].

UDA transfers the knowledge learned in a label-rich domain to a label-lacking domain, bridging the domain gap. Currently, there are two main streams for UDA. One is image-level adaption [4,10,21], which aims to make the images of different domains appear similar, so that the label-lacking target domain can

S. Albarqouni et al. (Eds.): DART 2021/FAIR 2021, LNCS 12968, pp. 133–143, 2021.
https://doi.org/10.1007/978-3-030-87722-4_13

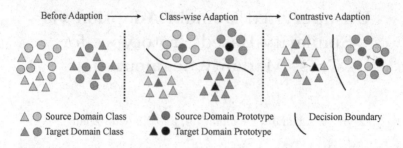

Fig. 1. Illustration of our adaption procedure. On the one hand, our method performs class-wise adaption to align semantic features to their prototypes, on the other hand, we align class-wise prototypes across domains using the contrastive loss.

learn from the transferred source domain. The other stream focuses on feature-level adaption [16,22], which aims to match the feature distributions with adversarial learning or contrastive learning. Besides, DualHierNet [1] also uses edges as self-supervision for the target domain, and EntMin [5] uses entropy minimization to narrow domain gaps.

For methods based on image adaption, most works only conduct the two-direction images translation between source and target domains separately, which may be insufficient to eliminate the domain gap. To this end, we propose a unified framework to fully exploit the two-direction translation results, and our network can be trained end-to-end. For methods based on feature adaption, most works employ adversarial learning to make the semantic features indistinguishable to discriminators, which aligns features in an implicit way. In this paper, we explicitly align features to their prototypes using a class-wise similarity loss, which aims to minimize intra-class and maximize inter-class feature distribution difference. Then, with the help of feature dictionaries, we use the contrastive loss to align class-wise prototypes across domains, which further alleviates the domain shift problem. Figure 1 shows the illustration of our adaption procedure.

2   Methodology

Given a labeled source dataset $\mathbb{D}_s = \left\{ x_s^i, y_s^i \right\}_{i=1}^{N_s}$ and an unlabeled target dataset $\mathbb{D}_t = \left\{ x_t^i \right\}_{i=1}^{N_t}$, unsupervised domain adaption (UDA) for semantic segmentation aims to train a model with supervision from $\mathbb{D}_s$ and information from $\mathbb{D}_t$ to narrow domain gap and improve segmentation performance on $\mathbb{D}_t$.

2.1   Motivation

In our method, we utilize class-wise feature prototypes to perform explicit feature alignment. Firstly, we use a similarity-based loss to regularize the embedded space, and the purpose is to boost feature consistency. Features of the same class are encouraged to be closer to the prototype, and prototypes of different classes are encouraged to be separable. Secondly, we use dictionary to store prototypes from various images, and then contrastive learning is used to improve feature adaption across domains. We expect to adapt prototypes from target domain to source domain, so features of both domains are explicitly aligned.

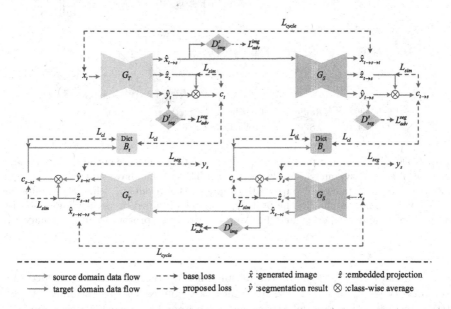

Fig. 2. Our framework has a cycle structure, and mainly consists of G_S and G_T. These modules have the same structure and output the translated image $\hat{x}$, segmentation result $\hat{y}$ and embedded projection $\hat{z}$. Prototypes c are obtained by performing a class-wise average operation on $\hat{z}$ under supervision from $\hat{y}$. During training, only c_s and $c_{s \to t}$ are stored into feature dictionaries B_s and B_t. Besides the widely used circle consistency loss L_{cycle}, segmentation loss L_{seg}, adversarial loss L_{adv}^{img} and L_{adv}^{seg}, we additionally use the proposed loss L_{sim} and L_{cl} to perform explicit feature alignment.

2.2 Proposed Framework

The overall framework is shown in Fig. 2. It has a cycle structure inspired from Cycle-GAN [12], and consists of two modules G_S and G_T, which have the same structures, but process images of different domains. Concretely, G_S processes source domains images, while G_T processes target domain images. Structurally, we input an image for G_S or G_T, and it will output the translated image $\hat{x}$, the segmentation result $\hat{y}$ and the embedded projection $\hat{z}$. During training, our framework is trained in a cycle manner. At each iteration, we calculate the class-wise prototypes c from $\hat{z}$ under the supervision of $\hat{y}$. Note that we only store c_s and $c_{s \to t}$ into feature dictionaries, since $\hat{y}_s$ and $\hat{y}_{s \to t}$ are trained under supervision of ground truth y_s, and we expect to adapt features from $c_{t \to s}$, c_t to c_s, $c_{s \to t}$. During inference, we get the final output by directly averaging the target segmentation result $\hat{y}_t$ and the target-to-source segmentation result $\hat{y}_{t \to s}$.

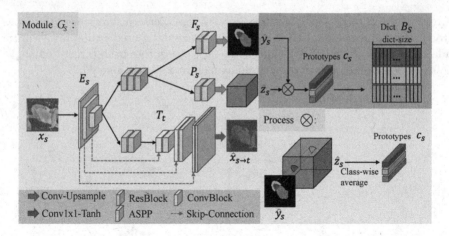

Fig. 3. Details of G_S and prototypes c_s. G_S consists of a feature encoder E_s, an image generator T_t, and symmetric heads F_s and P_s. These heads have same structures but different output channels. F_s outputs the segmentation results $\hat{y}_s$, while P_s outputs the embedded representation $\hat{z}_s$. The prototypes c_s are obtained by performing class-wise average on $\hat{z}_s$ under supervision of $\hat{y}_s$. And we store these prototypes in Dictionary B_s.

To make it clear, we show the detailed structure of G_S and the process to get prototypes c_s in Fig. 3. We introduce skip-connections to the image translation branch to help model convergence and image structure preservation [13]. The segmentation head C_s and projection head P_s have the same structure but different output channels. This symmetrical design is proved effective for semantic feature extraction and regularization [7]. We obtain prototypes c_s from projection z_s with the supervision from $\hat{y}_s$ by performing a class-wise average operation, and then we store c_s into feature dictionary B_s.

In Fig. 2, we denote the loss using the brown and red dash lines. Following GAN-based UDA methods [2,4,9], we use the cycle consistency loss L_{cycle}, segmentation loss L_{seg} and adversarial loss L_{adv}^{img}, L_{adv}^{seg} (see brown dash lines). Additionally, we design a class-wise similarity loss L_{sim} to promote intra-class consistency and inter-class discrepancy at feature level. L_{sim} is calculated between the projection $\hat{z}$ and the prototype c, which is calculated using $\hat{z}$ and $\hat{y}$. Besides, the contrastive loss L_{cl} is used to align prototypes across domains, further reducing the domain gap and improving the model performance. The calculation of L_{cl} needs the prototypes from feature dictionaries. At each iteration, prototypes c_s and $c_{s \to t}$ are first used to calculate L_{sim} and L_{cl}, then stored into Dict B_s and B_t, respectively.

2.3   Feature Prototypes and Class-Wise Similarity Loss

It is observed that the features of the same category tend to be clustered together [19], but the features across different domains have significant discrepancies.

To solve this issue, we regard class-wise prototypes as centers, and explicitly align features to their prototypes. As a result, prototypes of target domain are aligned to those of source domain.

Feature Prototypes. Following [6,7], we calculate the class-wise prototypes in a similar way, and the difference is that we use network segmentation result $\hat{y}$ instead of ground truth as supervision. As shown in Fig. 3, we get prototypes c_s by performing a class-wise average operation on $\hat{z}_s$ under supervision of $\hat{y}_s$. This procedure can be formulated as:

$$c_s^m = \frac{1}{N_m} \sum_{i=1}^{H \times W} \delta\left(\hat{y}_s[i], m\right) \hat{z}_{s,i}, \tag{1}$$

where c_s^m denotes the prototype of the m-th category from x_s, N_m denotes the total pixels of the m-th category, $\delta\left(\hat{y}_s[i], m\right) = 1$ if the i-th pixel of $\hat{y}_s$ belongs to category m, and $\hat{z}_s$ is the embedded representation.

Class-Wise Similarity Loss. We propose a cosine similarity based loss L_{sim} to explicitly regularize features in the embedding space, and we impose the constraint to $\hat{z}$. Taking $\hat{z}_s$ for example, the proposed loss L_{sim} is the summation of the following L_{sc} and L_{dc}.

$$L_{sc} = \frac{1}{C} \sum_{m=1}^{C} \frac{1}{N_m} \sum_{i=1}^{H \times W} \delta(\hat{y}_s[i], m)\left(1 - \cos sim\left(c_s^m, \hat{z}_{s,i}\right)\right), \tag{2}$$

$$L_{dc} = \frac{1}{N_c} \sum_{m=1}^{C} \sum_{n=m+1}^{C} \left(1 + \cos sim(c_s^m, c_s^n)\right), \tag{3}$$

where $\cos sim(u, v) = \frac{u^T v}{\|u\|\|v\|}$ denotes the cosine similarity, C denotes the number of categories in the image, $N_c = \frac{C!}{2!(C-2)!}$ denotes the number of combinations. L_{sc} becomes minimal when the similarity between category prototype c_s^m and representation $\hat{z}_{s,i}$ is maximal. This aims to minimize the intra-class feature discrepancy. In L_{dc}, the similarity is calculated between prototypes of different classes, and it becomes minimal when these prototypes are as dissimilar as possible. This aims to maximize inter-class variance.

For target domain images, since the $\hat{y}_t$ and $\hat{y}_{t \to s}$ are trained without ground truth supervision, we use the pixel-wise predictions of high confidence to supervise $\hat{z}_t$ and $\hat{z}_{t \to s}$, and get the prototypes c_t and $c_{t \to s}$. The similar idea has already been proven successful in pseudo-labeling [20].

2.4   Contrastive Loss via Feature Dictionaries

To align features across domains and boost representative embedded projection, we use dictionaries to store class-wise prototypes from various images, which avoids the category missing problem and enables contrastive learning.

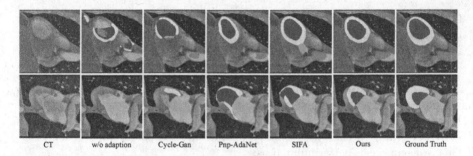

Fig. 4. Visual comparison of representative methods. The structures of MYO, LAC, LVC, AA are denoted by   , green, red, blue colors, respectively. (Color figure online)

Feature Dictionaries. In our framework, Dict B_s and B_t are used to store prototypes from x_s and $x_{s \to t}$, respectively. Following [8], each dictionary has category labels as the keys and the values of each key are prototypes. We denote B_s^c as the source domain dictionary accessed with category key c, and its shape is $[depth \times dict\_size]$. B_s and B_t are updated at every iteration, and old prototypes will be de-queued if the dictionary is full.

Contrastive Loss of Prototypes. Taking c_s^m (category m from source image x_s) as an example, we first calculate the cosine similarity between c_s^m and all prototypes stored in the dictionary B_s. Then for each category, we calculate the average of the highest k similarity values. And the contrastive loss can be formulated as:

$$[v_1^{m,c}, v_2^{m,c}, ..., v_L^{m,c}] = \cos sim(c_s^m, [d_{s,1}^c, d_{s,2}^c, ..., d_{s,L}^c]), \tag{4}$$

$$v^{m,c} = \frac{1}{k} \sum_{i=1}^{k} topk(v_1^{m,c}, v_2^{m,c}, ..., v_L^{m,c}), \tag{5}$$

$$L_{cl} = -\frac{1}{C} \sum_{m=1}^{C} \log \frac{\exp(v^{m,m}/\tau)}{\sum\limits_{i=1, i \neq m}^{C} \exp(v^{m,i}/\tau) + \exp(v^{m,m}/\tau)}, \tag{6}$$

where $v_i^{m,c}$ denotes the cosine similarity between c_s^m and $d_{s,i}^c$, $d_{s,i}^c$ denotes i-th value from category c of dictionary B_s, and τ is the temperature factor.

The contrastive loss not only makes the representation discriminative in embedding space, but also pulls target features closer to the source. Thus both domains are explicitly aligned at the feature level.

Overall Objectives. Following [2,4,9], the widely used cycle consistency loss L_{cycle}, segmentation loss L_{seg} and adversarial loss L_{adv}^{img}, L_{adv}^{seg} are also employed in our training process, we denote them as L_{base}. By adding our proposed similarity-base losses, and the overall objectives can be formulated as:

$$L_{all} = L_{base} + \lambda_1 L_{sim} + \lambda_2 L_{cl},\qquad(7)$$

where λ_1, λ_2 are balance parameters.

Table 1. Results of the MRI → CT task for four cardiac structures on *MMWHS*.

Methods	Volumetric dice↑					Volumetric ASD↓				
	AA	LAC	LVC	MYO	Avg.	AA	LAC	LVC	MYO	Avg.
Supervised training	92.7	91.1	91.9	87.7	90.9	1.5	3.5	1.7	2.1	2.2
W/o adaptation	28.4	27.7	4.0	8.7	17.2	20.6	16.2	-	48.4	-
PnP-AdaNet [16]	74.0	68.9	61.9	50.8	63.9	12.8	6.3	17.4	14.7	12.8
SynSeg-Net [23]	71.6	69.0	51.6	40.8	58.2	11.7	7.8	7.0	9.2	8.9
AdaOutput [22]	65.2	76.6	54.4	43.6	59.9	17.9	5.5	5.9	8.9	9.6
CycleGAN [12]	73.8	75.7	52.3	28.7	57.6	11.5	13.6	9.2	8.8	10.8
CyCADA [21]	72.9	77.0	62.4	45.3	64.4	9.6	8.0	9.6	10.5	9.4
EntMin [5]	83.0	**81.3**	67.2	58.4	72.5	**2.9**	**2.7**	6.3	6.4	**4.6**
SIFAv1 [3]	81.1	76.4	75.7	58.7	73.0	10.6	7.4	6.7	7.8	8.1
SIFAv2 [2]	81.3	79.5	73.8	61.6	74.1	7.9	6.2	5.5	8.5	7.0
DSFN [4]	**84.7**	76.9	79.1	62.4	75.8	-	-	-	-	-
Ours	82.6	**81.3**	**81.7**	**64.3**	**77.5**	6.1	6.0	**3.6**	**4.8**	5.1

Fig. 5. t-SNE visualization of foreground features in Fig. 4. **Left:** The results without the proposed losses, **Right:** The results with the proposed losses.

3 Experiments

3.1 Datasets and Details

The proposed method is validated on the *Multi-Modality Heart Segmentation Challenge 2017* (MMWHS) dataset [15], which consists 20 unpaired MR and CT volumes data with their pixel-level ground truth of heart structures. The left ventricle blood cavity (LVC), the left atrium blood cavity (LAC), the myocardium of the left ventricle (MYO) and the ascending aorta (AA) are usually selected to evaluate the model segmentation performance. For a fair comparison, we use

Table 2. Ablation study of the proposed losses

Components		Dice↑					ASD↓				
L_{sim}	L_{cl}	AA	LAC	LVC	MYO	Avg.	AA	LAC	LVC	MYO	Avg.
×	×	79.7	82.9	77.5	58.1	74.5	7.4	**4.3**	4.1	**4.3**	**5.0**
✓	×	77.7	**83.0**	78.0	61.8	75.2	9.5	5.0	4.8	4.7	6.0
✓	✓	**82.6**	81.3	**81.7**	**64.3**	**77.5**	**6.1**	6.0	**3.6**	4.8	5.1

Table 3. Ablation study of three methods to utilize feature dictionaries

Methods	Dice↑				
	AA	LAC	LVC	MYO	Avg.
Max Similarity	79.1	84.1	78.3	61.2	75.7
Mean All	80.5	**82.1**	79.1	62.7	76.1
Mean Top-k	**82.6**	81.3	**81.7**	**64.3**	**77.5**

Table 4. Ablation study of dict sizes S

S	Dice↑				
	AA	LAC	LVC	MYO	Avg.
w/o	77.7	83.0	78.0	61.8	75.2
200	81.1	**84.8**	81.6	58.4	76.5
400	**82.6**	81.3	**81.7**	64.3	**77.5**
600	80.3	82.8	80.3	63.6	76.8

the preprocessed data released by [2,16], which contains randomly selected 16 volumes for training and 4 volumes for testing for both modalities. All data were first normalized to zero-mean and unit standard deviation, and then switched to $[-1, 1]$. Each slice was cropped and resized to the size of 256×256. These data were also augmented by rotation, scaling, and affine transformations.

Implementations. The discriminators follow patchGAN [17], except that we replace log objective with least-squares loss for a stable training [14]. We empirically set $\lambda_1 = 0.05$, $\lambda_2 = 0.02$, and the dictionary size S is set to 400, temperature τ is set to 1, while top-k is set to 20. Batch size and training epoch are set to 4 and 35, respectively. Besides, we use Adam optimizer [18] with weight decay of 1×10^{-4}, and the learning rate for discriminators is set to 2×10^{-4}, while 3×10^{-4} for G_S and G_T. To warm up training, we apply our proposed loss after the first epoch, and our model is trained on a NVIDIA Tesla V100 with PyTorch.

3.2 Results and Analysis

Quantitative and Qualitative Analysis. Table 1 shows the MRI→CT adaption performance comparison with other methods. Since our experiment is conducted under the same setting as [2,16], we directly refer to their paper results. As shown in Table 1, the model without adaption gets a poor performance on the unseen target domain. Methods based on image-alignment [12,21,23] and methods based on feature-alignment [5,16,22] can significantly improve the model results by narrowing the domain gap. [2,4] further improve the performance by taking both perspectives into account. Our proposed method outperforms these methods in terms of dice, and achieves an average result of 77.5%, besides, we achieve an average ASD of 5.1, which is slightly worse than EntMin [5]. This indicates that our generated results may be not smooth on the boundary regions, while EntMin [5] conduct the entropy minimization to deal with high uncertainty of the boundary. Figure 4 shows the visual comparison, and we choose [2,12,16] as the representative methods from different alignment perspectives. We visualize the feature distribution of Fig. 4 using t-SNE [25] in Fig. 5.

Ablation Study. Firstly, we evaluate the effectiveness of the proposed losses. As shown in Table 2, when neither of the proposed losses were used, our method can be seen as a variant of [4], except that we redesign network structure to utilize low-level features to help image-translation and we do not use auxiliary task for feature adaption. In this case, our method achieves an average Dice of 74.5%. When only the class-wise similarity loss is used, the result gets a gain of 0.7%. When both losses are applied, our model achieves an average dice of 77.5%, surpassing other methods by a large margin.

Secondly, we test on several strategies to utilize prototypes in dictionaries. **Mean Top-k** means taking average of the top-k similarity values. **Mean All** means taking average of all similarity values. **Max Similarity** means only use the largest similarity value. Table 3 shows the results, and we can find that **Mean Top-k** achieves the best performance. This may due to the fact that sampling average can improve the robustness of similarity calculation, and the similarity will not be generalized to much.

Thirdly, we consider different dictionary sizes S. Table 4 shows the results, we can achieve the best performance when $S = 400$, which indicates that an appropriate dictionary size is necessary. A small dictionary may not have sufficient feature diversity, while a big dictionary may induce a slow updating of L_{cl}, as we calculate the average similarity using the top-k strategy.

4 Conclusion

This paper proposes a novel unsupervised domain adaption framework for medical image segmentation. The framework is a unified network that can be trained end-to-end. We propose an innovative class-wise loss (calculated within a single sample) to boost feature consistency and learn representative prototype. Moreover, we conduct contrastive learning of prototypes (calculated with prototypes of multiple samples) to further improve feature adaption across domains. Compared with existing adversarial learning based methods, we explicitly align features. Extensive experiments prove the effectiveness of our method, and show the superiority of the class-wise similarity loss and prototype contrastive learning via dictionary. In the future, we will test our method with different datasets and explore to apply it to domain generalization task.

References

1. Xue, Y., Feng, S., Zhang, Y., Zhang, X., Wang, Y.: Dual-task self-supervision for cross-modality domain adaptation. In: Martel, A.L., et al. (eds.) MICCAI 2020. LNCS, vol. 12261, pp. 408–417. Springer, Cham (2020). https://doi.org/10.1007/978-3-030-59710-8_40
2. Chen, C., Dou, Q., Chen, H., et al.: Unsupervised bidirectional cross-modality adaptation via deeply synergistic image and feature alignment for medical image segmentation. IEEE Trans. Med. Imag. **39**(7), 2494–2505 (2020)
3. Chen, C., Dou, Q., Chen, H., et al.: Synergistic image and feature adaptation: towards cross-modality domain adaptation for medical image segmentation. In: AAAI, vol. 33 no. 01, pp. 865–872 (2019)
4. Zou, D., Zhu, Q., Yan, P.: Unsupervised domain adaptation with dual-scheme fusion network for medical image segmentation. In: IJCAI, pp. 3291–3298 (2020)
5. Vesal, S., Gu, M., Kosti, R., et al.: Adapt everywhere: unsupervised adaptation of point-clouds and entropy minimisation for multi-modal cardiac image segmentation. IEEE Trans. Med. Imag. (2021). https://ieeexplore.ieee.org/document/9380742
6. Liu, Z., Zhu, Z., Zheng, S., et al.: Margin Preserving Self-paced Contrastive Learning Towards Domain Adaptation for Medical Image Segmentation. arXiv preprint arXiv:2103.08454 (2021)
7. Marsden, R.A., Bartler, A., Döbler, M., et al.: Contrastive Learning and Self-Training for Unsupervised Domain Adaptation in Semantic Segmentation. arXiv preprint arXiv:2105.02001 (2021)
8. Chung, I., Kim, D., Kwak, N.: Maximizing Cosine Similarity Between Spatial Features for Unsupervised Domain Adaptation in Semantic Segmentation. arXiv preprint arXiv:2102.13002 (2021)
9. Tomar, D., Lortkipanidze, M., Vray, G., et al.: Self-attentive spatial adaptive normalization for cross-modality domain adaptation. IEEE Trans. Med. Imag. (2021). https://ieeexplore.ieee.org/document/9354186
10. Chen, Y.C., Lin, Y.Y., Yang, M.H., et al.: Crdoco: pixel-level domain transfer with cross-domain consistency. In: CVPR, pp. 1791–1800 (2019)
11. Wang, J., Huang, H., Chen, C., Ma, W., Huang, Y., Ding, X.: Multi-sequence cardiac MR segmentation with adversarial domain adaptation network. In: Pop, M., et al. (eds.) STACOM 2019. LNCS, vol. 12009, pp. 254–262. Springer, Cham (2020). https://doi.org/10.1007/978-3-030-39074-7_27
12. Zhu, J.Y., Park, T., Isola, P., et al.: Unpaired image-to-image translation using cycle-consistent adversarial networks. In: ICCV., pp. 2223–2232 (2017)
13. Ronneberger, O., Fischer, P., Brox, T.: U-Net: convolutional networks for biomedical image segmentation. In: Navab, N., Hornegger, J., Wells, W.M., Frangi, A.F. (eds.) MICCAI 2015. LNCS, vol. 9351, pp. 234–241. Springer, Cham (2015). https://doi.org/10.1007/978-3-319-24574-4_28
14. Mao, X., Li, Q., Xie, H., et al.: Least squares generative adversarial networks. In: CVPR, pp. 2794–2802 (2017)
15. Zhuang, X., Shen, J.: Multi-scale patch and multi-modality atlases for whole heart segmentation of MRI. Med. Image Anal. **31**, 77–87 (2016)
16. Dou, Q., Ouyang, C., Chen, C., et al.: PnP-AdaNet: plug-and-play adversarial domain adaptation network at unpaired cross-modality cardiac segmentation. IEEE Access **7**, 99065–99076 (2019)

17. Isola, P., Zhu, J.Y., Zhou, T., et al.: Image-to-image translation with conditional adversarial networks. In: CVPR, pp. 1125–1134 (2017)
18. Kingma, D.P., Ba, J.: Adam: a method for stochastic optimization. arXiv preprint arXiv:1412.6980 (2014)
19. Zhang, Q., Zhang, J., Liu, W., et al.: Category anchor-guided unsupervised domain adaptation for semantic segmentation. arXiv preprint arXiv:1910.13049 (2019)
20. Li, Y., Yuan, L., Vasconcelos, N.: Bidirectional learning for domain adaptation of semantic segmentation. In: CVPR, pp. 6936–6945 (2019)
21. Cycada: cycle-consistent adversarial domain adaptation. In: ICML, pp. 1994–2003 (2018)
22. Tsai, Y.H., Hung, W.C., Schulter, S., et al.: Learning to adapt structured output space for semantic segmentation. In: CVPR, pp. 7472–7481 (2018)
23. Huo, Y., Xu, Z., Moon, H., et al.: Synseg-Net: synthetic segmentation without target modality ground truth. IEEE Trans. Med. Imag. 38(4), 1016–1025 (2018)
24. Shen, D., Wu, G., Suk, H.I.: Deep learning in medical image analysis. Ann. Rev. Biomed. Eng. 19, 221–248 (2017)
25. Van der Maaten, L., Hinton, G.: Visualizing data using t-SNE. J. Mach. Learn. Res. 9(11) (2008). https://www.jmlr.org/papers/v9/vandermaaten08a.html

Affordable AI and Healthcare

Classification and Generation of Microscopy Images with Plasmodium Falciparum via Artificial Neural Networks Using Low Cost Settings

Rija Tonny Christian Ramarolahy, Esther Opoku Gyasi,
and Alessandro Crimi[✉]

African Institute for Mathematical Sciences - Ghana, Accra, Ghana
{rija,eopoku,alessandro}@aims.edu.gh

Abstract. Recent studies use machine-learning techniques to automatically detect parasites in microscopy images. However, these tools are trained and tested in specific datasets. Indeed, even if over-fitting is avoided during the improvements of computer vision applications, large differences are expected. Differences might be related to settings of camera (exposure, white balance settings, etc.) and different blood film slides preparation. Generative adversial networks offer new opportunities in microscopy: data homogenization, and increase of images in case of imbalanced or small sample size. Taking into consideration all those aspects, in this paper, we describe a more complete view including both detection and generating synthetic images: i) an automated detection used to detect malaria parasites on stained blood smear images using machine learning techniques testing several datasets. ii) a generative approach to create synthetic images which can deceive experts, and provide new data for data transfer or augmentation. The tested architecture achieved 0.98 and 0.95 area under the ROC curve in classifying images with respectively thin and thick smear, in leave-one-out cross-validation settings. The generated images proved to be very similar to the original and difficult to be distinguished by an expert microscopist, which identified correctly the real data for one dataset but had 50% misclassification for another dataset of images. Moveover, the trained model was also tested on a low cost device as RaspberryPI 4 with display.

1   Introduction

Malaria is an infective disease caused by a blood parasite of the genus Plasmodium. According to the latest World malaria report, there were 228 million cases of malaria in 2018 compared to 231 million cases in 2017 [15]. Microscopy investigation remains the major form of diagnosis in malaria management in low- and middle-income countries in daily practice. Although, it has been shown that imprecision might occur due to non-expert microscopists especially in rural areas where malaria is endemic [1]. Microscopy diagnosis initially requires determining

© Springer Nature Switzerland AG 2021
S. Albarqouni et al. (Eds.): DART 2021/FAIR 2021, LNCS 12968, pp. 147–157, 2021.
https://doi.org/10.1007/978-3-030-87722-4_14

the presence of malarial parasites in the specimen in exam. Then, if parasites are detected, species identification and calculation of the degree of infection (number of parasites) is performed. Those tasks are currently still performed by human operators, leading to variability and possible imprecision [1]. Recently, advancements in computer vision are allowing malaria diagnosis in microscopy in an automated manner by using image processing and machine learning. Indeed there are two steps related to each of them. The image processing step has the purpose of detecting/segmenting the region of interest (ROI) inside blood smears from a microscope by using image processing techniques. The second step is given by a classifier which discriminates whether the detected elements in the ROI is an infected red blood cell or not. Before the deep-learning era, classifiers such as *support vector machine (SVM)* [2], *logistic regression* [9], *Bayesian learning* [2] or other machine learning techniques were used, and an extra step for feature extraction and selection were needed. This feature selection required an engineering expertise to define a specific feature to be extracted as color, textural and morphological features to distinguish the parasitized and healthy cells [10]. More recently, by using deep-learning models such as *convolutional neural network (CNN)*, the step of hand-engineered feature selection is not needed, CNN training performs feature extraction jointly to the training. Moreover, as in many fields, deep-learning methods outperformed other machine learning methods [3]. Also, deep-learning architectures introduced the concept of *transfer learning*, where pre-trained models are either fine-tuned on the underlying data or used as feature extractors to aid in visual recognition tasks. These models transfer their knowledge gained while learning generic features from large-scale datasets. In this study the architecture was a VGG-16 trained for the purpose [8]. Lastly, some studies included these image processing approaches in mobile phone applications with the goal of making the analysis accessible in remote settings without access to computers [11,16]. When using a machine learning based classifier, the number of available samples for training is an important key to high accuracy. However, obtaining images may be expensive and time consuming, and often the obtained image datasets are imbalanced. Namely, many sample of one specific plasmoid or uninfected are available and very few of a specific instance are available [4]. In this context and in other computational pathology contexts, data augmentation is rising as a promising solution thanks to *Generative Adversarial Networks (GANs)* [5]. The use of GANs in pathology is motivated by major factors related to data augmentation: 1) to offer variability to a pathologist learning possible variation of the images in study. 2) to have more balanced data (as usually control images are more available than specific case images). 3) data translation and homogenization. Therefore, we present here a comprehensive analysis on malaria images both in terms of automated classification and generation for thick and thin smear. Our goal is not to outperform any existing detection algorithm for malaria, but to show the an comprehensive system which integrates detection and generation.

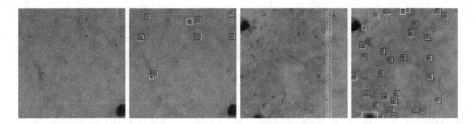

Fig. 1. Parasites detection on a thick smear image (original images and after the parasite detection). The yellow rectangles show the candidate parasites, the red boxes show the detected parasites from the model and the blue boxes show the ground-truth parasites annotated by a microscopist. (Color figure online)

2   Methods and Materials

Data and Code: In this paper, we used three different datasets with films stained according to the Giemsa protocol and all focused on Plasmodium falciparum, which is the deadliest [15]. The three datasets are publicly available. All data have been collected according to ethical standards following the he World Medical Association declaration of Helsinki, and informed consent was obtained from all the subjects. Ethical approval has been previously granted by local committees for the individual studies though they are independent from the study reported in this paper. i) A cell dataset containing 27,558 individual images of red blood cells, taken by a smartphone mounted on a microscope at Chittagong Medical College Hospital, Bangladesh [12]. The images comprise equal instances of infected and uninfected cells. The individual cell images are taken from stained thin blood smears of 150 infected and 50 healthy patients. The images were manually annotated by an expert slide reader at the Mahidol-Oxford Tropical Medicine Research Unit in Bangkok, Thailand. This dataset appears to have different pictures taken from the same patients (though information on how to separate patients is not available). ii) A dataset including 1182 pictures of stained thick blood smear taken from a smartphone attached to a microscope all presenting at least 1 infected cell, using field stain at x1000 magnification. The images were acquired and annotated by experts of the Makerere University, Uganda [11]. iii) A dataset comprising 655 images of thin smears acquired with a magnification of $100X$ by an immersion objective of Leica DLMA microscope [14], all with at least 1 infected cell and reported to be taken from different patients. The produced software tools are publicly available at the URL https:// github.com/alecrimi/malaria_detection_and_generation.

Parasites Detection: The detection experiments are performed on a trained CNN models which are fed with uniformly scaled patches of candidate regions potentially containing parasites. In this work we also adopt a ROI detection approach to identify the patches which are fed into the CNN as done in most of the aforementioned studies, as depicted in Fig. 2. More specifically, images are

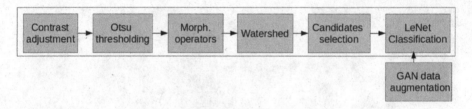

Fig. 2. The overall pipeline which comprises 4 steps of preprocessing to identify the candidates patches. The candidate ROIs are then tested by the CNN. In the expanded scenario (considered outside the main box) the CNN is trained with both originally annotated data and additional synthetic images generated by the GAN.

adjusted by using the contrast. Except for the single cell dataset, we use the Otsu automatic thresholding to separate background from foreground. Then, noise is removed by using the morphological operator. Finally, watershed algorithm is applied. Candidate ROIs to be checked are scaled to 100 × 100 pixels for consistency to the classifier. The used CNN model is inspired by the LeNet model [7], and it comprises three hidden convolutional layers. We consider architectures like LeNet as the state of art, and in this context the experiments are more confirmation than novel results, as LeNet for malaria detection has already been used [3]. Each fully-connected layer is followed by a dropout with probability of 0.3 and the output of the last dropout feeds into a sigmoid classifier. Giemsa-stained microscope images provide good contrast between deep purple nuclei and background. Hence, both in classification and generation, color information is used.

In the following section we report the results using the trained CNN for the thick and thin smear dataset. Moreover, we further test the trained thin smear model on an unseen dataset, and we repeat the experiments by carrying out transfer learning on a VGG architecture as used in previous literature [8,12].

Image Generation: For the image generation, deep convolutional networks are also used, but as mentioned earlier, the architecture is more complex and divided into generator and discriminator as shown in Fig. 3, rather than just a discriminator. Being two sub-systems, they can have independent sets of hyper-parameters. A relevant aspect is that the generator should have a faster convergence than the discriminator, otherwise the discriminator might always understand that the images are fake hence not allowing proper image generation. A solution is to give smaller learning rate. In our experiments, learning rates of 0.00001 and 0.0002 were respectively used for the discriminator and the generator. Moreover, 1000 iterations were used to allow the generator to create realistic images with an output size of 256 × 256 pixels. To assess the quality of the generated images, an experienced microscopist has been involved, being unaware of the process and of the number of generated images, only aware that some images were synthetic and some real. The micscopist received jointly all images. More specifically, he received 10 images comprising thin smear single infected cells, and 6 images of

Table 1. First 2 rows are the performance of the model described. Third row reports the performance of the mixed model: the described CNN described trained on the single cell dataset and tested on the thin smear dataset.

	Accuracy	Recall	F_1 score	Sensitivity	Specificity
Individual cells	0.953	0.953	0.953	0.947	0.959
Thick patches	0.989	0.989	0.989	0.993	0.982
Mixed model	0.895	0.895	0.895	0.894	0.896

thick staining. Morover, 5.000 synthetic generated individual cells were used to train the model once again and validate if improvements was occurring.

3 Results

Parasites Detection: The parasite detection achieved satisfactory results in line with previous literature with an AUC is equal to 0.98 and 0.95 respectively for the thick and thin staining. The performance metric for the models trained and tested on the two datasets are shown in the Table 1. It is worthwhile to mention that sometimes the selected ROIs were containing white blood cells which were then discarded by the neural networks (Table 2). As shown in Table 1, our models performed a satisfactory classification for the two datasets. For the first model, the accuracy value shows that 95% of the first dataset were correctly classified and for the second model, 98% of the second dataset were correctly classified. The high values of the precision, recall and F_1 score for the two models give information about their performances. Despite the difference in size between the patches and the input of the model, Table 1 also shows acceptable performance of the mixed test because from the accuracy value, 89% of the second dataset were correctly classified by the first model. And also, the value of precision, recall and F_1 score are high. The proposed model with fine-tuning outperformed the pre-trained VGG architecture with transfer-learning especially in the mixed model experiment. As expected, fine tuning performs better than transfer-learning though acceptable results are also obtained by the latter. Figure 1 shows the detection of parasites in a thick blood smear image.

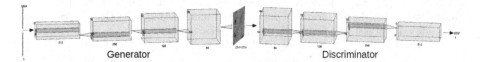

Generator   Discriminator

Fig. 3. GAN architecture, each block represent a convolutional filters bank or a fully connected layer. The pyramids represent the either the upsampling or downsampling. The inner parallelepiped represent the size of the convolution filter. In figure a raw image is depicted but the framework is also meant for patches containing only one cell.

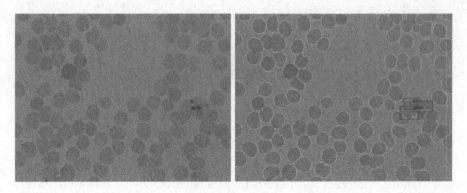

Fig. 4. Cells segmentation and infected cells detection on a thin smear image. On the left original image, on the right the results of the detection. The yellow lines show the candidate parasites, the red boxes show the detected parasites from the model and the blue boxes show the ground-truth parasites annotated by a microscopist. (Color figure online)

The first figure is the original image and the second is the result of our algorithm. The yellow rectangles show the candidate parasites, the red boxes show the detected parasites from the model and the blue boxes show the parasites from the annotations data. Figure 4 shows the result of the segmentation using watershed transform and detection of the infected cells. The first figure (left) is the original thin smear image, the second figure (right) shows the result of our segmentation which gives satisfactory results and the third figure gives the infected cells predict by our model. The cells marked in red are those detected as infected by the model. The training was performed on a Google Colab environment using $1\times$ GPU Tesla K80, having 2496 CUDA cores, and 12 GB VRAM. The training time was 2.4 min. While the testing time was less than a second per image on a Raspberry PI 4 (Cambridge, UK) with 8 GB RAM.

Image Generation: The computational time to produce a single image was 3 h on a single node cluster on a Openstack based system with 125 GB RAM and 32 VGPU and CUDA enabled. The resulting images given to the microscopist

Table 2. First 2 rows are performance of transfer learning using a pre-trained VGG-19. Third row reports the performance of the mixed model using the VGG-19, trained on the single cell dataset and tested on the thin smear dataset.

	Accuracy	Recall	F_1 score	Sensitivity	Specificity
Individual cells	0.940	0.940	0.940	0.932	0.947
Thick patches	0.985	0.985	0.985	0.987	0.982
Mixed model	0.697	0.697	0.620	0.682	0.920

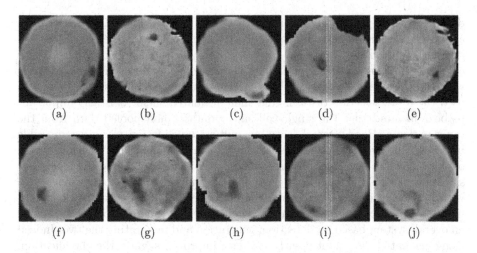

(a)   (b)   (c)   (d)   (e)

(f)   (g)   (h)   (i)   (j)

Fig. 5. Thin smear images: single cells infected by parasite showing different staining. Real: f, h, j. Synthetic: a, b, c, d, e, g, i.

during the test are reported in Fig. 5 and 6. The microscopist (EKN) misclassified 5 out of 10 images in the thin smear single cell test, but identified correctly the real images with thick staining. However, the microscopist described having difficulties in taking a decision with the thick smear image depicted in Fig. 6(a). The experiment confirmed the need of using slower convergence of the discriminator compared to the generator. If this approach was not used, the generator was never going to be able to produce plausible images. The generated data for the single cells dataset represent an already segmented and labeled dataset. Therefore, it is straightforward to use them in the context of data augmentation. Indeed, we repeated the detection experiment for this dataset introducing 5.000 new synthetic images for the uninfected cells to be used jointly to the real ones to balance the dataset, and we obtained slightly better performances, as accuracy of 0.989.

4   Discussions

Detection of malaria infected erythrocytes from peripheral blood smear samples using microscopy is a challenging and subjective task for humans. In this study we investigated both classification of plasmodium in microscopy images, as well as generation of new data for data augmentation.

We used thick and thin Giemsa smears containing Plasmodium falciparum previously annotated by experts. In contrast to legacy machine learning models which need further feature extractions, deep-learning models required less work and achieve more accuracy. Dong et al. [3] reached an accuracy of 0.92 using SVM, and Das et al. [2] achieved an accuracy of 0.84 and 0.83 with Bayesian learning and SVM respectively for malaria parasites classification. The proposed deep-learning approach gave satisfactory results with AUCs of 0.98 and 0.95 in a cross-validation setting. In the used CNN architecture inspired by LeNet [7], implicit regularization imposed by batch normalization, and dropouts in the convolutional and dense layers led to improved generalization. To be in line with the aforementioned research using deep-learning on microscopy images we used the a pre-selection of ROIs and then evaluated them. It is also important to bear in mind that the single-cell smear dataset had pooled data from the same patients. Here, even if leave-one-out cross-validation was used, possible bias might be present in this reported result, in this paper and previous works on the same dataset. The purpose of the paper is not to provide the ideal classification methods outperforming any existing methods, therefore being in line with previously proposed method is sufficient, especially as often works speculate on a specific dataset to advocate their methods. The purpose is to provide an overall system based on detection, generation and integrating the two. Indeed using generated data jointly with real data improved slightly the classification, and showed the usefulness of augmented datasets. We envision the translation to images from further datasets validated directly from rural clinics as a future

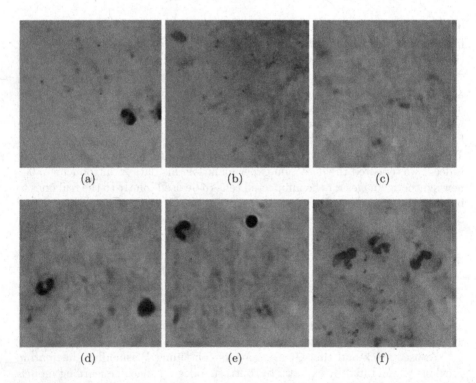

Fig. 6. Thick staining images. Real: a, b, and f. Synthetic: c, d, and e. In those images, it seems visible that GANs can easily generate white blood cells given their size, but has some challenges in creating sharp edges for the smaller red blood cells. (Color figure online)

work. Besides, recent architectures tested in other object detection tasks can also perform detection while performing classification, in this way avoiding also the initial step of ROI detection. The most successful technique is called *You-Only-Look-Once* (YOLO) [13]. Given the excellent results obtained with the proposed approach, we consider simultaneous detection/classification via CNN also as a future improvement. The transfer learning experiment gave also very promising results, with the VGG architecture having almost the same accuracy of our proposed CNN model, except for the mixed model experiment. Automated detection for malaria can help the technician to reduce their workload. Indeed it has been reported that a microscopist need an average of 50 min to evaluate a smear [18], while given the need of few seconds to detect infected cells with the proposed tools, the time can be reduced to few minutes plus the time of the staining. The algorithm focuses mostly on detection of infected cells, as the count of those cells can give an indication of the level of infection. Indeed, the number of cells needed to diagnose is defined as the number of red blood cells to be tested to yield a correct positive test [12]. It was also shown that GANs can be used to generate synthetic infected blood smears which look realistic, and can be used to increase human training on how differently a pathogen can appear and for data augmentation to further data analysis. The performance were increased by using the generated data, though the slight increase is most likely non-statistically significant. This can be related to the already a high accuracy score achieved before the use of the augmented data.

To our knowledge, this is the first study comprising both detection and generation of synthetic microscopy images with malaria parasites. The experienced microscopist involved in the study, reported anecdotally that the discrimination between real and synthetic images was challenging. He had 50% of misclassification with the thin smear single cells images. Instead, he successfully identified all real and synthetic images for the thick smear. A possible explanation is given by the fact that the for the thick smear data GANs were able to generate plausible white blood cells and artifacts, but the small erythrocytes were blurred in the synthetic images and sharper in the real one, and this led the microscopist to identify correctly the images. Nevertheless, data augmentation has become common tools in several digital pathology context [6], so it will be for malaria studies. Moreover, GANs are often used to translate data from one domain to another [4]. Indeed, there has been a recent growing interest in applying GANs to solve several tasks related to digital pathology, including staining normalization and staining transformation [17].

5   Conclusions

In this study we proposed a neural network architecture able to classify infected cells in microscopy images with high accuracy and in different datasets. This shows the potential of deep-learning architecture to be adopted in point of care settings to accelerate malaria diagnosis replicating human thought processes.

Lastly, we demonstrated the ability of neural networks in generating synthetic images which are hard to be distinguished from real ones by experienced microscopists, and can be used for data augmentation. This paves the way to the use of generated images to normalize and translate data from different centers or stainings, offering novel valuable solutions against malaria.

References

1. Bates, I., Bekoe, V., Asamoa-Adu, A.: Improving the accuracy of malaria-related laboratory tests in Ghana. Malaria J. **3**(1), 38 (2004)
2. Das, D., et al.: Machine learning approach for automated screening of malaria parasite using light microscopic images. Micron **45**, 97–106 (2013)
3. Dong, Y., et al.: Evaluations of deep convolutional neural networks for automatic identification of malaria infected cells. In: 2017 IEEE EMBS International Conference on Biomedical and Health Informatics (BHI), pp. 101–104. IEEE (2017)
4. Gadermayr, M., Appel, V., Klinkhammer, B.M., Boor, P., Merhof, D.: Which way round? a study on the performance of stain-translation for segmenting arbitrarily dyed histological images. In: Frangi, A.F., Schnabel, J.A., Davatzikos, C., Alberola-López, C., Fichtinger, G. (eds.) MICCAI 2018. LNCS, vol. 11071, pp. 165–173. Springer, Cham (2018). https://doi.org/10.1007/978-3-030-00934-2_19
5. Goodfellow, I., et al.: Generative adversarial nets. In: Advances in Neural Information Processing Systems, pp. 2672–2680 (2014)
6. Gupta, L., Klinkhammer, B.M., Boor, P., Merhof, D., Gadermayr, M.: GAN-based image enrichment in digital pathology boosts segmentation accuracy. In: Shen, D., et al. (eds.) MICCAI 2019. LNCS, vol. 11764, pp. 631–639. Springer, Cham (2019). https://doi.org/10.1007/978-3-030-32239-7_70
7. LeCun, Y., et al.: Gradient-based learning applied to document recognition. Proc. IEEE **86**(11), 2278–2324 (1998)
8. Mehanian, C., et al.: Computer-automated malaria diagnosis and quantitation using convolutional neural networks. In: Proceedings of the IEEE International Conference on Computer Vision Workshops, pp. 116–125 (2017)
9. Park, H.S., et al.: Automated detection of p. falciparum using machine learning algorithms with quantitative phase images of unstained cells. PloS One **11**(9), e0163045 (2016)
10. Poostchi, M., et al.: Image analysis and machine learning for detecting malaria. Translational Res. **194**, 36–55 (2018)
11. Quinn, J.A., et al.: Deep convolutional neural networks for microscopy-based point of care diagnostics. In: Machine Learning for Healthcare, pp. 271–281 (2016)
12. Rajaraman, S., et al.: Pre-trained convolutional neural networks as feature extractors toward improved malaria parasite detection in thin blood smear images. PeerJ **6**, e4568 (2018)
13. Redmon, J., Farhadi, A.: Yolov3: an incremental improvement. arXiv preprint arXiv:1804.02767 (2018)
14. Tek, F.B., et al.: Parasite detection and identification for automated thin blood film malaria diagnosis. Comput. Vis. Image Understand. **114**(1), 21–32 (2010)
15. World Health Organization : World malaria report 2019. World health organization 2019, Geneva (2020)
16. Yang, F., et al.: Deep learning for smartphone-based malaria parasite detection in thick blood smears. IEEE J. Biomed. Health Inform. **24**, 1427 (2019)

17. Zanjani, F.G., et al.: Stain normalization of histopathology images using generative adversarial networks. In: 2018 IEEE 15th International Symposium on Biomedical Imaging (ISBI 2018), pp. 573–577. IEEE (2018)
18. Zimmerman, P.A., Howes, R.E.: Malaria diagnosis for malaria elimination. Curr. Opinion Infect. Diseases **28**(5), 446–454 (2015)

Contrast and Resolution Improvement of POCUS Using Self-consistent CycleGAN

Shujaat Khan$^{(\boxtimes)}$ (ID), Jaeyoung Huh (ID), and Jong Chul Ye (ID)

Department of Bio and Brain Engineering, Korea Advanced Institute of Science and Technology (KAIST), 335 Gwahangno, Yuseong-gu, Daejeon 305-701, Korea
{shujaat,woori93,jong.ye}@kaist.ac.kr

Abstract. Point-of-Care Ultrasound (POCUS) imaging can help efficient resource utilization by reducing the secondary care referrals, and work as an extension in physical examination. Recently, many methods were proposed to reduce the size and power consumption of the system while improving the visual quality, but hand-held POCUS devices still have inferior image contrast and spatial resolution compared to the high-end ultrasound systems. To address this, here we propose an efficient solution for contrast and resolution enhancement of hand-held POCUS images using unsupervised deep learning. In contrast to the existing CycleGAN approaches that have difficulty in improving both contrast and image resolutions, the proposed method mitigate the problem by decomposing the contrast transfer and resolution improvement through CycleGAN and self-supervised learning. Experimental results confirmed that our method is superior than the conventional approaches.

Keywords: Ultrasound imaging · Hand-held ultrasound · POCUS · Unsupervised learning · CycleGAN

1 Introduction

Point-of-Care Ultrasound (POCUS) imaging is one of the most rapidly adapted technologies in medical imaging [21]. One major advantage of POCUS is that it can be used as for general physical examination to help assist primary care. According to a study [6], the POCUS can save the treatment cost about 33% by reducing the referral to secondary care. Furthermore, with the advent of artificial intelligence it is expected to become an in-demand electric gadget for personal and tele-health monitoring [18], and to be multi-billion dollar industry by the year 2025 [26]. Recently, to deal with the COVID-19 pandemic, a number of AI-integrated POCUS solutions have also been developed [4,23,27].

Unfortunately, currently available POCUS systems usually generate images with low contrast and resolution that are not suitable for clinical applications.

This work is supported by National Research Foundation of Korea, Grant Number: NRF-2020R1A2B5B03001980.

S. Albarqouni et al. (Eds.): DART 2021/FAIR 2021, LNCS 12968, pp. 158–167, 2021.
https://doi.org/10.1007/978-3-030-87722-4_15

Classical model-based iterative methods are usually computationally expensive, and require many hyper-parameter tuning [29]. Inspired by the success of deep learning approach in other medical imaging area, several AI-based US image improvement methods have been studied [2,3,24]. Unfortunately, supervised learning methods [8,10–12,14,20,25,28] are not effective in the case of POCUS as they require paired high quality data.

To deal with this, Zhou *et al.* [29] proposed a multi-stage GAN model to mimic the high quality images from a low quality portable ultrasound system. To improve the training using relatively small dataset, they utilize transfer learning and data augmentation approaches. However, the method requires paired data acquired from same anatomical region and carefully alignment. In this sense, CycleGAN approaches [19,30] may be ideally suitable for POCUS by training of neural network models in the absence of matched target samples. Accordingly, a variety of CycleGAN based ultrasound image enhancement methods have been developed [5,7,9,13,15,16]. For example, in [13,15] two retrospective studies are presented in which CycleGAN model is utilized for the accelerated high-quality US imaging. Similarly, in [9] a constrained CycleGAN model is proposed to improve the quality of POCUS images for the application of cardiac imaging by introducing the segmentation regularization. Although the study provides impressive results, the major limitation is that the model required large number of supervised labels for segmentation. In fact, one of the main reasons that the existing CycleGAN approaches [9,13,15] require additional data and regularization constrain is that the direct style transfer of images between POCUS and high-end system often results in artifacts and insufficient quality improvement. This is because the image quality differences are originated from two tightly entangled factors such as contrast and spatial resolution, which are often difficult to learn simultaneously.

Therefore, one of the main contributions of this paper is a novel self-consistent CycleGAN architecture that decomposes the contrast transfer and resolution improvement in two steps. Specifically, CycleGAN is trained between the POCUS image and down-sampled images from a high-end system, after which the resolution improvement is achieved using an additional super-resolution network. In particular, the super-resolution network can be trained in a supervised manner, as down-sampled images from high-end system have matched high-resolution labels. Once the neural networks are trained, the images from POCUS can be enhanced by applying the two networks by cascaded applications.

We organized this paper as follows. The proposed method is described in Sect. 2. In Sect. 3, experimental results are discussed followed by the conclusion in Sect. 4.

2   Method

The task of POCUS image quality enhancement require the reconstruction of multiple features of image, including contrast and resolution enhancement. To simplify the learning task, we design a novel CycleGAN approach shown in Fig. 1.

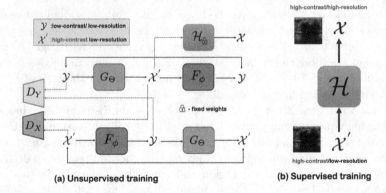

Fig. 1. (a) Self-consistent CycleGAN model for contrast enhancement, which is trained in an unsupervised manner. Here, $\mathcal{H}$ is the super-resolution network for intermediate quality image in $\mathcal{X}'$ to the target quality image in $\mathcal{X}$, which is trained a priori. (b) Super-resolution network trained in a supervised manner.

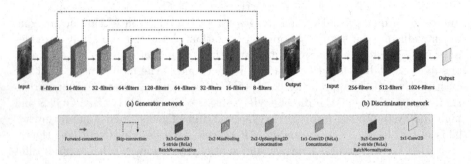

Fig. 2. Proposed neural network architecture: (a) generator network, and (b) discriminator network.

Specifically, to decouple the burden of resolution enhancement from the contrast transfer, we trained our generators using images in the intermediate domain $\mathcal{X}'$ that have contrast from the high-end system but with reduced spatial resolution. Using the intermediate domain images, our objective is to learn mapping functions G_θ that can convert the input POCUS image in $\mathcal{Y}$ to the target images in the intermediate domain $\mathcal{X}'$. For the backward path, F_ϕ is trained to convert the intermediate domain images to the input POCUS domain. Both networks are trained as shown in Fig. 1(a). Since intermediate domain images were generated using high quality images, they have matched high-resolution images for supervised training: so we can additionally train a super-resolution model in a supervised fashion as shown in Fig. 1(b). At the training phase, the super-resolution network $\mathcal{H}$ is separately trained using a combined loss of mean-absolute-error (MAE) and structural-similarity (SSIM) defined as $\ell(\mathcal{X}, \mathcal{X}') = 1 + MAE(\mathcal{X}, \mathcal{X}') - SSIM(\mathcal{X}, \mathcal{X}')$, after which CycleGAN network is trained. The loss function for our CycleGAN training is given by $\ell(G_\theta, F_\phi, D_\mathcal{Y}, D_{\mathcal{X}'}) :=$

$\mathcal{E}_{GAN} + \alpha\mathcal{E}_{cycle}$. where α is a hyper-parameter which is set to be 10, $G_\theta, D_{\mathcal{X}'}$ and $F_\phi, D_{\mathcal{Y}}$ are the generator and discriminator networks for domain $\mathcal{X}'$ and $\mathcal{Y}$ respectively. Whereas $\mathcal{E}_{cycle}$ is the cycle consistency loss function defined as $\mathcal{E}_{cycle} = \mathbb{E}_{\mathcal{X}'}[||D_{\mathcal{X}}(\mathcal{X}') - D_{\mathcal{X}}(F_\phi(G_\theta(\mathcal{X}')))||_1] + \mathbb{E}_{\mathcal{Y}}[||D_{\mathcal{Y}}(\mathcal{Y}) - D_{\mathcal{Y}}(G_\theta(F_\phi(\mathcal{Y})))||_1]$, and $\mathcal{E}_{GAN}$ is the adversarial-loss given by $\mathcal{E}_{GAN} = \mathbb{E}[\log D_{\mathcal{X}}(\mathcal{Y})] + \mathbb{E}[\log(1 - D_{\mathcal{X}}(F_\phi(\mathcal{X}')))] + \mathbb{E}[\log D_{\mathcal{Y}}(\mathcal{X}')] + \mathbb{E}[\log(1 - D_{\mathcal{Y}}(G_\theta(\mathcal{Y})))]$. In regard to the discriminator for enhanced images, there are potentially two feasible implementation: one in the high-end domain $\mathcal{X}$, and the other in the intermediate domain $\mathcal{X}'$. Through experiments, we found that the difference are minimal although the discriminator in $\mathcal{X}$ domain requires more memory. Therefore, we use the CycleGAN architecture in Fig. 1(a), where the discriminator is implemented in the intermediate domain. Once the networks are trained, during inference phase a target domain image can be generated by the composite mapping $y \mapsto \mathcal{H}(G_\theta(y))$, which can be realized by cascade applications of contrast enhancement network and super-resolution network (see Fig. 3).

The model is designed to process an input image of size 256×256 pixels to produce an enhanced quality image of the same size. The architecture for the generator and discriminators are shown in Fig. 2. Note that both contrast enhancement and super-resolution network have same architecture as shown in Fig. 2(a).

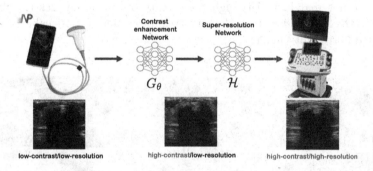

Fig. 3. Our self-consistent CycleGAN approach for contrast and resolution enhancement.

To design an image translation model we required two datasets representing low and high quality images distributions. To generate a high quality dataset, 512 images were acquired using the E-CUBE 12R US systems (Alpinion Co., Korea). The images were scanned in focused mode with (L3–12) linear array and filtered by NLLR [31] and Deep deconvolution [13] algorithms. The dataset of 512 images comprises of 320 *in vivo* images acquired from the carotid and thyroid regions of 8 healthy volunteers and 192 ATS-539 phantom images.

Similarly, 400 images scanned using NPUS050 (NOPROBLEM MEDICAL Co., China) portable US system were used as low-quality input. The low quality dataset consist of 200 *in vivo* and 200 phantom images. The *in vivo* dataset acquired from the carotid, trachea/thyroid, artery, forearm, and calf muscle areas of 2 volunteers, and 100 temporal frames were scanned from each subject. For phantom images ATS-539 phantom was scanned from different views angles. For model training 157 *in vivo* and 168 phantom images were used, while remaining 43 *in vivo* images from unseen region and 32 images from different regions of phantom were used for testing. For the intermediate domain images, an additional dataset of 512 simulated low-high quality image was generated by down grading the quality of high quality images acquired from and high-end system.

The proposed method was implemented using Python with TensorFlow [1] on an Nvidia GEFORCE GTX 1080 Ti GPU. For network training Adam optimizer [17] was used and learning rate linearly changed from 5×10^{-4} to 1×10^{-4} in 200 epochs.

3   Experimental Results

3.1   Qualitative Evaluation

Figures 4 show the sample results from different regions of phantom and anatomical regions. As for reference, high quality images from the similar region/anatomy scanned using high-end system are also shown. Readers are reminded that due to the nature of unsupervised learning, the improved images and the high-end images are not matched. Although the quality of output images is still quite inferior to the high-end system, the quality improvement is noticeable compared to original low-quality images.

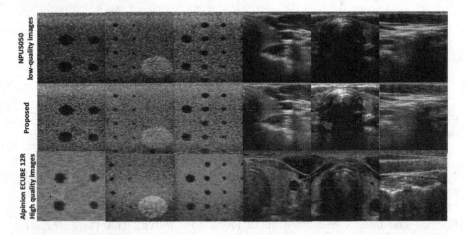

Fig. 4. Sample of (first row) original low qualtiy images from NPUS050 hand-held POCUS device, (second row) improved quality images using proposed method, and (third row) high quality images from the similar region/anatomy using high-end system.

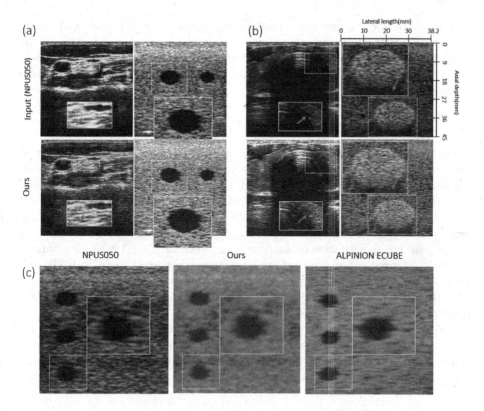

Fig. 5. Reconstruction results from *in vivo* scans of carotid, and trachea/thyroid regions, and the ATS 539 tissue mimicking phantom. B-Mode images highlighting (a) resolution improvement, (b) contrast enhancement, and (c) comparison of image quality using same region scanned using high and low-end machines.

To show the effect of the quality enhancement using proposed method, four examples are shown in Fig. 5. Interesting regions are highlighted and zoomed in inset figures. In Fig. 5(a), it can be clearly seen that the resolution in carotid and phantom images has been improved, the red arrow showing the spots where proposed method recover the missing details and produced sharped results. Similarly, to highlight the gain in contrast, trachea/thyroid and hyper-echoic phantom scans are shown in Figs. 5(b). Note that the details of the trachea's surrounding regions are virtually in-visible in the original input image, whereas in the output image it is quite prominent. A similar contrast enhancement effect can be seen in the phantom example where shadow artifacts around the hyperechoic region are noticeably reduced.

To confirm whether the reconstructed features truly depicting the anatomical details or not, we acquired two scans from a similar region of ATS-539 tissue mimicking phantom using NPUS050 and ALPINION ECUBE systems and compared the results. In Fig. 5(c), input and output as well as the target image of a tissue mimicking anechoic phantom are shown. From the results it can be clearly seen that compared to the input image, the contrast and texture of the output image are quite close to those of the target image. This direct comparison with high-end system validate image enhancement performance in an objective manner.

3.2   Quantitative Evaluation

Unfortunately, conventional reconstruction metrics such as PSNR and SSIM require paired data and generating exactly matched pair of low and high quality data samples is not possible, therefore PSNR and SSIM are only used to measure the fitness of super-resolution network. In particular, for the simulated low quality images that are generated from $2\times$ axially and $2\times$ laterally sub-sampled images of high-end images, the PSNR and SSIM metrics show 13.58 dB and 0.63 units gain. For quantitative evaluation of proposed $\mathcal{H}(G_\theta(y))$ network, contrast statistics of input (y) and output (x) images are compared using CR, CNR, and GCNR [22] metrics. On average, the proposed method achieves 14.96 dB, 2.38, 0.8604 units CR, CNR and GCNR, which is 21.77%, 30.06%, and 44.42% higher, respectively, than those of the input images.

Apart from reconstruction quality improvement, one major advantage of our method is the fast reconstruction time. This is significantly important for real time imaging, on average the reconstruction time for a single image is around 13.18 (milliseconds), and it could further reduce by optimized implementation.

3.3   Comparative Evaluation

In Fig. 6, we compare the results of three different approaches. Specifically, the input images in (a) are processed using (b) histogram equalization, (c) conventional CycleGAN, and (d) the proposed method. The histogram equalization is a popular method used to improve the visual quality of low-contrast images. It is a passive filtration technique which doesn't require model training. From the results in Fig. 6(b) it can be easily seen that holistic contrast enhancement can only increase the contrast difference and it cannot recover the anatomical details which is only possible with learning based methods.

Figure 6(c) shows the results by the conventional CycleGAN-based method that was employed in the existing approach [9] for a similar task, whereas the results by the proposed method are shown in Fig. 6(d). Here the objective is to highlight the importance of the intermediate domain $\mathcal{X}'$ learning in Fig. 1. Both models were trained using same training data-samples, the conventional CycleGAN was trained to directly maps $\mathcal{Y}$ to $\mathcal{X}$, and the proposed method with intermediate domain conversion. On careful observation, it can be easily seen that the contrast and texture of phantom and *in vivo* images using proposed method is relatively better than the conventional method.

To highlight aforementioned claim, in Fig. 1(c) three interesting spots $\{A, B,$ and $C\}$ of *in vivo* are marked with orange arrows. A and B show that the conventional approach only manages to mimic the contrast and smooth-out the

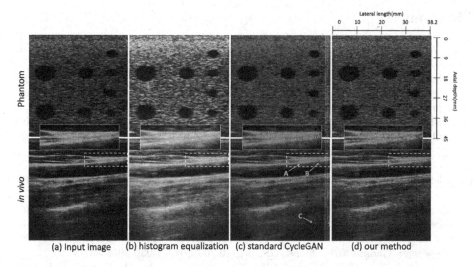

Fig. 6. Comparison of different image enhancement approaches using phantom and *in vivo* data. (a) original low-quality image acquired using NPUS050, (b) results of histogram equalization, (c) directly mapping of $\mathcal{Y}$ to $\mathcal{X}$ using conventional CycleGAN, and (d) the proposed CycleGAN in Fig. 1.

anatomical fine details. On the other hand, our model is designed based on the prior knowledge of the source of the image degradation therefore it not only improved contrast but also improve the resolution without dramatically increasing the noise components which is the case in conventional CycleGAN (see spot C a noticeable effect of noise enhancement can be easily seen).

4   Conclusion

To overcome the limitation of visual quality in hand-held POCUS imaging, herein we proposed an unsupervised learning based image enhancement method. Unlike contemporary supervised learning methods which requires large training data from identical anatomical regions, the proposed method can directly process unpaired image domain data to generate high-quality noise and artifact free images from low-quality images. Furthermore, in contrast to the existing CycleGAN approach, we decomposed the contrast enhancement and resolution improvement into two steps - CycleGAN based unsupervised learning step and supervised resolution enhancement step - so that the overall image quality improvement were stably obtained without spurious artifacts. Even though proposed method shows noticeable gain in image quality, the results are still not ideal and require a comprehensive clinical comparison. However, the design was evaluated on *in vivo* and phantom data for both qualitatively and quantitatively, using the standard quality measures such as contrast recovery and reconstruction loss. Which suggest that the proposed schemes may substantially help in designing low-powered, high quality POCUS systems.

References

1. Abadi, M., et al.: TensorFlow: a system for large-scale machine learning. OSDI **16**, 265–283 (2016)
2. Blaivas, M., Arntfield, R., White, M.: DIY AI, deep learning network development for automated image classification in a point-of-care ultrasound quality assurance program. J. Am. Coll. Emergency Phys. Open **1**(2), 124–131 (2020)
3. Blaivas, M., Blaivas, L.: Are all deep learning architectures alike for point-of-care ultrasound?: evidence from a cardiac image classification model suggests otherwise. J. Ultrasound Med. **39**(6), 1187–1194 (2020)
4. Cristiana, B., et al.: Automated lung ultrasound B-line assessment using a deep learning algorithm. IEEE Trans. Ultrasonics Ferroelectrics Frequency Control **67**, 2312 (2020)
5. Escobar, M., Castillo, A., Romero, A., Arbeláez, P.: UltraGAN: ultrasound enhancement through adversarial generation. In: Burgos, N., Svoboda, D., Wolterink, J.M., Zhao, C. (eds.) SASHIMI 2020. LNCS, vol. 12417, pp. 120–130. Springer, Cham (2020). https://doi.org/10.1007/978-3-030-59520-3_13
6. Han, P.J., Tsai, B.T., Martin, J.W., Keen, W.D., Waalen, J., Kimura, B.J.: Evidence basis for a point-of-care ultrasound examination to refine referral for outpatient echocardiography. Am. J. Med. **132**(2), 227–233 (2019)
7. Huang, O., et al.: Mimicknet, mimicking clinical image post-processing under black-box constraints. IEEE Trans. Med. Imag. **39**(6), 2277–2286 (2020)
8. Hyun, D., Brickson, L.L., Looby, K.T., Dahl, J.J.: Beamforming and speckle reduction using neural networks. IEEE Trans. Ultrasonics Ferroelectrics Frequency Control **66**(5), 898–910 (2019)
9. Jafari, M.H., et al.: Cardiac point-of-care to cart-based ultrasound translation using constrained CycleGAN. Int. J. Comput. Assist. Radiol. Surg. **15**(5), 877–886 (2020)
10. Khan, S., Huh, J., Ye, J.C.: Deep learning-based universal beamformer for ultrasound imaging. In: Shen, D., et al. (eds.) MICCAI 2019. LNCS, vol. 11768, pp. 619–627. Springer, Cham (2019). https://doi.org/10.1007/978-3-030-32254-0_69
11. Khan, S., Huh, J., Ye, J.C.: Universal plane-wave compounding for high quality us imaging using deep learning. In: 2019 IEEE International Ultrasonics Symposium (IUS), pp. 2345–2347. IEEE (2019)
12. Khan, S., Huh, J., Ye, J.C.: Adaptive and compressive beamforming using deep learning for medical ultrasound. IEEE Trans. Ultrasonics Ferroelectrics Frequency Control **67**, 1558 (2020)
13. Khan, S., Huh, J., Ye, J.C.: Unsupervised deconvolution neural network for high quality ultrasound imaging. In: 2020 IEEE International Ultrasonics Symposium (IUS), pp. 1–4. IEEE (2020)
14. Khan, S., Huh, J., Ye, J.C.: Switchable deep beamformer for ultrasound imaging using Adain. In: 2021 IEEE 18th International Symposium on Biomedical Imaging (ISBI), pp. 677–680. IEEE (2021)
15. Khan, S., Huh, J., Ye, J.C.: Unsupervised deep learning for accelerated high quality echocardiography. In: 2021 IEEE 18th International Symposium on Biomedical Imaging (ISBI), pp. 1738–1741. IEEE (2021)
16. Khan, S., Huh, J., Ye, J.C.: Variational formulation of unsupervised deep learning for ultrasound image artifact removal. IEEE Trans. Ultrasonics Ferroelectrics Frequency Control **68**, 2086 (2021)
17. Kingma, D., Ba, J.: Adam: a method for stochastic optimization. arXiv preprint arXiv:1412.6980 (2014)

18. Lee, L., DeCara, J.M.: Point-of-care ultrasound. Curr. Cardiol. Rep. **22**(11), 1–10 (2020)
19. Lim, S., Ye, J.C.: Blind deconvolution microscopy using cycle consistent CNN with explicit PSF layer. In: Knoll, F., Maier, A., Rueckert, D., Ye, J.C. (eds.) MLMIR 2019. LNCS, vol. 11905, pp. 173–180. Springer, Cham (2019). https://doi.org/10.1007/978-3-030-33843-5_16
20. Luchies, A.C., Byram, B.C.: Deep neural networks for ultrasound beamforming. IEEE Trans. Med. Imag. **37**(9), 2010–2021 (2018)
21. Ramsingh, D., Bronshteyn, Y.S., Haskins, S., Zimmerman, J.: Perioperative point-of-care ultrasound: from concept to application. Anesthesiology **132**(4), 908–916 (2020)
22. Rodriguez-Molares, A., Rindal, O.M.H., D'hooge, J., Måsøy, S.E., Austeng, A., Torp, H.: The generalized contrast-to-noise ratio. In: 2018 IEEE International Ultrasonics Symposium (IUS), pp. 1–4. IEEE (2018)
23. Roy, S., et al.: Deep learning for classification and localization of COVID-19 markers in point-of-care lung ultrasound. IEEE Trans. Med. Imag. **39**(8), 2676–2687 (2020)
24. Shokoohi, H., LeSaux, M.A., Roohani, Y.H., Liteplo, A., Huang, C., Blaivas, M.: Enhanced point-of-care ultrasound applications by integrating automated feature-learning systems using deep learning. J. Ultrasound Med. **38**(7), 1887–1897 (2019)
25. Solomon, O., et al.: Deep unfolded robust PCA with application to clutter suppression in ultrasound. IEEE Trans. Med. Imag. **39**(4), 1051–1063 (2019)
26. Toth, J.: Utility of Point-of-Care Ultrasound Across Clinical Applications Spurs Continued Growth (2021). https://www.itnonline.com/article/utility-point-care-ultrasound-across-clinical-applications-spurs-continued-growth-0
27. Wynants, L., et al.: Prediction models for diagnosis and prognosis of COVID-19: systematic review and critical appraisal. BMJ **369**, m1328 (2020). https://doi.org/10.1136/bmj.m1328
28. Yoon, Y.H., Khan, S., Huh, J., Ye, J.C.: Efficient b-mode ultrasound image reconstruction from sub-sampled RF data using deep learning. IEEE Trans. Med. Imag. **38**(2), 325–336 (2018)
29. Zhou, Z., Wang, Y., Guo, Y., Qi, Y., Yu, J.: Image quality improvement of hand-held ultrasound devices with a two-stage generative adversarial network. IEEE Trans. Biomed. Eng. **67**(1), 298–311 (2019)
30. Zhu, J.Y., Park, T., Isola, P., Efros, A.A.: Unpaired image-to-image translation using cycle-consistent adversarial networks. In: Proceedings of the IEEE International Conference on Computer Vision, pp. 2223–2232 (2017)
31. Zhu, L., Fu, C.W., Brown, M.S., Heng, P.A.: A non-local low-rank framework for ultrasound speckle reduction. In: Proceedings of the IEEE Conference on Computer Vision and Pattern Recognition, pp. 5650–5658 (2017)

Low-Dose Dynamic CT Perfusion Denoising Without Training Data

Viswanath P. Sudarshan[1]([⊠]), R. AthulKumar[2], Pavan Kumar Reddy[1],
Jayavardhana Gubbi[1], and Balamuralidhar Purushothaman[1]

[1] Embedded Devices and Intelligent Systems, TCS Research, Bangalore, India
viswanath.pamulakantysudarshan@tcs.com
[2] Computer Science and Engineering, IIT Gandhinagar, Gujarat, India

Abstract. Cerebral perfusion maps derived from low-dose computed
tomography (CT) data typically suffer from low signal-to-noise ratio
(SNR). Obtaining denoised perfusion maps is critical for improved quantitative accuracy and clinical decision making. Several prior works focus
on denoising the low-dose CT (LD-CT) images followed by perfusion
map generation via regularized deconvolution. Recently, supervised deep
neural networks (DNN) have been employed for learning a mapping
between the perfusion maps obtained at low-dose and corresponding
maps obtained at standard dose. Supervised learning-based methods rely
on large amount of training data for improved accuracy. However, they
suffer from changes in acquisition protocol and risk missing patient specific information, which is critical for applications such as stroke imaging. In this work, we address the problem of handling changes in signal characteristics while retaining patient-specific information. For this
purpose, we explore a combination of self-supervised and unsupervised
deep neural networks to obtain denoised perfusion maps from low-dose
noisy CT data. We propose a two-stage sequential approach for obtaining
improved perfusion maps to leverage statistical independence of noise in
the measurement space. First, we denoise the low-dose CT projection
space data using a self-supervised method. Subsequently, we reconstruct
the CT images from the denoised sinograms using fast reconstruction
methods such as filtered backprojection. Second, we use the improved
CT images as anatomical prior to refine the noisy CBF maps obtained
directly from the LD-CT data. Through empirical experiments, we show
that our model is robust both qualitatively and quantitatively to changes
in signal characteristics of the acquired dynamic CT data.

Keywords: CT perfusion · Low-dose CT · Unsupervised learning ·
Self-supervised learning

1 Introduction and Related Work

Dynamic X-ray computed tomography perfusion (CTP) imaging provides quantitative analysis of the blood flow parameters. Fast and accurate assessment of

© Springer Nature Switzerland AG 2021
S. Albarqouni et al. (Eds.): DART 2021/FAIR 2021, LNCS 12968, pp. 168–179, 2021.
https://doi.org/10.1007/978-3-030-87722-4_16

hemodynamic parameters is imperative during time-critical situations such as ischemic stroke. Compared to diffusion-weighted imaging (DWI) based on magnetic resonance imaging (DWI), CTP offers faster and quantitative imaging capability. Typically, CTP imaging involves acquisition of multiple CT volumes after a bolus injection of the contrast agent. The amount of radiation accumulated during a stroke scanning session (about 10 mSv) is a cause of concern [18] and does not align with the advocated principle of *as low as reasonably achievable* (ALARA) [21]. Hence, recently, works on low-dose CTP imaging have been an active area of interest [19]. However, lowering the dose by means of lowering the X-ray tube current, reduces the signal-to-noise ratio (SNR) in the observed data (sinogram) and consequently affects the quality of the reconstructed CT images and hence, the estimated perfusion maps. Works attempting enhancement of low-dose CT (LD-CT) perfusion maps rely on denoising the CT perfusion images, followed by conventional methods such as truncated singular value decomposition (TSVD) [12] for improved CBF map estimation [23]. A few other influential works [5,6] focused on estimating the perfusion maps directly from the 4-D spatiotemporal contrast-enhanced CT data using model-based approaches. Recently, methods based on deep neural network (DNN) methods learn a mapping between the perfusion maps obtained from LD-CT and those obtained using standard-dose CT (SD-CT) images [9]. In these above-mentioned works, the LD-CT images are retrospectively simulated by adding noise in the image domain of SD-CT images with the noise model described in [9]. Typically, corruption in the signal occurs in the measurement domain (sinogram) where the statistical properties of the noise can be well characterized such as independently identically distributed (i.i.d) Gaussian or Poisson noise. Thus, the various approaches to estimate perfusion maps from LD-CT data can be categorized as: (i) regularized deconvolution to obtain cerebral blood flow (CBF) maps from noisy LD-CT data; (ii) denoising/enhancing the LD-CT images to obtain an estimate of the SD-CT images followed by application of TSVD algorithm to obtain improved CBF maps; and (iii) post-deconvolution denoising/enhancement of the CBF maps obtained from LD-CT images to estimate perfusion maps at standard dose. The aforementioned approaches involved in CT perfusion imaging are summarized in Fig. 1.

The CT perfusion imaging scenario is analogous to several multimodal imaging scenarios where both anatomical (here contrast-enhanced CT) and functional (perfusion maps) images are used for clinical diagnosis. While supervised learning based methods have shown substantial improvement for LD-CT imaging, they are prone to missing patient-specific information and in certain extreme cases, lead to hallucinating artifacts [2]. Several works have studied the degradation in performance of DNN models when presented with testing data that deviates from training data [8,13]. Recently, several self-supervised as well as unsupervised DNN have been proposed e.g., [14] for the denoising of natural images which do not necessitate the presence of training data, but require multiple instances of the noisy image for training the DNN. More recently, the work in [1] showed that denoising can be achieved with just the noisy image alone using the elegant theory of $\mathcal{J}$-invariant functions. These methods assume that the noise is element-wise statistically independent in the image domain. However, this does not hold for tomographic inverse

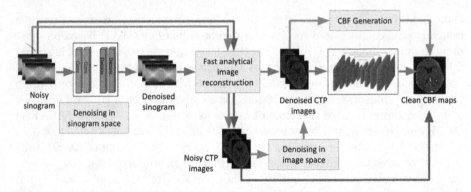

Fig. 1. Summary of the various approaches for perfusion map generation using low-dose CT data. The proposed framework is highlighted using green arrows. Methods focusing on improving CT image quality via image space denoising (orange arrows) typically employ TSVD for CBF map generation. Alternative approaches include generating CBF maps directly from low-quality CT data (blue arrow) like tensor total variation (TTV) [6] or sparse perfusion deconvolution (SPD) [5]. (Color figure online)

problems. The property of element-wise statistical independence can be assumed in the projection domain. Denoising the sinogram to enable low-dose imaging has been explored in several prior works [11,16,20]. On the other hand, works such as [3] leverage prior information from noise-free anatomical images to improve noisy images from another (functional) modality. The effect of noise on the anatomical image is not studied. This work tries to address the problem of denoising both the anatomical and functional images using non-supervised DNN methods. Specifically, this work proposes a novel two-stage framework that (i) performs self-supervised denoising in the projection domain (sinogram) that leverages the statistical independence of noise in the measurement space, and (ii) uses the cleaner (or denoised) anatomical images within an unsupervised denoising framework for further refinement of the LD-CBF images (obtained using LD-CTP images). Because our proposed approach does not involve learning a mapping between low and high dose images, and rather relies only on patient-specific information, the framework is agnostic to changes in the data domain such as the injected X-ray dose and acquisition site, etc. Using empirical experiments on simulated and *in vivo* data, we demonstrate the efficacy of the proposed framework across *three* levels of noise variations in the CT data.

2   Methods

We describe the statistical model for CT perfusion imaging followed by the application of a self-supervised learning scheme in the sinogram space. Subsequently, the CT images with reduced noise (reconstructed from denoised sinograms) are used to denoise the CBF maps using an unsupervised method.

2.1   Problem Formulation and Strategy

Let $\{Y^t\}_{t=1}^T$ represent the set of T observed dynamic CT sinograms, each containing K elements. For each CT sinogram Y^t, we represent the corresponding reconstructed CT image as X^t containing N elements. The corrupted (by noise) sinogram domain measurements are modeled as $Y_i^t \sim \text{Poisson}((\mathcal{R}X^t)_i)$, where $\mathcal{R}$ represents the CT forward operator and Y_i^t denotes the i-th element of the sinogram Y^t. Here, the noise in the sinogram domain is assumed to be i.i.d. Many denoising algorithms that have been proposed in the literature to denoise a signal in a self-supervised or unsupervised manner. These algorithms assume element-wise statistical independence for the noise added in the image space. In the general context of tomographic inverse problems, although the noise in the sinogram domain is i.i.d, the reconstruction of CT images using techniques such as filtered backprojection (FBP), causes the noise to be correlated across pixels. Thus, noise in the image domain for tomographic inverse problems is no longer i.i.d. Thus, to obtaining a set of denoised CT images, we perform self-supervised denoising in the sinogram domain. Once the set of T sinograms have been denoised, we obtain the corresponding reconstructed CT images using fast analytical methods such as filtered backprojection (FBP). Subsequently, we use the cleaner CT images to further refine the perfusion maps obtained as described below.

Notations. Generally, we use X to refer to a single CT image with appropriate spatial and temporal index subscripts. Further, let X_{noisy} represent the CT image reconstructed from noisy sinograms and $X_{denoised}$ represent the CT images reconstructed from the denoised sinograms. Lastly, we use the prefix LD and SD to denote signals obtained from low-dose and standard-dose CT data, respectively.

Forward Model for Perfusion Maps Generation. Let $A(t)$ represent the arterial input function (AIF), determined from the set of T dynamic LD-CT data, $\{X_{noisy}^n\}_{n=1}^N$. If $R_v(t)$ denotes the unobserved and unknown tissue residue function within a local region v, $R_v(t)$ denotes the residual contrast in the region v. The time-concentration curve $C_v(t)$ is given by $C_v(t) = \int_0^t A(\tau)R_v(t-\tau)\,d\tau$ [7]. In the discrete setting, if column vectors c_v and r_v, each containing P elements obtained at equally spaced intervals, denote the measured contrast and the unknown residue function, the forward model becomes $c_v = \mathbf{A}r_v$. Here, matrix $\mathbf{A}$ is block-circulant with columns representing $A(t)$ as described in [22]. We obtain r_v as the maximum-a-posteriori (MAP) estimate by solving the following optimization problem $\widehat{r_v} = \arg\min_{r_v} \|\mathbf{A}r_v - c_v\|^2 + \lambda\mathcal{D}(r_v)$, where $\mathcal{D}(\cdot)$ denotes a suitable regularizer of choice like total variation (TV) and $\lambda > 0$ is the prior weight. The CBF maps are obtained by suitably scaling the image r_v as in [6]. Typically, the CBF maps are obtained by computing the singular value decomposition of the matrix $\mathbf{A}$. Because the smaller singular values can lead to instable solutions, truncated singular value decomposition (TSVD) algorithm removes smaller singular values (chosen heuristically) to obtain a regularized inverse solution. In this paper, we obtain both LD-CBF (CBF corresponding to low-dose data) and SD-CBF (CBF corresponding to standard-dose data) using the TSVD algorithm.

Thus, given a set of LD noisy measurement data $\{Y^t\}_{t=1}^T$, we estimate the CT reconstructions $\{X_{noisy}^t\}_{t=1}^T$ using FBP. Subsequently, the reconstructed CT images are used to compute the LD-CBF maps using the TSVD algorithm. The corresponding SD counterparts are used for evaluating the performance of the proposed framework and other baseline methods (detailed in Sect. 4). In this paper, we propose the following sequential approach: First, denoise the LD sinograms using a self-supervised DNN to obtain a set of denoised sinograms. Secondly, reconstruct the CT images corresponding to the denoised dynamic sinogram data. Finally, use the reconstructed CT images (obtained in the previous stage) as anatomical prior information to refine the LD-CBF maps. Note that the LD-CBF maps were obtained via TSVD applied on the noisy CT images $\{X_{noisy}^t\}_{t=1}^T$.

2.2   Self-supervised Low-Dose Sinogram-Space Denoising

The work in Noise2Self [1] showed that by leveraging $\mathcal{J}$-invariance, self-supervised denoising can be achieved. A function $f(\cdot)$ is said to be $\mathcal{J}$-invariant if the value of $f(\cdot)$ restricted within a subset $J \in \mathcal{J}$ does not depend on the elements within that subset. The Noise2Self algorithm relies only on a single instance of the image and does not necessitate any additional data. The fundamental assumption underlying the Noise2Self method are: (i) element-wise statistical independence of noise and (ii) spatial correlation of the true underlying signal in a neighborhood. We leverage the above strategy to obtain a $\mathcal{J}$ invariant version of a denoising DNN. We call this network SS-DNN (self-supervised DNN). To train the SS-DNN, we employ a mask M that replaces the pixels within a subset region J with the mean of the neighborhood pixels. For every iteration, a different masked image is used as the input to the SS-DNN and the noisy image is used as the reference image.

2.3   Unsupervised CBF Map Denoising Using CTP Information

We now describe the method used to obtain the denoised CBF maps. We employ a modified network proposed in [3]. We call the unsupervised denoising approach as CBF-CDIP. The CBF-CDIP takes the set of reconstructed CT images obtained from the denoised sinograms from the SS-DNN. At the input layer, we use a 1×1 convolution to fuse the T channels. Given the noisy LD-CBF image r_{noisy}, the estimated denoised LD-CBF image $\hat{r}$, is given by $\hat{r} := \Phi(\widehat{\Theta}|\{X_{denoised}\}_{t=1}^T)$, where Φ represents the DNN and $\widehat{\Theta}$ represents the optimal DNN parameters. The optimal DNN parameters are obtained by solving the optimization problem

$$\arg \min_{\Theta} \|\Phi(\Theta|\{X_{denoised}\}_{t=1}^T) - r_{noisy}\|. \tag{1}$$

2.4   Low-Dose Simulation

We consider the provided CT data as standard-dose data, *i.e.*, SD-CT images. We simulate LD-CT data by applying noise in the measurement domain unlike prior works that apply correlated noise in the image space (CTP) [5, 6, 9]. That is, we generate the LD-CT images by retrospectively projecting the SD-CT images and applying Poisson noise in the sinogram space. Specifically, we apply a scaled Poisson noise to obtain the noisy LD-CT data Y_{noisy} as $Y_{noisy} = (\text{Poisson}(Y_{true} * S))/S$, where Y_{true} represents the uncorrupted (clean) sinogram and $S \in \mathbb{R}^+$ is a scalar value that determines the amount of noise in the data.

2.5   DNN Architectures

For the self-supervised SS-DNN network, we employ an architecture similar to the convolutional DNN [24]. On the other hand, the CBF-CDIP uses a modified U-Net architecture consisting of 3 double convolution layers, with max-pooling layers in between them. We employ a dropout rate of $1/\text{channels}$ ($= 0.004$) at the bottleneck layer. Each convolutional layer consists of a 2D convolution, followed by batch normalization, application of ReLU activation function. For both SS-DNN and CBF-CDIP, we use ℓ_2 loss between the DNN output and the label/reference image. As mentioned before, the reference image for the SS-DNN is the noisy sinogram and that for the CBF-CDIP DNN is the noisy LD-CBF.

3   Data and Experiments

Simulated Data for Workflow Demonstration. We use a generic multi-modal 2D *in silico* data to demonstrate the steps of the proposed framework. The simulated data is obtained from [4] and consists of two images corresponding to an anatomical (MRI) and functional image (PET). Henceforth, we refer to the anatomical image and the functional image as AI and FI, respectively. Compared to the correlation between the CTP and CBF images, the 2D simulated data is a slightly exaggerated case where the AI and FI exhibit negative correlation of edges at specific locations and show modality-specific features. We generate the noisy data for AI by forward projecting the image (using a parallel-beam geometry) followed by application of independently identically distributed Gaussian noise in the projection (sinogram) domain. To simulate the noisy FI, we add i.i.d Gaussian noise in the image domain.

***In vivo* Data.** We use the dynamic CT perfusion images provided as part of the ischemic stroke lesion (ISLES) challenge 2018 [15]. The dataset consists of several volumes of dynamic CT images each containing varying number of axial slices (two to eight) and time-frames (40 to 50). We tune the *SF* to generate data (sinograms) corresponding to three different noise levels, arranged in ascending levels of noise as NL1, NL2, and NL3. The LD-CT images corresponding to each NL were reconstructed using filtered backprojection (FBP) with Hann filter. The PSNR between the set of SD-CT and the set of LD-CT images at NL1 averaged over the entire dataset was around 38. Similarly, the mean PSNR for NL2 and NL3 were around 36 and 35, respectively. To obtain the perfusion maps at SD and LD, first, we remove (replace with zero) the bone regions with a threshold of 120 HU. Secondly, we computed the AIF and venous output function (VOF) as detailed in [10,17]. We obtain corresponding perfusion (CBF) maps at SD and LD using the TSVD-based deconvolution algorithm [22]. The LD-CBF maps obtained using the reconstructed LD-CT images is used as the noisy reference image for the CBF-CDIP DNN.

4   Evaluation, Results, and Discussion

We compare our method with three other approaches to generate perfusion maps from CT data: (i) TSVD applied to LD-CT images [22], (ii) tensor total variation (TTV) which is an edge-preserving prior and penalizes the gradients along both

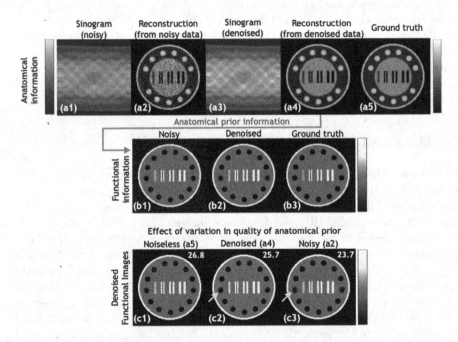

Fig. 2. Results on Simulated Toy Data. (**a1**) Noisy sinogram of anatomical image (AI), (**a2**) reconstructed AI from noisy sinogram, (**a3**) denoised sinogram of AI using SS-DNN, (**a4**) reconstructed AI from denoised sinogram, (**a5**) ground truth AI, (**b1**) Noisy functional image (FI), (**b2**) denoised FI using AI (a3) as anatomical information using CBF-CDIP, (**b3**) ground truth FI. **Effect of varying quality of input data (anatomical information).** Denoised functional image produced from: (**c1**) noiseless AI as in (a5), (**c2**) denoised AI reconstructed from denoising the noisy measurement domain data as in (a4), (**c3**) noisy AI image reconstructed directly from the noisy measurement domain data. The PSNR values of the denoised images are embedded within each panel in row (c)

spatial and temporal dimensions [6], (iii) Noise2Self based self-supervised CT image denoising in the image space (N2S-IS), which is a modified version of the method proposed in [23]. For a fair comparison, none of the methods use the bias compensation loss proposed in [23]. For quantitative evaluation of the CBF maps, we use PSNR as the image quality metric.

We use *one* randomly chosen subject's data from the ISLES dataset as the validation data to tune all the methods for optimal performance. For both TSVD and TTV methods, we tune the underlying free parameters to give the best PSNR based on the validation data. Similar strategy was employed to determine the stopping criteria for the DNNs N2S-IS, SS-DNN, and CBF-CDIP.

Figure 2 demonstrates the proposed framework on simulated data representing a generic scenario consisting of multimodal (anatomical and functional) data. The SS-DNN network is used to denoise the sinogram corresponding to AI and subsequently used to reconstruct a cleaner AI image with reduced noise (Fig. 2 row (a)). The reconstructed image is subsequently used as input to the CBF-CDIP DNN (Fig. 2 row (b)). The noisy FI acts as the reference image and the output of the CDIP CNN is the denoised FI with reduced noise. We performed

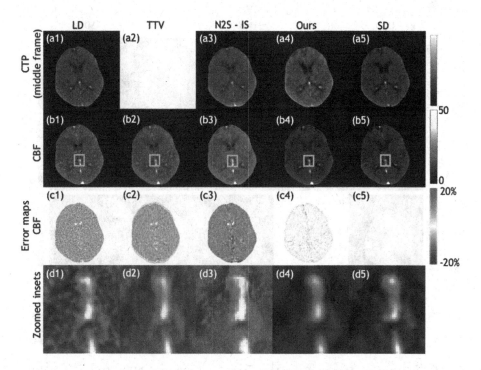

Fig. 3. Qualitative Results on Real Data at NL1. Row (a) shows a representative slice from the dynamic set of CTP images. Row (b) shows corresponding CBF maps generated from data at NL1. Row (c) shows the residual error images between the generated CBF images (in row (b)) and SD-CBF (b5). Row (d) shows the zoomed region of interest depicted by the green rectangle in row (b). (Color figure online)

an experiment to determine the effect of variation in input image quality on the quality of the output image (Fig. 2 row (c)). Clearly, an improvement in the quality of the input image leads to improved denoised FI images. The best output is obtained when the ground-truth AI is used as input, followed by the output when the denoised AI image is used (Fig. 2(a4)), which is very close in structure and contrast to the output with the ground-truth AI image. On the other hand, the output FI image obtained with the noisy AI as input shows structural distortions as highlighted by the green arrows.

Figure 3 shows the qualitative results of the estimated CTP images and the corresponding CBF maps using LD-CT data at NL1, for various methods. The CT frame corresponding to $ceil(T/2)$ is shown in row (a). The TTV method estimates the CBF image directly from the LD-CT images and and does not produce an intermediate denoised CT slice; hence, it is shown as an empty image. Among the CTP images (row (a)), the N2S-IS method shows a slight improvement over the LD-CTP image but fails to remove majority of the noise. On the other hand, our image reconstructed from the denoised sinogram (using SS-DNN) shows smoother reconstructions and a superior recovery of structure and

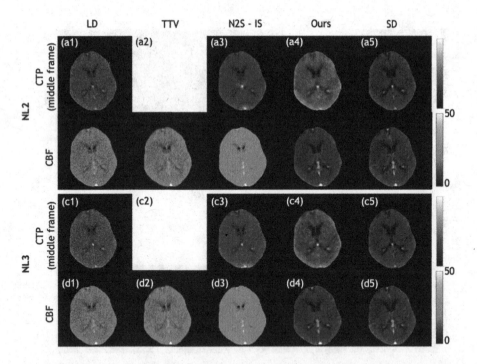

Fig. 4. Qualitative Results on Real Data at NL2 and NL3. Rows (a) and (b) correspond to CTP images (representative middle frame slice) and corresponding CBF maps generated from data at NL2, respectively. Similarly, rows (c) and (d) correspond to CTP images and corresponding CBF maps generated from data at NL2, respectively.

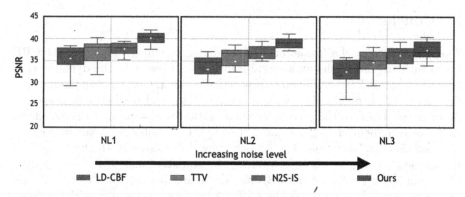

Fig. 5. Quantitative performance of various methods across multiple subjects. PSNR values with SD-CBF as reference across 30 subjects randomly selected from the ISLES data, evaluated at various noise levels: (a1) NL1; (a2) NL2; (a3) NL3.

contrast while retaining important anatomical edges with substantially reduced noise compared to Fig. 3(a3). For the generated CBF images (row (b)), the TTV method (Fig. 3(b2)) suffers from patchy artifacts that is characteristic of image-gradient based methods. The N2S-IS method (Fig. 3(b3)) uses the TSVD deconvolution method and improves over the TTV method. Our proposed method that uses the CTP images (as in Fig. 3(a4)) appears closest to the SD-CBF maps (Fig. 3 (b5)) and shows a quantitative improvement of ≥ 3.2 in terms of PSNR compared to other methods. The performance of the proposed approach is also reflected in the (i) residual CBF error maps with our approach yielding the least residual (Fig. 3(c4)) and (ii) zoomed insets showing the improved restoration of the high intensity regions (Fig. 3(d4)).

Figure 4 shows CTP and CBF images from CT data at higher noise levels than NL1, namely NL2 and NL3. Compared to Fig. 3, for all the methods, there is a slight degradation in the quality of the generated images. Nevertheless, our method produces CBF map that is comparable to the CBF map image quality obtained using NL1 data. That is, even at higher noise levels, our method is able to restore the perfusion maps more accurately compared to other non-supervised methods.

Figure 5 shows the PSNR values of the generated CBF maps with respect to the SD-CBF image as reference, for all the methods at the three different noise levels discussed in this paper: NL1, NL2, and NL3. The methods are evaluated on a randomly chosen set of 30 subjects from the ISLES data. Clearly, as the noise in the data increases, all the methods show a dip in performance. For all the levels, the TTV method demonstrates larger variance in the performance metrics compared to N2S-IS and the proposed framework. Our method shows significant improvement over other methods at all noise levels with relatively smaller variance.

5 Conclusion

This paper presented a framework for improving the image quality of the CBF maps with the aid of the anatomical information present in the dynamic contrast enhanced CT images, in an unsupervised manner. Our denoising framework does not rely on previously acquired training data and retains patient-specific information. Because the anatomical information is also prone to noise effects in its measurement space, it is natural to aim to denoise both the anatomical and functional information. The self-supervised denoising method was applied in the measurement domain to leverage the statistical independence property of the noise. For denoising the CBF maps, the unsupervised DNN method uses patient's own CTP images as the CBF-CDIP network input and the original noisy LD-CBF image was treated as the training label. Our method demonstrated improved robustness to practical perturbations in the data across different noise levels in the CTP data. This work showed that sinogram denoising is effective in improving the quality of low-dose CT images which subsequently aid in improved CBF map generation. Our results on simulated data showed that in the context of multimodal imaging, our framework can be applied to scenarios where both anatomical and functional images are used for diagnostic purposes e.g., MRI and PET. Future work calls for simultaneous denoising of both these images rather than sequential denoising.

References

1. Batson, J., Royer, L.: Noise2self: blind denoising by self-supervision. In: International Conference on Machine Learning, pp. 524–533. PMLR (2019)
2. Cohen, J.P., Luck, M., Honari, S.: Distribution matching losses can hallucinate features in medical image translation. In: International Conference on Medical Image Computing and Computer-assisted Intervention. pp. 529–536. Springer (2018)
3. Cui, J., et al.: Pet image denoising using unsupervised deep learning. Eur. J. Nucl. Med. Mole. Imag. **46**(13), 2780–2789 (2019)
4. Ehrhardt, M., et al.: Pet reconstruction with an anatomical MRI prior using parallel level sets. IEEE Trans. Med. Imaging **35**(9), 2189–2199 (2016)
5. Fang, R., Chen, T., Sanelli, P.: Towards robust deconvolution of low-dose perfusion CT: sparse perfusion deconvolution using online dictionary learning. Med. Imaging Anal. **17**(4), 417–428 (2013)
6. Fang, R., Zhang, S., Chen, T., Sanelli, P.: Robust low-dose CT perfusion deconvolution via tensor total-variation regularization. IEEE Trans. Med. Imag. **34**(7), 1533–1548 (2015)
7. Fieselmann, A., Kowarschik, M., Ganguly, A., Hornegger, J., Fahrig, R.: Deconvolution-based CT and MR brain perfusion measurement: theoretical model revisited and practical implementation details. Int. J. Biomed. Imag. **2011**, 467563 (2011)
8. Hsu, Y., Shen, Y., Jin, H., Kira, Z.: Generalized ODIN: Detecting out-of-distribution image without learning from out-of-distribution data. In: CVPR (2020)

9. Kadimesetty, V., Gutta, S., Ganapathy, S., Yalavarthy, P.: Convolutional neural network-based robust denoising of low-dose computed tomography perfusion maps. IEEE Trans. Radiat. Plasma Med. Sci. **3**(2), 137–152 (2018)

10. Kao, Y., et al.: Automatic measurements of arterial input and venous output functions on cerebral computed tomography perfusion images: a preliminary study. Comp. Biol. Med. **51**, 51–60 (2014)

11. Karimi, D., Deman, P., Ward, R., Ford, N.: A sinogram denoising algorithm for low-dose computed tomography. BMC Med. Imag. **16**(1), 1–14 (2016)

12. Koh, T., Wu, X., Cheong, L., Lim, C.: Assessment of perfusion by dynamic contrast-enhanced imaging using a deconvolution approach based on regression and singular value decomposition. IEEE Trans. Med. Imag. **23**(12), 1532–1542 (2004)

13. Lee, K., Lee, K., Lee, H., Shin, J.: A simple unified framework for detecting out-of-distribution samples and adversarial attacks. In: Proceedings of the 32nd International Conference on Neural Information Processing System, pp. 7167–7177 (2018)

14. Lehtinen, J., et al.: Noise2noise: learning image restoration without clean data. arXiv preprint arXiv:1803.04189 (2018)

15. Maier, O., Menze, B., von der Gablentz, J., et al.: ISLES 2015-A public evaluation benchmark for ischemic stroke lesion segmentation from multispectral MRI. Med. Imag. Anal. **35**, 250–269 (2017)

16. Manduca, A.: Projection space denoising with bilateral filtering and CT noise modeling for dose reduction in CT. Med. Phys. **36**(11), 4911–4919 (2009)

17. Mouridsen, K., Christensen, S., Gyldensted, L., Østergaard, L.: Automatic selection of arterial input function using cluster analysis. Magn. Reson. Med. **55**(3), 524–531 (2006)

18. Niesten, J., van der Schaaf, I., Riordan, A., et al.: Radiation dose reduction in cerebral CT perfusion imaging using iterative reconstruction. Eur. Radiol. **24**(2), 484–493 (2014)

19. Othman, A., et al.: Radiation dose reduction in perfusion CT imaging of the brain: a review of the literature. J. Neuroradiol. **43**(1), 1–5 (2016)

20. Shtok, J., Elad, M., Zibulevsky, M.: Sparsity-based sinogram denoising for low-dose computed tomography. In: 2011 IEEE International Conference on Acoustics, Speech and Signal Processing (ICASSP), pp. 569–572. IEEE (2011)

21. Voss, S., Reaman, G., Kaste, S., Slovis, T.: The ALARA concept in pediatric oncology. Pediat. Radiol. **39**, 1142–1146 (2009)

22. Wittsack, H., et al.: CT-perfusion imaging of the human brain: advanced deconvolution analysis using circulant singular value decomposition. Comp. Med. Imag. Graph. **32**(1), 67–77 (2008)

23. Wu, D., Ren, H., Li, Q.: Self-supervised dynamic CT perfusion image denoising with deep neural networks. IEEE Trans. Radiat. Plasma Med. Sci. (2020)

24. Zhang, K., Zuo, W., Chen, Y., Meng, D., Zhang, L.: Beyond a gaussian denoiser: Residual learning of deep CRN for image denoising. IEEE Trans. Image Proces. **26**(7), 3142–3155 (2017)

Recurrent Brain Graph Mapper for Predicting Time-Dependent Brain Graph Evaluation Trajectory

Alpay Tekin[1], Ahmed Nebli[1,2] (iD), and Islem Rekik[1(✉)] (iD)

[1] BASIRA Lab, Faculty of Computer and Informatics, Istanbul Technical University, Istanbul, Turkey
irekik@itu.edu.tr
[2] National School of Computer Science (ENSI), University of Manouba, Manouba, Tunisia
http://basira-lab.com

Abstract. Several brain disorders can be detected by observing alterations in the brain's structural and functional connectivities. Neurological findings suggest that early diagnosis of brain disorders, such as mild cognitive impairment (MCI), can prevent and even reverse its development into Alzheimer's disease (AD). In this context, recent studies aimed to predict the evolution of brain connectivities over time by proposing machine learning models that work on brain images. However, such an approach is costly and time-consuming. Here, we propose to use brain connectivities as a more efficient alternative for time-dependent brain disorder diagnosis by regarding the brain as instead a large interconnected graph characterizing the interconnectivity scheme between several brain regions. We term our proposed method Recurrent Brain Graph Mapper (RBGM), a novel efficient edge-based **recurrent graph neural network** that predicts the time-dependent evaluation trajectory of a brain graph from a single baseline. Our RBGM contains a set of *recurrent neural network*-inspired mappers for each time point, where each mapper aims to project the ground-truth brain graph onto its next time point. We leverage the teacher forcing method to boost training and improve the evolved brain graph quality. To maintain the topological consistency between the predicted brain graphs and their corresponding ground-truth brain graphs at each time point, we further integrate a topological loss. We also use $l1$ loss to capture time-dependency and minimize the distance between the brain graph at consecutive time points for regularization. Benchmarks against several variants of RBGM and state-of-the-art methods prove that we can achieve the same accuracy in predicting brain graph evolution more efficiently, paving the way for novel graph neural network architecture and a highly efficient training scheme. Our RBGM code is available at https://github.com/basiralab/RBGM.

A. Tekin and A. Nebli—Co-first authors.

© Springer Nature Switzerland AG 2021
S. Albarqouni et al. (Eds.): DART 2021/FAIR 2021, LNCS 12968, pp. 180–190, 2021.
https://doi.org/10.1007/978-3-030-87722-4_17

Keywords: Recurrent graph convolution · Transformation layer · Topological loss · Time-dependent graph evolution prediction

1   Introduction

Latest neuroscience studies have emphasized the importance of personalized treatments for brain disorders that can significantly improve patient's recovery [1]. Brain disorders such as mild cognitive impairment (MCI) can be easily reversed if diagnosed at an early stage before evolving into irreversible Alzheimer's disease (AD) [2]. As such, recent landmark studies [3–6] proposed using the breadth of machine learning to predict brain connectome evolution trajectory. For instance, [3] has proposed a learning-based framework to predict the longitudinal development of cortical surface and white matter fibers. To do so, first, they used multiple atlases to generate a spatially heterogeneous atlas that mimics the cortical surface of the target subject. Second, they predicted spatio-temporal connectivity features from neonatal brains using low-rank tensor completion. Furthermore, [4] proposed a deep learning model that jointly classifies and predicts the evolution trajectory of the brain from a single acquisition point. At the baseline, they identified the landmarks, namely regions of interest (ROIs), and utilized supervised and unsupervised learning to predict the evolution trajectory at each landmark.

However, such studies were conducted only on brain images that solely consider local neighborhood connectivities undeniably overseeing the brain global connectivity pattern. To properly diagnose brain connectivity disorders, the brain must be viewed as a large interconnected graph [7] where each ROI represents a node, and each pairwise connectivity between two ROIs represents an edge. To address this limitation, we set out a more challenging problem, which is predicting the evolution of brain graphs from a baseline observation. In this context, recent works [5,6] leveraged generative adversarial networks (GANs) [8], where they proposed the first graph-based GAN specialized in AD evolution trajectory prediction. Namely, GANs [8] address the problem of unsupervised learning by training two neural networks: generator and discriminator. The generator takes randomly distributed data as input and generates synthetic data that mimics a real distribution. On the other hand, the discriminator inputs the generated data and predicts whether it is real or fake. Both generator and discriminator compete in an adversarial way. Hence, the overall quality of the generated data improves with training epochs. [5] used gGAN to learn how to normalize a brain graph with respect to a fixed connectional brain template (CBT). Their gGAN architecture is made up of a graph normalizer network that learns a high-order representation of each brain in a graph and produces a CBT-based normalized brain graph. Furthermore, [6] proposed EvoGraphNet that connects series of gGANs each specialized in how to construct a brain graph from the predicted brain graph of the previous gGANS in the time-dependent cascade.

Although gGANs were proven to be successful for predicting brain connectome evolution trajectory given a single observation, there is still a considerable amount of computational complexity. Since gGANs contain two graph neural networks (i.e., generator and discriminator), they require a significant amount of computational power and training time. Also, these architectures require a sequential framework where each gGAN has to generate a time point t_i to get the following time point prediction t_{i+1}.

This challenge raises the following question: *can we mimic the prediction power of the above-mentioned sequential gGANs without the need to use highly complex sequential networks?*

According to [9], brain connectivities do not evolve randomly. Their evolution follows a temporal scheme that aims to satisfy the patient's needs for each given age and health condition. In that regard, recurrent neural networks (RNNs) [10] are known for their temporal pattern recognition. Therefore, in this paper, we propose to power our model with RNNs. We term our proposed model Recurrent Brain Graph Mapper (RBGM), the *first* framework to predict brain connectome evolution while efficiently reducing complexity and training time consumption. Our model uses a shallow graph convolutional neural network architecture to predict brain disease's evolution trajectory given a baseline graph. To do so, for a given time point, t_i each mapper uses the ground-truth from the previous time point t_{i-1} to predict the brain connectivity scheme at the time point t_i instead of starting from the initial time point. To do so, we apply the teacher forcing method [11], which is known for its quick and efficient training recurrent-based models.

We propose preserving the topological consistency between the predicted and ground-truth brain graphs at each time point. To do so, we integrate a topological loss measuring the topological discrepancy between the predicted brain graph and its corresponding ground-truth brain graph. Furthermore, we leverage a $l1$ loss to minimize the sparse distance between two serialized brain graphs to capture time-dependency between two consecutive observations. We also investigate the effect of Kullback-Leibler divergence (i.e., KL-divergence), where we enforce the preservation of node distribution between predicted and ground-truth brain graphs over time. We articulate the main contributions of our work as follows:

1. *On a methodological level.* Our proposed RGBM is the first RNN-based geometrical deep learning framework that predicts the time-dependent brain graph evolution trajectory from a single observation.
2. *On a conceptual level.* Our model reduces the complexity of the geometrical deep learning framework by speeding up the training while maintaining similar performances to the state-of-the-art methods.
3. *On clinical level.* Our RGBM can be used to prevent and reverse the onset of neurological diseases.

2   Proposed Method

This section introduces the key steps of our RBGM for predicting brain graph evolution from a single observation. Table 1 displays the mathematical notations

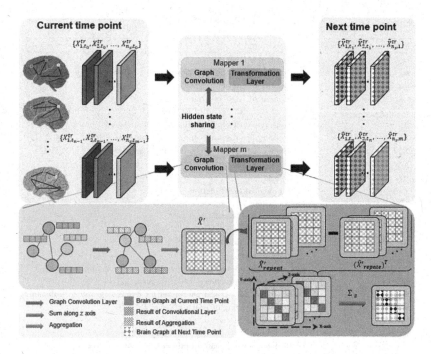

Fig. 1. *Proposed Recurrent Brain Graph Mapper architecture (RBGM) for predicting the evaluation trajectory of brain disease given a single time point.* We develop a mapper that learns how to morph an input at time point t_{i-1} to its next time point t_i. Given m mappers for m time points, each mapper contains a graph neural network and transformation layer. The graph neural network takes the input brain graph $\mathbf{X}_{t_{i-1}}^{tr} \in \mathbb{R}^{n_r \times n_r}$ for a given time point t_i, where n_r is the number of ROIs to learn node embedding $\mathbf{V}^l = [\mathbf{v}_1^l, \mathbf{v}_2^l, \ldots, \mathbf{v}_{n_r}^l]^T$ that captures the node-to-node relation and visualizes it in vector form. Then a transformation layer takes these node embeddings $\mathbf{V}^l$ and computes the pairwise absolute difference to predict the brain graph at the time point t_i given by $\hat{\mathbf{X}}_{t_i}^{tr} \in \mathbb{R}^{n_r \times n_r}$. First $\mathbf{V}^l$ is repeated horizontally n_r times to obtain $\mathcal{R} \in \mathbb{R}^{n_r \times n_r \times n_r}$. Next, we compute the absolute difference between $\mathcal{R}$ and its transpose. Finally, the resulting tensor is sum along z-axis to obtain the predicted brain graph $\hat{\mathbf{X}}_{t_i}^{tr}$ for the time point t_i.

that we use throughout our paper. We denote the matrices as boldface capital letters (e.g., $\mathbf{X}$) and scalars as lowercase letters (e.g., m). The transpose operator is denoted as $\mathbf{X}^T$.

Overview of Recurrent Brain Graph Mapper for Predicting Brain Graph Evolution Trajectory from a Single Baseline. Our proposed RBGM is composed of m mappers for m time points, as shown in **Fig. 1**. Each mapper can predict a brain graph for a given time point t_i using its corresponding ground-truth brain graph at the time point t_{i-1} as an input. The recurrent graph convolution enables each mapper to capture temporal changes in the brain connectivity pattern between consecutive time points. Also, it increases

Table 1. Mathematical definitions.

Mathematical notation	Definition
m	Number of time points
n_s	Number of training subjects
m_r	Number of edges
n_r	Number of ROIs in brain
$\mathbf{s}_i^{tr}$	Node strength vector of ROI i in the ground-truth brain graph
$\hat{\mathbf{s}}_i^{tr}$	Node strength vector of ROI i in the predicted brain graph
$\mathbf{x}_{t_i}^{tr}$	Training brain graph connectivity matrices $\in \mathbb{R}^{n_r \times n_r}$ at t_i
$\hat{\mathbf{x}}_{t_i}^{tr}$	Predicted brain graph connectivity matrices $\in \mathbb{R}^{n_r \times n_r}$ at t_i
M_i	Mapper at time point t_i
$\mathcal{L}_{l1}$	l_1 loss
$\mathcal{L}_{TP}$	Topological loss function
λ_1	Coefficient of l_1 loss
λ_2	Coefficient of topological loss
V	a set of n_r nodes
E	a set of m_r undirected or directed edges
l	index of layer
$\mathcal{N}(i)$	The neighborhood containing all the adjacent nodes of node i
F^l	edge-conditioned filter
$\mathbf{\Theta}^l$	Learnable edge-based parameter for dynamic graph convolution
$\mathbf{v}_i{}^l$	Node embedding of ROI i at layer $l \in \mathbb{R}^{d_t}$
$\mathbf{W}^l$	Weight parameter
$\mathbf{b}^l$	Bias term
$\mathcal{R}$	Horizontally replicated brain connectivity matrix $\in \mathbb{R}^{n_r \times n_r \times n_r}$
$\mathcal{R}^T$	Transpose of horizontally replicated brain connectivity matrix $\in \mathbb{R}^{n_r \times n_r \times n_r}$
$\mathbf{h}_i$	Hidden state matrix at $t_i \in \mathbb{R}^{m_r \times m_r}$
$\mathbf{W}_{ih}$	Input to hidden weight for recurrent filter $\in \mathbb{R}^{1 \times m_r}$
$\mathbf{W}_{hh}$	Hidden to hidden weight for recurrent filter $\in \mathbb{R}^{m_r \times m_r}$
$\mathbf{b}_h$	Bias term for recurrent filter $\in \mathbb{R}^{m_r \times m_r}$

the prediction power of each mapper, hence using fewer convolutional layers. Furthermore, we apply the teacher forcing method [11] to quickly and efficiently train our RGBM.

To enhance our method's robustness, we propose using $l1$ loss thanks to its resilience against outliers to enforce the connectivity consistency across time points. Thus, we express the $l1$ loss for each subject tr using the predicted brain graph $\hat{\mathbf{X}}_{t_{i-1}}^{tr}$ from the mapper M_{i-1}, and its corresponding ground-truth brain graph at t_i as follows:

$$\mathcal{L}_{l1}(M_{i-1}) = ||\hat{\mathbf{X}}_{t_{i-1}}^{tr} - \mathbf{X}_{t_i}^{tr}||_1 \tag{1}$$

This acts as a regularizer over time and aligns with the sparse nature of brain connectivity evolution. In addition to the $l1$ loss, we propose a second loss to preserve the topological consistency between predicted brain graphs and their corresponding ground-truth at each time point. To frame the topological loss, we define a node strength vector measuring the topological strength for each node in a given graph. Since our brain graph is fully-connected (each node has a same number of edges), we chose node strength as a centrality measures. We compute the node strength vector by adding the weights of all edges connected to a node of interest. As such, $\mathbf{S} = [\mathbf{S}_1, \mathbf{S}_2, ..., \mathbf{S}_{n_r}]^T$ represents the node strengths for all ROIs where n_r is the number of ROIs. The following equation gives the topological loss:

$$\mathcal{L}_{TP}(\mathbf{S}^{tr}, \hat{\mathbf{S}}^{tr}) = \frac{1}{n_r} \sum_{i=1}^{n_r} \left(\mathbf{S}_i^{tr} - \hat{\mathbf{S}}_i^{tr} \right)^2 \tag{2}$$

The Full Loss. We combine the previous losses to train our RGBM as follows:

$$\mathcal{L}_{Full} = \sum_{i=1}^{m} \left(\lambda_1 \mathcal{L}_{l1}(M_{i-1}) + \lambda_2 \mathcal{L}_{TP}(\mathbf{S}^{tr}, \hat{\mathbf{S}}^{tr}) \right) \tag{3}$$

where λ_1, and λ_2 are hyperparameters adjusting each corresponding loss.

The Mapper Network Architecture. Each mapper m uses our proposed recurrent graph convolution (RGC) function. We leverage the teacher forcing method [11] to speed up training and increase the overall performance in training. Namely, the teacher forcing is a common method that speeds up training and improves the quality of recurrent-based models. It enforces the recurrent model to use ground-truth samples to predict the brain graph at the following time point. According to this method, for a given time point, t_i the mapper takes the ground-truth $\mathbf{X}_{t_{i-1}}^{tr}$ from the time point t_{i-1} instead of taking the predicted brain graph $\hat{\mathbf{X}}_{t_{i-1}}^{tr}$ from the preceding mapper $M_{t_{i-1}}$ to make prediction for the time point t_i in training phase. Therefore, the need to start from the initial time point is eliminated during training.

Dynamic Edge-filtered Convolution. Each mapper in our RBGM uses the dynamic graph convolution with edge-conditioned filter proposed by [12]. Let $G = (V, E)$ is a directed or undirected graph, where V is the set of n_r ROIs and $E \in V \times V$ is a set of m_r edges between each ROI. Let l be the layer index. For each layer $l \in \{1, 2, ..., L\}$, $F^l : \mathbb{R}^{d_m} \mapsto \mathbb{R}^{d_l \times d_{l-1}}$ represents a filter-generating network that generate edge weights for the message passing between ROIs i and j given features of e_{ij}. d_m and d_l are dimensionality indexes. This operation is expressed as follows:

$$\mathbf{v}_i^l = \mathbf{\Theta}^l \mathbf{v}_i^{l-1} + \frac{1}{|\mathcal{N}(i)|} \left(\sum_{j \in \mathcal{N}(i)} F^l(\mathbf{e}_{ij}; \mathbf{W}^l) \mathbf{v}_j^{l-1} + \mathbf{b}^l \right), \tag{4}$$

where $\mathbf{v}_i^l$ is the node embedding for the ROI i at layer l. $\mathcal{N}(i)$ denotes the neighbors of ROI i. F^l is the neural network that maps $\mathbb{R}^{d_m}$ to $\mathbb{R}^{d_l \times d_{l-1}}$ with weights $\mathbf{W}^l$. $\mathbf{\Theta}^l$ is the dynamically generated edge specific weights by F^l. The $\mathbf{b}^l \in \mathbb{R}^{d_l}$ is the bias term. We note that F^l can be any type of neural network.

We draw inspiration from the image-based recurrent neural network architecture, which shows outstanding performances on time-series data prediction

[13, 14]. This type of network can remember the former information and process new events accordingly thanks to its hidden state, which holds the former information (i.e., learned information in the previous layer). Each RNN cells takes two distinct inputs: (i) the input brain graphs from the current time point, and (ii) hidden state value from the brain graphs at the previous time point, then updates the hidden state, which holds the representation of the knowledge from the prior time point.

Proposed Graph Recurrent-Filter. We propose *the first edge-based recurrent graph neural network* Fig. 2 by re-designing the edge-conditioned filter [12] in the graph convolution layer as a graph recurrent-filter so that each mapper can capture temporal changes on brain connections over time, as shown in Fig. 2. Therefore, unlike [6], the need to start from the initial time point is eliminated during training. Our proposed graph recurrent filter can process past information when generating messages between each ROI to capture temporal changes of brain connectivity. To do so, it takes the set of edges $\mathbf{e} \in \mathbb{R}^{m_r \times 1}$ for a given time point t_i and the hidden state matrix from the previous time point t_{i-1} given by $\mathbf{h}^{t_{i-1}} \in \mathbb{R}^{m_r \times m_r}$, which acts as a memory and processes past information. Then it updates the hidden state matrix $\mathbf{h}^{t_i} \in \mathbb{R}^{m_r \times m_r}$ for the current time point t_i. In order to avoid the vanishing gradient problem which makes the network's gradients tend to zero (i.e., hard to learn parameters), we need a function which can bound the gradient and eliminate the risk of divergence during the training. To do so, we use tanh [15] as an activation function in our graph-recurrent filter since it allows the state values to update by bounding in the range of $[-1, 1]$ compared to other activation functions such as sigmoid. The equation for our recurrent edge-filtering function $F^l(\mathbf{e}^{t_i}, \mathbf{h}^{t_{i-1}})$ is expressed as follows:

$$\mathbf{h}^{t_i} = \tanh\left([\mathbf{e}^{t_i}, \mathbf{h}^{t_{i-1}}] \odot [\mathbf{W}_{ih}, \mathbf{W}_{hh}]^T + \mathbf{b}_h\right) \tag{5}$$

where $\mathbf{W_{ih}} \in \mathbb{R}^{1 \times m_r}$ and $\mathbf{W_{hh}} \in \mathbb{R}^{m_r \times m_r}$ are learnable parameters for input-to-hidden weight and hidden-to-hidden weight respectively. $\mathbf{b_h} \in \mathbb{R}^{m_r \times m_r}$ is bias term.

The Transformation Layer Architecture. Let $\mathbf{X}_{t_{i-1}}^{tr} \in \mathbb{R}^{n_r \times n_r}$ be the input brain connectivity matrix at a given time point, t_i where n_r is the number of ROIs. After obtaining the output node embeddings $\mathbf{V}^l = [\mathbf{v}_1^l, \mathbf{v}_2^l, ..., \mathbf{v}_n^l]^T$ of RGC layer from a given input $\mathbf{X}_{t_{i-1}}^{tr}$, we construct predicted brain graph $\hat{\mathbf{X}}_{t_i}^{tr}$ at t_i by computing pairwise absolute difference of learned embeddings [16]. To do so, first $\mathbf{V}^l$ is replicated with respect to the horizontal axis n_r times to obtain $\mathcal{R} \in \mathbb{R}^{n_r \times n_r \times n_r}$. Then, we compute the absolute difference between $\mathcal{R}$ and its transpose $\mathcal{R}^T$. Finally, the resulting tensor is the sum along z-axis to obtain the predicted brain graph $\hat{\mathbf{X}}_{t_i}^{tr} \in \mathbb{R}^{n_r \times n_r}$ for the time point t_i.

3   Results and Discussion

Evaluation Dataset. We conducted experiments on OASIS-21 longitudinal dataset with 113 subjects [17]. This set contains longitudinal collection of 150

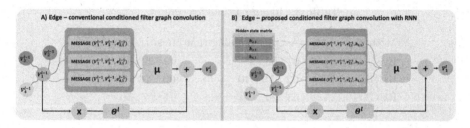

Fig. 2. *Illustration of the key differences between the conventional edge-conditioned filter graph generation and our proposed recurrent graph convolution.* (**A**) *Conventional edge filter for graph convolution.* First, messages are created between ROIs i and its neighbors $\mathcal{N}(i)$. Then, the average of the messages is computed by the mean operation. To inherit the previous layer embedding $\mathbf{V}^{l-1} \in \mathbb{R}^{n_r \times n_r}$, we multiply $\mathbf{V}^{l-1}$ by dynamically generated edge specific weight Θ^l. Finally, $\mathbf{V}^l$ is computed by combining the previous layer embedding and the average message passing between ROIs. (**B**) *Recurrent graph convolution.* First, the graph recurrent-filter network creates the message between ROIs i and $\mathcal{N}(i)$ by taking hidden state value h_{ij} in contradiction to the conventional edge-conditioned filter graph generation. To inherit previous layer embedding $\mathbf{V}^{l-1} \in \mathbb{R}^{n_r \times n_r}$, we multiply $\mathbf{V}^{l-1}$ by dynamically generated edge specific weight Θ^l. Finally, $\mathbf{V}^l$ is computed by combining the previous layer embedding and average message passing between ROIs.

subjects aged between 60 to 96. Each subject's brain scans were acquired 3 times one year apart. For each subject, we construct a cortical morphological network derived from cortical thickness measure using structural T1-w MRI as proposed in [18]. Each cortical hemisphere is parcellated into 35 ROIs using the Desikan-Killiany cortical atlas. We construct our RBGM with PyTorch Geometric library [19].

Parameter Setting. In Table 2, we report the mean absolute error between ground-truth and synthesized brain graphs at follow-up time points t_1 and t_2. In Table 3, we publish the required training time for each comparison method respectively. We set hyperparameters of each mapper as follows: $\lambda_1 = 1$, $\lambda_2 = 10$. We used AdamW [20] optimizer and set the learning rate at 0.0001 for each mapper. Finally, we trained our model by using 3-fold cross-validation for 200 epochs using an NVIDIA Tesla V100 GPU.

Table 2. Prediction accuracy of compared methods using MAE at t_1 and t_2.

Method	t_1		t_2	
	Mean MAE ± std	Best MAE	Mean MAE ± std	Best MAE
EvoGraphNet[6]	0.05544 ± 0.01140	0.04555	**0.05991 ± 0.00937**	**0.05168**
RBGM (w/KL)	0.05585 ± 0.00349	0.05341	0.13509 ± 0.0098	0.12249
RBGM	**0.04465 ± 0.00473**	**0.03994**	0.06228 ± 0.00315	0.05870

Table 3. Required time for training.

Method	Average training time	Best training time
EvoGraphNet [6]	$07:08:33$	$02:22:35$
RBGM (w/KL)	$\mathbf{02:19:40}$	$\mathbf{00:46:70}$
RBGM	$03:24:08$	$01:07:50$

Comparison Method and Evaluation. Due to the lack of RNN-based comparison methods that consider time-dependency for brain graph prediction, we benchmarked our RBGM against some of its variants. We call the first benchmark method: RBGM (w/KL), where we replaced topological loss with KL-divergence loss between the predicted graph at t_i and its corresponding ground-truth brain graph. The KL-divergence minimizes the discrepancy between ground-truth and predicted connectivity weight distribution at each time point t_i. The second benchmarking method is against the current state-of-the-art EvoGraphNet [6] in order to assess the power of topological loss and the recurrent graph convolution. In Table 2 we report the mean absolute error (MAE) between ground-truth and predicted brain graphs for consecutive time points t_1 and t_2 for each comparison method. Our proposed RBGM outperformed baseline methods at t_1 by achieving both the lowest mean MAE (averaged across the 3 folds) and the overall best MAE as shown in Table 2. However, for t_2, EvoGraphNet achieved both the best MAE and mean MAE results. Notably, results show that our RBGM *closely* matches the best results at time point t_2 by an error difference of 7×10^{-3} in mean MAE yet outperformed EvoGraphNet in time consumption by achieving 46% less training time. To the best of our knowledge, such *time/complexity/error* is a delicate compromise to make. Under such a compromisation paradigm, we can fairly judge the *outperformance* of our proposed RBGM is matching state-of-the-art results given less complexity and time consumption.

Overall, our RBGM performs almost similar to the state-of-the-art method EvoGraphNet while speeding up the training and reducing the complexity Table 3 since it consists of fewer convolutional layers. However, our RBGM has a few limitations. So far, we have worked on brain graphs where the single edge connects only two ROIs. We aim to generalize our RGBM to handle brain hypergraphs where multiple edges can link two ROIs in our future work. This will enable us to better model and capture the complexity of the brain as a highly interactive network with different topological properties.

4   Conclusion

In this paper, we proposed the first edge-based recurrent graph neural network RBGM that uses a novel recurrent graph convolution to predict the brain connectivity evolution trajectory from a single time point. Our architecture contains m number of mappers for m time points. We proposed a time-dependency loss between consecutive time points and a topological loss to preserve topological consistency between predicted and ground-truth brain graphs at the same time point. The results showed that our time-dependent RBGM achieved a similar prediction accuracy compared to the state-of-the-art EvoGraphNet while reducing the training time and complexity. The RBGM is generic and can be used to predict brain graphs for any given time point. In future studies, we aim to generalize our RBGM to using hypergraphs and account for brain hyperconnectivity.

Acknowledgements. This work was funded by generous grants from the European H2020 Marie Sklodowska-Curie action (grant no. 101003403, http://basira-lab.com/ normnets/) to I.R. and the Scientific and Technological Research Council of Turkey to I.R. under the TUBITAK 2232 Fellowship for Outstanding Researchers (no. 118C288, http://basira-lab.com/reprime/). However, all scientific contributions made in this project are owned and approved solely by the authors.

References

1. Lohmeyer, J.L., Alpinar-Sencan, Z., Schicktanz, S.: Attitudes towards prediction and early diagnosis of late-onset dementia: a comparison of tested persons and family caregivers. Aging Ment. Health **25**, 1–12 (2020)
2. Stoessl, A.J.: Neuroimaging in the early diagnosis of neurodegenerative disease. Trans. Reurodegen. **1**, 1–6 (2012)
3. Rekik, I., Li, G., Yap, P.T., Chen, G., Lin, W., Shen, D.: Joint prediction of longitudinal development of cortical surfaces and white matter fibers from neonatal MRI. NeuroImage **152**, 411–424 (2017)
4. Gafuroglu, C., Rekik, I.: Image evolution trajectory prediction and classification from baseline using learning-based patch atlas selection for early diagnosis. arXiv preprint arXiv:1907.06064 (2019)
5. Gürler, Z., Nebli, A., Rekik, I.: Foreseeing brain graph evolution over time using deep adversarial network normalizer. In: Rekik, I., Adeli, E., Park, S.H., Valdés Hernández, M.C. (eds.) PRIME 2020. LNCS, vol. 12329, pp. 111–122. Springer, Cham (2020). https://doi.org/10.1007/978-3-030-59354-4_11
6. Nebli, A., Kaplan, U.A., Rekik, I.: Deep EvoGraphNet architecture for time-dependent brain graph data synthesis from a single timepoint. In: Rekik, I., Adeli, E., Park, S.H., Valdés Hernández, M.C. (eds.) PRIME 2020. LNCS, vol. 12329, pp. 144–155. Springer, Cham (2020). https://doi.org/10.1007/978-3-030-59354-4_14
7. van den Heuvel, M.P., Sporns, O.: A cross-disorder connectome landscape of brain dysconnectivity. Nat. Rev. Neurosci. **20**, 435–446 (2019)
8. Goodfellow, I.J., et al.: Generative adversarial networks. arXiv preprint arXiv:1406.2661 (2014)
9. McGrath, J.M., Cone, S., Samra, H.A.: Neuroprotection in the preterm infant: further understanding of the short-and long-term implications for brain development. Newborn Infant Nurs. Rev. **11**, 109–112 (2011)
10. Connor, J.T., Martin, R.D., Atlas, L.E.: Recurrent neural networks and robust time series prediction. IEEE Trans. Neural Netw. **5**, 240–254 (1994)
11. Drossos, K., Gharib, S., Magron, P., Virtanen, T.: Language modelling for sound event detection with teacher forcing and scheduled sampling. arXiv preprint arXiv:1907.08506 (2019)

12. Simonovsky, M., Komodakis, N.: Dynamic edge-conditioned filters in convolutional neural networks on graphs. In: Proceedings of the IEEE Conference on Computer Vision and Pattern Recognition, pp. 3693–3702 (2017)
13. Cui, R., Liu, M., Initiative, A.D.N., et al.: RNN-based longitudinal analysis for diagnosis of Alzheimer's disease. Comput. Med. Imag. Graph. **73**, 1–10 (2019)
14. Xu, X., Zhou, F., Liu, B.: Automatic bladder segmentation from CT images using deep CRM and 3d fully connected CRF-RRN. Int. J. Comput. Assist. Radiol. Surg. **13**, 967–975 (2018)
15. Shewalkar, A., Nyavanandi, D., Ludwig, S.A.: Performance evaluation of deep neural networks applied to speech recognition: RNN, LSTM and GRU. J. Artif. Intell. Soft Comput. Res. **9**, 235–245 (2019)
16. Gurbuz, M.B., Rekik, I.: Deep graph normalizer: a geometric deep learning approach for estimating connectional brain templates. In: Martel, A.L., et al. (eds.) MICCAI 2020. LNCS, vol. 12267, pp. 155–165. Springer, Cham (2020). https://doi.org/10.1007/978-3-030-59728-3_16
17. Marcus, D.S., Fotenos, A.F., Csernansky, J.G., Morris, J.C., Buckner, R.L.: Open access series of imaging studies: longitudinal MRI data in nondemented and demented older adults. J. Cogn. Neurosci. **22**, 2677–2684 (2010)
18. Mahjoub, I., Mahjoub, M.A., Rekik, I.: Brain multiplexes reveal morphological connectional biomarkers fingerprinting late brain dementia states. Sci. Rep. **8**, 1–14 (2018)
19. Fey, M., Lenssen, J.E.: Fast graph representation learning with pytorch geometric. arXiv preprint arXiv:1903.02428 (2019)
20. Loshchilov, I., Hutter, F.: Fixing weight decay regularization in Adam. (2018)

COVID-Net US: A Tailored, Highly Efficient, Self-attention Deep Convolutional Neural Network Design for Detection of COVID-19 Patient Cases from Point-of-Care Ultrasound Imaging

Alexander MacLean[1]([⊠]), Saad Abbasi[1], Ashkan Ebadi[2], Andy Zhao[1],
Maya Pavlova[1], Hayden Gunraj[1], Pengcheng Xi[3], Sonny Kohli[5],
and Alexander Wong[1,4]

[1] Department of Systems Design Engineering, University of Waterloo,
Waterloo N2L 3G1, Canada
alex.maclean@uwaterloo.ca
[2] National Research Council Canada, Montreal, QC H3T 1J4, Canada
[3] National Research Council Canada, Ottawa, ON K1K 2E1, Canada
[4] Waterloo Artificial Intelligence Institute, Waterloo, ON N2L 3G1, Canada
[5] Oakville Trafalgar Memorial Hospital, McMaster University,
Hamilton, ON, Canada

Abstract. The Coronavirus Disease 2019 (COVID-19) pandemic has impacted many aspects of life globally, and a critical factor in mitigating its effects is screening individuals for infections, thereby allowing for both proper treatment for those individuals as well as action to be taken to prevent further spread of the virus. Point-of-care ultrasound (POCUS) imaging has been proposed as a screening tool as it is a much cheaper and easier to apply imaging modality than others that are traditionally used for pulmonary examinations, namely chest x-ray and computed tomography. Given the scarcity of expert radiologists for interpreting POCUS examinations in many highly affected regions around the world, low-cost deep learning-driven clinical decision support solutions can have a large impact during the on-going pandemic. Motivated by this, we introduce COVID-Net US, a highly efficient, self-attention deep convolutional neural network design tailored for COVID-19 screening from lung POCUS images. Experimental results show that the proposed COVID-Net US can achieve an AUC of over 0.98 while achieving 353× lower architectural complexity, 62× lower computational complexity, and 14.3× faster inference times on a Raspberry Pi. Clinical validation was also conducted, where select cases were reviewed and reported on by a practicing clinician (20 years of clinical practice) specializing in intensive care (ICU) and 15 years of expertise in POCUS interpretation. To advocate

S. Albarqouni et al. (Eds.): DART 2021/FAIR 2021, LNCS 12968, pp. 191–202, 2021.
https://doi.org/10.1007/978-3-030-87722-4_18

affordable healthcare and artificial intelligence for resource-constrained environments, we have made COVID-Net US open source and publicly available (https://github.com/maclean-alexander/COVID-Net-US/) as part of the COVID-Net open source initiative.

Keywords: COVID-19 detection · Ultrasonic imaging · Deep learning

1   Introduction

Since the beginning of 2020, the Coronavirus Disease 2019 (COVID-19) pandemic has had an enormous impact on global healthcare systems, and there has not been a region or domain that has not felt its impact in one way or another. While vaccination efforts have certainly been shown as effective in mitigating further spread of COVID-19, screening of individuals to test for the disease is still necessary to ensure the safety of public health, and the gold-standard of this screening is reverse transcriptase-polymerase chain reaction (RT-PCR) [17]. With RT-PCR being laborious and time-consuming, much work has gone into exploring other possible screening tools, namely using chest x-ray (CXR) and computed tomography (CT) to observe abnormalities in images, and extending from needing to have radiologists analyze the images and make diagnoses to using automated models to screen for the presence of COVID-19 [8,16].

Another imaging modality that is gaining traction as a tool for treating lung related diseases is the lung point-of-care ultrasound (POCUS), and has been suggested as most useful in contexts that are resource limited, such as emergency settings or low-resource countries [1,6]. POCUS devices allow for easier and quicker applications than CXR and CT systems, and are much cheaper to acquire leading to more access in a variety of point-of-care locations, thus enhancing the ability for possible COVID-19 screening [1].

Despite the comparably superior ease-of-use and -access for POCUS devices, the interpretation of ultrasound images is far from simple, and protocols have been written providing instruction for radiologists to standardize such interpretation [15]. This need for expert radiologists creates a bottleneck in the COVID-19 screening pathway. Typically, POCUS devices are mobile and do not have large computational resources available on-board mandating computationally lightweight solutions. In addition, POCUS analysis is typically performed on video rather than still images. Thus any solution must also operate in real-time.

Motivated by this challenge, this study introduces COVID-Net US, a novel, highly customized self-attention deep neural network architecture tailored specifically for the detection of COVID-19 cases from POCUS images. The goal with COVID-Net US is to design a highly efficient yet high performing deep neural network architecture that is small enough to be implemented on low-cost devices allowing for limited additional resources needed when used with POCUS devices in low-resource environments. More specifically, we employ a collaborative human-machine design approach. We leverage human knowledge and experience to design an initial prototype network architecture. This prototype is subsequently leveraged alongside a set of operational requirements catered around low-power, real-time embedded scenarios to generate a unique, highly customized deep neural network architecture tailored for COVID-19 case detection from POCUS images via an automatic machine driven design exploration strategy. This combination of human and machine driven design strategy results in a highly accurate yet efficient network architecture. Finally, to demonstrate the efficacy of the proposed COVID-Net US for edge devices, we evaluate its practical performance by deploying it on a Raspberry Pi with a 1.5 GHz ARM Cortex 72 CPU with 4 GB memory. To advocate affordable healthcare and artificial intelligence for resource-constrained environments, we have made COVID-Net US open source and publicly available as part of the COVID-Net open source initiative[1] [7,8,16,21] for accelerating the advancement and adoption of deep learning for tackling this pandemic.

2   Related Work

There are a multitude of initiatives aiming to apply machine learning to classification of medical images for screening of COVID-19 infections. One such initiative is the COVID-Net initiative, which has shown success in curating open-source publicly available CXR and CT image datasets to act as training data for deep neural network models, and in using machine-driven design exploration to optimize both micro and macroarchitecture to build an architecture tailored to the problem at hand, improving both efficacy and efficiency [7,8,16,21].

There have been numerous other works developing deep neural network architectures for processing POCUS images, especially in the context of COVID-19 infections in recent times. The work of Arntfield et al. showed effectiveness in building deep neural network architectures, using Xception architectures, to identify the presence of COVID-19 in clinically captured lung ultrasound (LUS) images, reaching a higher area under the receiver operating characteristic curve (AUC) with the developed model than was able to be done by expert radiologists [2]. The authors have released the code used in this research, however neither the data used to train the model nor any resulting models are available, preventing complete reproducibility or even application by others at this time.

In another study [4], the authors constructed a deep neural network architecture trained using their own compiled datasets, which were a mix of public and private datasets, and were able to achieve relatively high performance (>0.94 AUC for each of the normal, pneumonia, and COVID-19 classes) with their final model based off of VGG-16. However, the authors did not appear to investigate how the resulting architecture fares on low-cost devices which could limit the contexts in which their work is applicable.

The study by Rojas-Azabache et al. [13] involved the development their own deep neural network architecture, trained on POCUS images collected from individuals in the city of Lima, Peru (and kept private by the authors). The authors

---

[1] http://www.covid-net.ml/.

demonstrate the efficacy of their VGG-16 inspired model not only in identifying COVID-19 but also by deploying it on a Raspberry Pi. Although the study does not report the memory footprint or latency of the proposed model, a standard VGG-16 architecture consists of 134 million parameters and requires 15 billion FLOPS for inference [14]. While it is possible to run this model on a Raspberry Pi, the inference will be far from real-time. The protocol for physician analysis of an ultrasound examination consists of them analyzing the video taken by the POCUS device, often from various views and positions [1]. It is likely that for effective automatic analysis, many frames from a single video, or possibly multiple, will need to be processed by a given architecture, so to be feasibly implemented it will need to perform this analysis in as close to real time as possible. Thus, a goal of this study is to develop the architecture such that not only is it effective in terms of the accuracy metrics but also efficient both in size of the resulting architecture and efficiency of model processing time to facilitate affordable use on low-cost embedded processors to better facilitate for widespread use in highly affected regions around the world.

3   Methods

This study introduces COVID-Net US, a tailored, highly efficient self-attention neural network design that aims to detect COVID-19 patient cases from POCUS images. As mentioned earlier, POCUS systems are typically resource constrained but also mandate real-time solutions. Thus any practical solution would need to exhibit low latency and high accuracy, simultaneously. To this end, this study leverages a machine-driven design exploration to automatically discover highly customized micro and macro architecture designs that result in a high-performing yet highly efficient deep neural network architecture catered around task and operational requirements at hand. The found macro-architecture design take advantage of self-attention via attention-condensers to yield an efficient and accurate network architecture [18].

3.1   COVIDx-US Dataset

We employ the COVIDx-US dataset [5] to train and evaluate our network. The COVIDx-US dataset contains a total of 173 ultrasound lung videos and 16,822 processed images, curated from multiple sources. The videos comprise of 60 COVID-19 positive cases, 15 normal cases, 41 patients with non-COVID related pneumonia, and 57 "other" non-COVID cases containing other pulmonary issues such as Chronic Obstructive Pulmonary Disorder (COPD), Pneumonthorax, and Pulmonary Oedema. Since the COVIDx-US dataset is compiled from various sources, many of which did not have extensive demographic data available, a complete summary of demographics for the overall dataset is not able to be provided at this time. Of those for which demographic information is known, there are POCUS videos from both male and female individuals with ages ranging from 7 to 72.

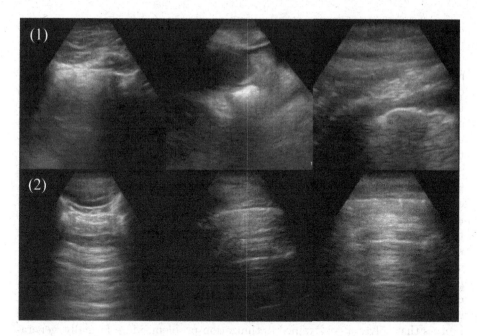

Fig. 1. Example POCUS images: (1) SARS-CoV-2 positive patient cases and (2) SARS-CoV-2 negative patient cases.

Due to the low number of normal cases and the heterogeneity of the "other" cases, initial attempts for 4-way multi-label classification were challenging. Instead, the dataset was treated as a binary classification problem, where the COVID-19 cases were labeled as positive and the normal, pneumonia, and "other" cases were labeled as negative. Additionally, due to very low numbers of positive examples taken with linear ultrasound probes, only cases of studies with a convex ultrasound probe were selected for this experiment. Selected example images are presented in Fig. 1. The resulting images were split into a training set with 7,644 images (3,947 positive and 3,697 negative), a validation set with 2,383 images (1,085 positive and 1,298 negative), and a test set with 2,233 images (1,052 positive and 1,181 negative). The splits were made such that all frames from each video were either in the training, validation, or testing set, and patient-level demographic information was leveraged in a similar manner, ensuring that obvious instances of data leakage across sets is avoided to the greatest possible extent.

3.2 Network Design

In this study, we construct an initial network design prototype by leveraging residual architecture design principles [9] given their ability to achieve high performance while maintaining ease of training by alleviating the issue of vanishing gradient present in conventional deep neural network architectures. More specifically, the initial network design prototype was designed to provide one of two predictions: a) SARS-CoV-2 negative c) SARS-CoV-2 positive.

Based on this initial network design prototype and a set of operational requirements catered around an embedded scenario, we leverage a machine-driven design exploration strategy to automatically determine the micro and macro-architecture for the final COVID-Net US deep neural network. This strategy allows us to automatically strike a balance between detection accuracy and network efficiency, yielding a network design that is tailored for on-device detection of COVID-19 from POCUS images.

In recent years, there has been increased interest in automatic exploration of the architectural space to yield tailor-made network designs that achieve both excellent accuracy while simultaneously having a small memory footprint and an efficient architecture. Methods such as MONAS [10] and DARTS [12] employ reinforcement learning or gradient descent to find an optimal set of parameters and thus yield an architecture that satisfies the given operational requirements and constraints. In this study, we leverage generative synthesis [19] to explore micro and macro-architecture designs in an efficient yet effective manner. In contrast to techniques that rely on gradient descent or reinforcement learning, generative synthesis finds the optimal network architecture through an iterative process that solves a constrained optimization problem. More formally, generative synthesis can be formulated as the following expression

$$\mathcal{G} = \max_{\mathcal{G}} \mathcal{U}(\mathcal{G}(s)) \text{ subject to } 1_r(G(s)) = 1, \ \forall \in \mathcal{S}. \tag{1}$$

where the goal is to find generator $\mathcal{G}$ that generates deep neural network architectures that maximize a performance function $\mathcal{U}$ (e.g. [20]) given the operational constraints $1_r(\cdot)$. Generative synthesis aims to find an approximate solution to the above expression through an iterative process. The ultimate goal of this study is to produce an efficient and accurate network architecture that can be employed on ultrasound machines and detect COVID-19 in real-time. To this end, we formulate $1_r(\cdot)$ in terms of parameter count (<1M), number of FLOPS (<1B) and area under the curve (AUC) (>0.9).

Figure 2 describes the resulting network. Several interesting observations can be made from the resulting architecture. First, we can note that that network design is comprised of a different number of macro-architectures including point-wise convolutions, attention condensers and standard convolutions. This high macro-architecture diversity is a direct result of using machine-driven design strategy to identify an architecture design specifically for identification of COVID-19 in ultrasound images. Secondly, we note that the use of depthwise convolutions leads to a very computationally light-weight design. This light-weight design is the result of imposing constraints on the parameter count and the number of FLOPS. Finally, we note that the architecture relies heavily on attention condensers. Attention condensers were originally introduced in [18] and have demonstrated themselves to be a highly efficient mechanism for self-attention [3]. Visual attention condensers produce a compressed embedding that represents the spatial and cross-channel activation relationships. This compressed representation yields

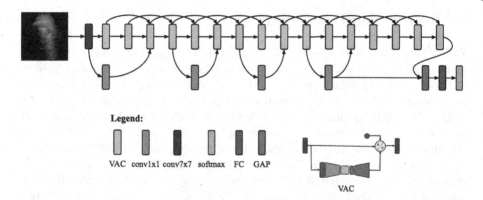

Fig. 2. The COVID-Net US architecture design. The COVID-Net US design exhibits high architectural heterogeneity and the use of visual attention condensers with macro-architecture and micro-architecture designs tailored specifically for the detection of COVID-19 from ultrasound images.

selective attention which improves representational capability while exhibiting low architectural complexity.

3.3 Explanation-Driven Performance Validation

To further validate the performance of the COVID-Net US, we examine the decision-making behaviour of the resulting architecture by employing GSInquire, a state-of-the-art explainability method which identifies critical factors in the test data which are quantifiably shown to be integral to the decisions made by the system [11]. In short, GSinquire makes use of an inquisitor $\mathcal{I}$ in a generator-inquisitor pair $\mathcal{G}, \mathcal{I}$ during the machine-driven exploration generative synthesis described above [19]. During generative synthesis, the inquisitor $\mathcal{I}$ probes the connections in the network $\mathcal{N}$ (here, COVID-Net US) iteratively with the generator $\mathcal{G}$ to identify the factors in the input to the network that are affecting the decision-making process of $\mathcal{N}$. The result of this iterative process is a highlighting of the regions or components of desired test inputs, in this case, test images, which if removed or modified would most drastically change the decision made by the network $\mathcal{N}$ and thus are critical factors.

In the context of COVID-Net US, the results of this explanation-driven performance validation is to identify whether the architecture's decision-making process is successfully using the factors in ultrasound examinations noted by radiologists as important in identifying the presence or absence of COVID-19 in individuals. To that end, the critical factors will be compared to those noted in the previously mentioned studies [1] and [15] to validate that the architecture makes use of clinically relevant features interpretable by radiologists.

Select patient cases from the benchmark dataset are also reviewed and reported on by a practicing clinician (20 years of clinical practice) specializing in intensive care (ICU) and 15 years of expertise in POCUS interpretation.

4   Results and Discussion

We evaluate the resulting network architecture through a number of important metrics. Historically, machine learning has been mostly concerned with providing accurate performance, which is still vital in our case, and to that end the architecture with the AUC score to keep consistent with previously mentioned studies [2,4]. Additionally, to ensure that the resulting architecture meets the requirements of effectiveness on low-resource, "edge" devices, the feasibility of this implementation will be evaluated by comparing the size and speed of processing on a test device, a Raspberry Pi. For comparison purposes, we also evaluate the performance of a ResNet-50 [9] architecture design. We also perform qualitative analysis using an explainability-driven performance validation method.

4.1   Quantitative Analysis

To further contextualize the balance of an architecture's performance compared to its computational and architectural complexities, which predict well its feasibility of implementation in realistic contexts, researchers have proposed many quantitative metrics, of which NetScore is notable [20]. NetScore is based on an analysis of many publicly available deep neural network architectures, and is described as useful in evaluating the effectiveness of architectures in practical contexts on edge devices, which describes well the use case of analyzing POCUS images in low-resource situations.

Table 1. Comparison of test AUC, FLOPS and model size of the proposed COVID-Net US architecture with the ResNet-50 architecture.

Model	NetScore	Params	FLOPS	Latency (ms)	Test AUC
COVID-Net US	82.53	65K	596M	216	0.9824
ResNet-50	49.08	23M	37B	3087	0.9196

Table 1 presents the accuracy results and the architecture and computation complexity results (NetScore, number of parameters, FLOPs, and Latency in ms on a Raspberry Pi). In terms of performance on the desired task, namely the AUC of the resulting model during testing, both the ResNet-50 and the COVID-Net US network architecture designs performed well with both achieving AUC values higher than 0.9, with COVID-Net US reaching above 0.98, comparable to the related works. Since the datasets used by each work are different, comparison of the exact performance values across studies is difficult, and in future work, studies that release their training code such as [2] and [4] will be tested to better compare the results of our training methods and analyze the sensitivity and positive predictive value of the architectures. As such, the effectiveness of the resulting architecture is consistent with the related works in suggesting that

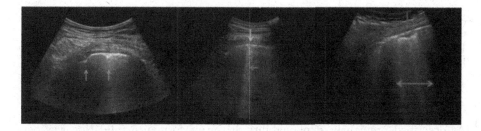

Fig. 3. Examples of ultrasound images with annotations of signs used by radiologists for scoring the examinations, on a scale from 0 to 3 with 0 being normal and 3 being the most severe [15]. A score 1 example is shown on the left, with red arrows pointing to indentations in the pleural line. The central image shows a score 2 example with breaking points in the pleural line annotated, as well as "white lung" artefacts extending further down. Score 3 is shown in the right image, with significantly discontinuous pleural lines with a large area of white lung.

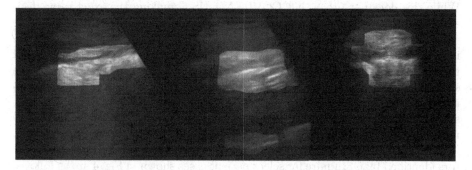

Fig. 4. Examples of COVIDx-US test images with GSInquire critical factor results overlaid, each for separated positive COVID-19 cases. Each image has the pleural line region identified as important to the decision making process, and many have regions further down the lung as highlighted as well, where white lung would be found.

automated detection systems can be successful in identifying cases of COVID-19 infections from ultrasound examination, and thus prove effective as a complementary COVID-19 screening tool, particularly where RT-PCR tests or CXR or CT examinations could be difficult due to resource or time constraints.

The second level of analysis concerns the measures of computational and architectural complexity. As desired, the utilization of a machine-driven design exploration strategy to determine the best micro and macro-architecture designs resulted in the proposed COVID-Net US outperforming ResNet-50 across all metrics in this area. The NetScore achieved by COVID-Net US was 82.53 compared to 49.03 as achieved by ResNet-50; in the original study, examined models ranged from around 30 to 80 on this scale, showing that this difference is significant. COVID-Net US exhibits various macro-architectures which work in tandem for greatly reduced architectural and computational complexity. The result is a highly efficient architecture that requires just 65K parameters (**353×** fewer than ResNet-50), 596M FLOPS (**62×** lower than ResNet-50) and exhibits an on-device latency of just 216 ms (**14.3×** lower than ResNet-50) on a Raspberry Pi. The small size and low latency of the proposed architecture enable them to be deployed on POCUS devices without significant hardware cost.

4.2 Qualitative Analysis

As described previously, explainability-driven performance validation was performed on COVID-Net US using GSInquire. Figure 3 shows examples of images corresponding to scores 1, 2, and 3 (out of 3, with 3 being more severe cases of lung involvement) based on the system described by [15]. Figure 4 then shows examples of the output from GSInquire for selected test images from COVIDx-US, with the critical factor regions highlighted. Note that the regions identified as quantifiably important for COVID-Net US when making classifications line up well with the areas annotated by radiologists based on their experiences. While this does not prove that COVID-Net US is making the exact same decisions as those radiologists, is does suggest two things; first, it confirms that the architecture is learning to examine features within the region of interest of the images, rather than unrelated regions or possible artefacts beyond the extent of the relevant content; and second, that since the architecture is focusing on the same regions as radiologists it is likely making clinically relevant decisions helping clinicians to trust the results of those decisions, supporting the notion that such models can be relied upon to automate the screening process for COVID-19 using ultrasound images.

The clinical expert findings and observations with regards to the critical factors identified by GSInquire for select patient cases shown in Fig. 4 are as follows. In all these cases, COVID-Net US detected them to be SARS-CoV-2 positive, which was clinically confirmed. It was observed in all three cases that the critical factors leveraged by COVID-Net US correspond to pleural line irregularities, specifically hypoechoic areas and breakages in the normally smooth contoured pleural surface (subpleural consolidations) associated with lung pathology consistent with SARS-CoV-2 induced lung inflammation and pneumonia. As such, it was demonstrated that the identified critical factors leveraged by COVID-Net US are consistent with radiologist interpretation.

5   Conclusions

In this study, we introduce COVID-Net US, a highly efficient, self-attention based deep convolutional neural network architecture tailored for screening of COVID-19 infections from lung ultrasound images in low-resource environments. A machine-driven exploration strategy was used to perform the architecture design, resulting in an architecture that made use of visual attention condensers highly customized to the task at hand. Experimental results show that COVID-Net US achieved a higher accuracy than the ResNet-50 in terms of AUC score, while having significantly lower architectural and computational complexity, making it much more feasible of an architecture in low-resource contexts. Additionally, explainability-driven performance validation was done to ensure that the decision-making process of COVID-Net US is consistent with that of COVID-19 lung ultrasound examination protocols published by radiologists.

With the expectation that COVID-19 will continue to have an impact on global health for the foreseeable future, it is important that work done to combat the pandemic is ongoing in all possible avenues. The ability for COVID-Net US to allow for point-of-care ultrasound devices to act as a COVID-19 screening tool with less of a need for radiologist oversight for analysis means that contexts where current commonly used tools, namely RT-PCR, CXR, and CT, are rare or unavailable due to resource constraints can hopefully improve their ability to screen for the disease. Whether that is in contexts with time constraints, such as emergency situations, or with resource constraints, such as developing countries, they will be able to benefit from this tool and protect individuals from further detrimental effects. We look forward to this work expanding the capabilities of the general COVID-Net initiative to provide support to clinicians and researchers alike.

Acknowledgements. We thank the Natural Sciences and Engineering Research Council of Canada (NSERC), the Canada Research Chairs program, DarwinAI Corp., and the National Research Council Canada (NRC). The study has received ethics clearance from the University of Waterloo (42235).

References

1. Amatya, Y., Rupp, J., Russell, F.M., Saunders, J., Bales, B., House, D.R.: Diagnostic use of lung ultrasound compared to chest radiograph for suspected pneumonia in a resource-limited setting. Int. J. Emerg. Med. **11**(1) (2018). https://doi.org/10.1186/s12245-018-0170-2
2. Arntfield, R., et al.: Development of a convolutional neural network to differentiate among the etiology of similar appearing pathological b lines on lung ultrasound: a deep learning study. BMJ Open **11**(3), e045120 (2021)
3. Bahdanau, D., Cho, K., Bengio, Y.: Neural machine translation by jointly learning to align and translate. CoRR abs/1409.0473 (2015)
4. Born, J., et al.: Pocovid-net: Automatic detection of Covid-19 from a new lung ultrasound imaging dataset (pocus). ArXiv abs/2004.12084 (2020)
5. Ebadi, A., Xi, P., MacLean, A., Tremblay, S., Kohli, S., Wong, A.: Covidx-us - an open-access benchmark dataset of ultrasound imaging data for AI-driven Covid-19 analytics (2021), https://arxiv.org/abs/2103.10003
6. Gehmacher, O., Mathis, G., Kopf, A., Scheier, M.: Ultrasound imaging of pneumonia. Ultrasound Med. Biol. **21**(9), 1119–1122 (1995)
7. Gunraj, H., Sabri, A., Koff, D., Wong, A.: Covid-net ct-2: Enhanced deep neural networks for detection of covid-19 from chest ct images through bigger, more diverse learning (2021)
8. Gunraj, H., Wang, L., Wong, A.: COVID Net-CT: a tailored deep convolutional neural network design for detection of COVID-19 cases from chest CT images. Front. Med. **7**, December 2020. https://doi.org/10.3389/fmed.2020.608525

9. He, K., Zhang, X., Ren, S., Sun, J.: Deep residual learning for image recognition. In: 2016 IEEE Conference on Computer Vision and Pattern Recognition (CVPR). IEEE (Jun 2016). DOI: 10.1109/cvpr.2016.90, https://doi.org/10.1109/cvpr.2016.90

10. Hsu, C.H., et al.: Monas: multi-objective neural architecture search using reinforcement learning. ArXiv abs/1806.10332 (2018)

11. Lin, Z.Q., Shafiee, M.J., Bochkarev, S., Jules, M.S., Wang, X.Y., Wong, A.: Do explanations reflect decisions? A machine-centric strategy to quantify the performance of explainability algorithms (2019)

12. Liu, H., Simonyan, K., Yang, Y.: DARTS: Differentiable architecture search. In: International Conference on Learning Representations (2019). https://openreview.net/forum?id=S1eYHoC5FX

13. Rojas-Azabache, C., Vilca-Janampa, K., Guerrero-Huayta, R., Núñez-Fernández, D.: Implementing a detection system for Covid-19 based on lung ultrasound imaging and deep learning (2021). https://arxiv.org/abs/2106.10651

14. Simonyan, K., Zisserman, A.: Very deep convolutional networks for large-scale image recognition. In: Bengio, Y., LeCun, Y. (eds.) 3rd International Conference on Learning Representations (ICLR 2015), San Diego, CA, USA, May 7–9, 2015, Conference Track Proceedings (2015). http://arxiv.org/abs/1409.1556

15. Soldati, G., et al.: Proposal for international standardization of the use of lung ultrasound for patients with COVID-19. J. Ultrasound Med. **39**(7), 1413–1419 (2020)

16. Wang, L., Lin, Z.Q., Wong, A.: COVID-net: a tailored deep convolutional neural network design for detection of COVID-19 cases from chest x-ray images. Sci. Rep. **10**(1) (2020). https://doi.org/10.1038/s41598-020-76550-z

17. Wang, W., et al.: Detection of SARS-CoV-2 in different types of clinical specimens. JAMA (2020). https://doi.org/10.1001/jama.2020.3786

18. Wong, A., Famouri, M., Shafiee, M.: Attendnets: tiny deep image recognition neural networks for the edge via visual attention condensers. ArXiv abs/2009.14385 (2020)

19. Wong, A., Shafiee, M., Chwyl, B., Li, F.: Ferminets: learning generative machines to generate efficient neural networks via generative synthesis. ArXiv abs/1809.05989 (2018)

20. Wong, A.: NetScore: towards universal metrics for large-scale performance analysis of deep neural networks for practical on-device edge usage. In: Karray, F., Campilho, A., Yu, A. (eds.) ICIAR 2019. LNCS, vol. 11663, pp. 15–26. Springer, Cham (2019). https://doi.org/10.1007/978-3-030-27272-2_2

21. Wong, A., et al.: Towards computer-aided severity assessment via training and validation of deep neural networks for geographic extent and opacity extent scoring of chest x-rays for SARS-COV-2 lung disease severity. Sci. Rep. **11**, 9315 (2021)

Inter-domain Alignment for Predicting High-Resolution Brain Networks Using Teacher-Student Learning

Başar Demir[1] , Alaa Bessadok[1,2,3] , and Islem Rekik[1]($\boxtimes$)

[1] BASIRA Lab, Faculty of Computer and Informatics,
Istanbul Technical University, Istanbul, Turkey
irekik@itu.edu.tr
[2] Higher Institute of Informatics and Communication Technologies,
University of Sousse, Sousse 4011, Tunisia
[3] National Engineering School of Sousse, LATIS-Laboratory of Advanced Technology
and Intelligent Systems, University of Sousse, Sousse 4023, Tunisia
http://basira-lab.com

Abstract. Accurate and automated super-resolution image synthesis is highly desired since it has the great potential to circumvent the need for acquiring high-cost medical scans and a time-consuming preprocessing pipeline of neuroimaging data. However, existing deep learning frameworks are solely designed to predict high-resolution (HR) image from a low-resolution (LR) one, which limits their generalization ability to brain graphs (i.e., connectomes). A small body of works has focused on superresolving brain graphs where the goal is to predict a HR graph from a single LR graph. Although promising, existing works mainly focus on superresolving graphs belonging to the same domain (e.g., functional), overlooking the domain fracture existing between multimodal brain data distributions (e.g., morphological and structural). To this aim, we propose a novel inter-domain adaptation framework namely, Learn to SuperResolve Brain Graphs with Knowledge Distillation Network (L2S-KDnet), which adopts a teacher-student paradigm to superresolve brain graphs. Our teacher network is a graph encoder-decoder that firstly learns the LR brain graph embeddings, and secondly learns how to align the resulting latent representations to the HR ground truth data distribution using an adversarial regularization. Ultimately, it decodes the HR graphs from the aligned embeddings. Next, our student network learns the knowledge of the aligned brain graphs as well as the topological structure of the predicted HR graphs transferred from the teacher. We further leverage the decoder of the teacher to optimize the student network. In such a way, we are not only bringing the learned embeddings from both networks closer to each other but also their predicted HR graphs. L2S-KDnet presents the first TS architecture tailored for brain graph super-resolution synthesis that is based on inter-domain alignment. Our experimental results demonstrate substantial performance gains over benchmark methods. Our code is available at https://github.com/basiralab/L2S-KDnet.

B. Demir and A. Bessadok—Co-first authors.

© Springer Nature Switzerland AG 2021
S. Albarqouni et al. (Eds.): DART 2021/FAIR 2021, LNCS 12968, pp. 203–215, 2021.
https://doi.org/10.1007/978-3-030-87722-4_19

1   Introduction

Multi-resolution neuroimaging has spanned several neuroscientific works thanks to the rich and complementary information that it provides [7,19]. Existing works showed that the diversity in resolution allows early disease diagnosis [14,21]. While super-resolution images provide more details about brain anatomy and function they correspondingly increase the scan time. Consequently, several deep learning-based cross-resolution image synthesis works have been proposed. For instance, [7] combined convolutional neural network with Generative Adversarial Network (GAN) [8] to superresolve magnetic resonance imaging (MRI). In addition, [18] proposed an autoencoder-based architecture for predicting 7T T1-weighted MRI from its 3T counterparts by leveraging both spatial and wavelet domains. While several multi-resolution image synthesis works have been proposed, the road to superresolving brain graphs (i.e., connectomes) is still less traveled [2]. In a brain graph, nodes denote the region of interest (i.e., ROI) and edges denote the connectivity between pairs of ROIs. Thus, each of the LR and HR brain graphs captures unique aspects of the brain to assess the connectivity patterns and functionalities of the brain regions. For instance, cortical brain graphs are generated using time-consuming image processing pipelines that include several steps such as reconstruction, segmentation and parcellation of cortical surfaces [13].

To circumvent the need for costly data acquisition and image processing pipelines, a few cross-resolution brain graph synthesis solutions have been proposed [6,10,11,14]. For instance, [6,14] predicted HR brain graph of a particular subject given a single LR graph by first selecting its neighboring LR training brain graphs, second performing a weighted averaging for the selected graphs. Specifically, [6] relied on manifold learning and correlation analysis techniques to learn how to generate the nearest graph embeddings for a given testing brain graph. With a different perspective, [14] estimated a connectional brain template for the LR domain and selected the most similar graphs by computing the residual distance to the estimated template. Although pioneering, these machine learning (ML) based works have two severe limitations: (i) they are designed in a dichotomized manner, with distinct sub-parts of the framework being trained separately. As a result, being agnostic to cumulative errors is obviously important for developing clinically interpretable forecasts, and (ii) they overlook the handling of *domain fracture* problem resulting in the difference in distribution between the LR and HR domains.

To alleviate these issues, two recent single modality geometric deep learning (GDL) works have been proposed [10,11], where graph U-net architectures were designed based on a novel graph superresolving propagation rule. In particular, [10] firstly learned the embeddings of the LR brain graphs by leveraging graph convolutional network (GCN) [12] and next superresolved the HR graphs using the proposed graph convolution operation. Later on, [11] was introduced as an extending work of [10], which proposed an adversarial loss to align the predicted HR brain graphs with the ground truth ones. Such framework mainly performed an *intra-domain alignment* as it is learned within the same domain (i.e., HR brain graphs). Besides, it exclusively superresolves unimodal brain graphs where both LR and HR brain graphs are derived from a single modality such as resting state functional MRI (rsfMRI). Therefore, both landmark works could not overcome the second limitation faced in ML-based frameworks either. In this context, several works have demonstrated the impact of domain alignment in boosting the medical image segmentation and reconstruction [20,23] which incited researchers in GDL to propose frameworks for aligning graphs [15,16]. However, neither the image-based works nor the graph-based ones can be generalized to brain graph super-resolution task. Specifically, to superresolve brain connectomes, one needs to align the LR brain graphs (e.g., morphological) to the HR brain graphs (e.g., functional). However, such an *inter-domain alignment* is strikingly lacking in the super-resolution brain graph synthesis task.

To solve the above challenges, we propose a 'Learn to SuperResolve Brain Graphs with Knowledge Distillation Network' (L2S-KDnet) architecture, an affordable artificial intelligence (AI) solution for the brain graph scarcity problem. Essentially, L2S-KDnet is the first inter-domain alignment learning framework designed for brain graph super-resolution rooted in a teacher-student (TS) paradigm [9]. The goal of a TS framework is to effectively train a simpler network called student, by transferring the knowledge of a pretrained complex network called teacher. It was originally proposed to reduce the computation cost of existing cumbersome deep models encompassing a large number of parameters. Therefore, a TS framework is deployable in real-time applications.

Motivated by such a design, our L2S-KDnet framework consists of two graph encoder-decoder models. The first one is a *domain-aware teacher* which is adversarially regularized using a discriminator, and the second one is a *topology-aware student* network which seeks to preserve the topological properties of the aligned LR to HR graphs and the ground truth graphs. Mainly, we aim to enforce the synthesized HR brain graphs to not only retain the global topological scale of a graph encoded in its connectivity structure but also the local topology of a graph encoded in the hubness of each brain region (i.e., node). Moreover, we propose to leverage the pretrained decoder of the teacher to further boost the learning representation of the student network thereby ensuring that the HR synthesis of the student preserves the topological properties learned by the teacher network.

Table 1. Major mathematical notations used in this paper

Notation	Dimension	Definition
n_s	$\mathbb{N}$	Number of training subjects
n_l	$\mathbb{N}$	Number of brain regions (i.e. ROIs) in LR brain graph
n_h	$\mathbb{N}$	Number of brain regions (i.e. ROIs) in HR brain graph
n'_f	$\mathbb{N}$	Number of features extracted from the HR brain graph
n_f	$\mathbb{N}$	Number of features extracted from the LR brain graph
n_z	$\mathbb{N}$	Number of features of the embedded graphs
X_l	$\mathbb{N}^{n_l \times n_l}$	Symmetric connectivity matrix representing the LR brain graph
X_h	$\mathbb{N}^{n_h \times n_h}$	Symmetric connectivity matrix representing the HR brain graph
$Z_{l \to h}^T$	$\mathbb{N}^{n_s \times n_z}$	Embedding of the aligned LR to HR brain graphs learned by the teacher's encoder
Z_l^S	$\mathbb{N}^{n_s \times n_z}$	Embedding of the LR brain graphs learned by the student's encoder
F_l	$\mathbb{N}^{n_s \times n_f}$	Feature matrix stacking n_s feature vectors extracted from LR brain graphs
F_h	$\mathbb{N}^{n_s \times n'_f}$	Feature matrix stacking n_s feature vectors extracted from HR brain graphs
$\hat{F}_h^T$	$\mathbb{R}^{n_s \times n'_f}$	HR feature matrix predicted by the teacher network given the teacher's embedding
$\hat{F}_h^{T'}$	$\mathbb{R}^{n_s \times n'_f}$	HR feature matrix predicted by the teacher network given the student's embedding
$\hat{F}_h^S$	$\mathbb{R}^{n_s \times n'_f}$	HR feature matrix predicted by the student network
D_{align}	–	Discriminator used for inter-domain alignment taking as inputs the real HR distribution extracted from F_l matrix and the embedded LR brain graphs learned by teacher encoder E^T

2 Methodology

Problem Definition. A brain graph is a fully connected, weighted and undirected graph that can be represented as $\mathcal{G} = \{\mathbf{V}, \mathbf{E}, \mathbf{X}\}$ where $\mathbf{V}$ is set of vertices denoting the brain regions, $\mathbf{E}$ is set of connectivity strengths existing between each pair of vertices and $\mathbf{X}_r$ denotes a connectivity matrix with a specific resolution r and capturing either functional, structural or morphological relationship between two ROIs. Given a LR connectivity matrix $\mathbf{X}_l \in \mathbb{R}^{n_l \times n_l}$ derived from a particular modality (e.g., structural connectivity) we aim to predict a HR $\mathbf{X}_h \in \mathbb{R}^{n_h \times n_h}$ where $n_h > n_l$ that represents another modality (e.g., functional connectivity). We refer the readers to Table 1 which summarizes the major mathematical notations we used in this paper.

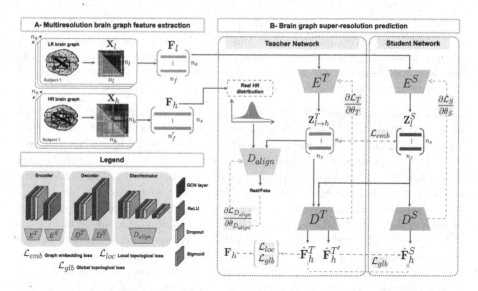

Fig. 1. *Proposed pipeline for predicting high-resolution brain graph from a low-resolution one using teacher-student paradigm.* (**A**) **Multiresolution brain graph feature extraction.** Each subject is represented by a low-resolution (LR) connectivity matrix $\mathbf{X}_l \in \mathbb{R}^{n_l \times n_l}$ and a high-resolution (HR) connectivity matrix $\mathbf{X}_h \in \mathbb{R}^{n_h \times n_h}$ where n_l and n_h represent the resolution of each graph (i.e., number of ROIs) and $n_h > n_l$. Knowing that each brain graph is encoded in a symmetric matrix, we propose to vectorize the upper-triangular part and stack the resulted feature vectors in a matrix $\mathbf{F}_r$ for each brain resolution r. (**B**) **Brain graph super-resolution prediction.** Firstly, we train a teacher network to align the LR domain to the SR domain in the low-dimensional latent space and we regularize it by a discriminator D_{align}. To decode the aligned embeddings, we train the teacher network with a *topology-aware adversarial loss* that so that we enforce the teacher decoder D^T to capture both global and local node properties of the real SR graphs. Secondly, we train a student network that aims to accurately predict the SR brain graphs given an LR graph with the knowledge transferred by the teacher. We regularize it with a *topology-aware distillation loss* and we further use the decoder of the teacher to enforce the decoder of the student to give the same prediction results as the teacher network.

A-Multiresolution Brain Graph Feature Extraction. Since each brain graph of resolution r is conventionally encoded in a symmetric connectivity matrix, we propose to reduce the redundancy of edge weights by firstly vectorizing the off-diagonal upper-triangular part using following equation: $n = \frac{r \times (r-1)}{2}$ where n is the dimension of the feature vector and r is the number of nodes (i.e., r can represent n_l or n_h). Second, we stack the resulting feature vectors of all subjects into a feature matrix $\mathbf{F}_r$. In that way, we get a reduced representation of the connectivity matrices denoting the connectivity strength between pairs of ROIs. Hence, we create two feature matrices representing our graph population with n_s subjects: $\mathbf{F}_l \in \mathbb{R}^{n_s \times n_f}$ and $\mathbf{F}_h \in \mathbb{R}^{n_s \times n'_f}$ representing the LR and HR brain graph domains, respectively (Fig. 1-A). In other words, we represent the whole population by a single subject-based graph where nodes denote the extracted feature vectors and edges linking pairwise nodes are not weighted. Since our goal is to reduce the redundancy of the weights in the connectivity matrices, we choose to not focus on edges relating nodes thereby defining the adjacency matrix of the graph population as an identity matrix.

B-Brain Graph Super-Resolution Prediction. To map the LR brain graph of a subject to the HR graph, we need to first solve the problem of domain fracture resulting in the difference in distribution between both LR and HR domains. Therefore, we propose a TS scheme where we first train a *domain-aware* teacher network aiming to reduce the discrepancy between both domains by learning an *inter-domain alignment* in the low-dimensional space, second, we train a *topology-aware* student network that seeks to preserve the topological properties of both ground truth HR graphs and the aligned brain graph embeddings using the knowledge distilled by the trained teacher. We design our teacher network as a graph GAN composed of a generator designed as a graph encoder-decoder and a single discriminator. On the other hand, we design our student network as a graph encoder-decoder regularized using three losses which we will explain in detail in the following section. Specifically, each graph encoder-decoder network is composed of an encoder that learns the LR brain graph embeddings, and a decoder that predicts the HR brain graphs. Specifically, we use GCN layers [12] for building all components of our pipeline.

- **Teacher network.** To align the LR brain graphs to the HR graphs, our teacher network firstly maps the LR graphs into a low-dimensional space using the encoder $E^T(\mathbf{F}_l, \mathbf{A})$. To eliminate redundancy of edge weights, we set the second input to the GCN layer which is the adjacency matrix to an identity matrix $\mathbf{I}$. We build both encoder and decoder with the same sequence of layers, we stack two GCN layers followed by Rectified Linear Unit (ReLU) and dropout function. We define the graph convolution operation proposed by [12] and employed by the GCN layers is defined as follows:

$$\mathbf{Z}^{(l)} = f_{ReLU}(\mathbf{F}^{(l)}, \mathbf{A}|\mathbf{W}^{(l)}) = ReLU(\widetilde{\mathbf{D}}^{-\frac{1}{2}}\widetilde{\mathbf{A}}\widetilde{\mathbf{D}}^{-\frac{1}{2}}\mathbf{F}^{(l)}\mathbf{W}^{(l)}) \qquad (1)$$

$\mathbf{Z}^{(l)}$ is the learned graph embedding resulting from the layer l which will be considered as the aligned LR to HR embeddings when training our discriminator. In the first layer of our encoder $E^T(\mathbf{F}_l, \mathbf{I})$ we define $\mathbf{F}^{(l)}$ as the feature matrix of the LR domain $\mathbf{F}_l$ while we define it in the second layer as the learned embeddings $\mathbf{Z}$. We denote by $\mathbf{W}^{(l)}$ the learnable parameter for each layer l. We define the graph convolution function as $f_{(.)}$ where $\widetilde{\mathbf{A}} = \mathbf{A} + \mathbf{I}$ with $\mathbf{I}$ being an identity matrix, and $\widetilde{\mathbf{D}}_{ii} = \sum_j \widetilde{\mathbf{A}}_{ij}$ is a diagonal matrix (Fig. 1-B). Since our goal is to enforce the encoder E^T to learn an *inter-domain* alignment, we design a discriminator D_{align} to bridge the gap between the LR brain graph embedding distribution and $\mathbf{Z}$ the original HR brain graph distribution. We optimize it using the following loss function:

$$\max_{D_{align}} \mathbb{E}_{\mathbf{F}'_h \sim \mathbb{P}_{\mathbf{F}_h}} [\log D_{align}(\mathbf{F}'_h)] + \mathbb{E}_{\mathbf{Z}_{l \to h} \sim \mathbb{P}_{\mathbf{Z}}} [\log(1 - D_{align}(\mathbf{Z}_{l \to h}))] \quad (2)$$

where $\mathbb{P}_{\mathbf{F}_h}$ is the real HR graph distribution, and the distribution $\mathbb{P}_{\mathbf{Z}}$ is the generated distribution by our encoder E^T representing the learned aligned LR to HR. On the other hand, our decoder $D^T(\mathbf{Z}^{(l)}, \mathbf{I})$ which has a reversed structure of the encoder takes the learned embedding as input and superresolves it to get the predicted feature matrix of the HR domain $\hat{\mathbf{F}}_h^T$. Particularly, both first and second GCN layers play a scaling role since they map the initial dimension of the graph embedding domain into the HR domain. In another word, our decoder can superresolve the brain graphs given the aligned LR brain graph embedding to the HR ground truth data distribution.

Owing to prior works [1,4] showing the unique topological properties that have a brain connectome such as hubness and modularity, it is important to preserve such properties when superresolving the brain graphs. To achieve this goal, we proposed to regularize the graph encoder-decoder using a novel *topology loss function* $\mathcal{L}_{top}$ that is composed by both local and global topology losses. Specifically, the global topology loss aims to learn the *global graph structure* such as the number of vertices and edges. So, we define it as the mean absolute difference (MAE) between the predicted and ground truth brain features. Moreover, our local topology loss aims to preserve the *local graph structure* which reflects the node importance in the graph. Hence, we define it as the absolute difference in the centrality metrics of nodes. Specifically, we consider the node strength distance between the ground truth and the predicted connectivity matrix $\mathbf{X}_h$ and $\hat{\mathbf{X}}_h$, respectively. We generate the predicted HR matrix by antivectorizing the feature matrix $\hat{\mathbf{F}}_h^T$ generated by the decoder D^T. Thus, we compute the node strength vector for each subject by summing and normalizing over the rows of the corresponding brain graph. Then, we stack them vertically to obtain $\mathbf{S}_h$ and $\hat{\mathbf{S}}_h$ by summing and normalizing over the rows of $\mathbf{X}_s^{m,r_k}$ and $\hat{\mathbf{X}}_s^{m,r_k}$, respectively. Ultimately, we define the *topology loss function* $\mathcal{L}_{top}$ to train the teacher network as follows:

$$\mathcal{L}_{top} = \sigma \underbrace{\ell_{MAE}(\mathbf{S}_h^T, \hat{\mathbf{S}}_h^T)}_{\text{local topology loss}} + \underbrace{\ell_{MAE}(\mathbf{F}_h^T, \hat{\mathbf{F}}_h^T)}_{\text{global topology loss}} \quad (3)$$

Note that our graph encoder-decoder acts as a generator in a GAN, thus this boils down to computing the adversarial loss function as follows:

$$\mathcal{L}_{adv} = \min_{D_{align}} \mathbb{E}_{\mathbf{Z}_{l \to h} \sim \mathbb{P}_{\mathbf{Z}}} [\log(1 - D_{align}(\mathbf{Z}_{l \to h}))] \quad (4)$$

Given the above definitions of the topology loss Eq. 3, the adversarial Eq. 4 loss in addition to a hyper-parameter λ that controls the relative importance of the topology loss, we introduce the overall *domain-aware adversarial loss* function of our teacher's generator as follows:

$$\mathcal{L}_T = \mathcal{L}_{adv} + \lambda \mathcal{L}_{top} \quad (5)$$

In that way, we unprecedentedly enforce our teacher network to superresolve brain graphs while jointly solving two critical problems [2]: the *inter-domain alignment* and the *topological property preservation* of brain graphs (Fig. 1-B).

- **Student network.** Recall that the ultimate goal of this work is to predict a HR brain graph from a LR graph each derived from a specific modality (e.g., functional and structural). However, there exists an obvious domain gap between the LR and HR domains, which raises the question of whether we can have a model performing an *inter-domain alignment* in that way we ensure that the superresolved graph distribution matches the distribution of the ground truth HR graph. To approach this, we propose a TS learning-based framework where the student network takes advantage of a pretrained *domain-aware* teacher network's knowledge. In that way, our efficient and effective lightweight student model can be easily deployed in real-world clinical applications [9]. Therefore, we propose to freeze the teacher network to distill its knowledge that will be used to train the student model. Knowing that our teacher is trained using the topology loss function, we assume that the knowledge transferred to the student already encompasses the information related to the local topological properties of the HR brain graphs. Thus, we propose to solely train the student to preserve the global structure of the teacher's predicted graphs $\hat{\mathbf{F}}_h^T$. Additionally, since the teacher and the student take the same input data that is the LR feature matrix $\mathbf{F}_l$ and perform a graph embedding using their GCN encoders, we make the assumption that their latent space should be similar. To ensure that, we further train the student network with additional information of the teacher. More specifically, we propose to regularize it with two global topology-based losses $\mathcal{L}_{emb}$ and $\mathcal{L}_{glb}$ related to the aligned brain graph embeddings and the predicted HR brain graphs, respectively. We define both loss functions optimizing the student network as follows:

$$\mathcal{L}_{emb} = \ell_{MAE}(\mathbf{Z}_{l \to h}^T, \mathbf{Z}_l^S); \quad \mathcal{L}_{glb} = \ell_{MAE}(\hat{\mathbf{F}}_h^S, \hat{\mathbf{F}}_h^T) \tag{6}$$

where $\mathbf{Z}_{l \to h}^T$ and $\mathbf{Z}_l^S$ denotes the aligned LR to HR graph embeddings learned by the teacher's encoder E^T and the learned LR graph embeddings learned by the student's encoder E^S. $\hat{\mathbf{F}}_h^S$ and $\hat{\mathbf{F}}_h^T$ are the predicted HR brain feature matrices of the student and teacher, respectively. By doing so, we enforce the student to preserve both global and local topological properties of not only the synthesized HR graphs from the teacher but also those of the aligned brain graphs.

Although we want the embedding of the teacher and student to be similar as well as their predicted graphs, still their decoders are not the same. There is no communication between the prediction result between the teacher and student. Therefore, training our student network only with the two losses defined in Eq. 6 means that we are bringing the lower representations of both teacher and student networks closer to each other independently of the decoder outputs of both networks. However, our goal is to obtain the same HR graph prediction from both networks during the testing. Therefore, inspired by a recent work [22], we propose to bridge the gap between the teacher and the student embeddings in such a way this increases the HR brain graph prediction result of the student. Hence, we propose to leverage the pretrained decoder D^T of the trained teacher network to further enhance the student loss function. We pass the learned graph embeddings of the student $\mathbf{Z}_i^S$ to the decoder of the teacher D^T which will generate the HR predicted graphs $\hat{\mathbf{F}}_h^{T'}$. Next, we compute an additional global topology loss between the prediction of the student and teacher decoders and we express our full *topology-aware distillation loss* function as follows:

$$\mathcal{L}_S = mean[\lambda_1 \ell_{MAE}(\hat{\mathbf{F}}_h^S, \hat{\mathbf{F}}_h^{T'}) + \lambda_2 \mathcal{L}_{glb} + \lambda_3 \mathcal{L}_{emb}] \tag{7}$$

3 Results and Discussion

Connectomic Dataset and Parameter Setting. We evaluated our framework on a connectomic dataset derived from the Southwest University Longitudinal Imaging Multimodal (SLIM) Brain Data Repository [17]. It consists of 276 subjects each represented by a morphological LR brain graph of size 35×35 derived from T1-weighted MRIs and a functional HR brain graph of size 160×160 derived from resting-state functional MRI. Our encoder comprises two hidden layers that contains 100 and 50 neurons, respectively. We construct discriminator with 3 hidden layers and each contains 32, 16, 1 neurons. We train our model with 150 iterations using 0.0001 as learning rate and $\beta_1 = 0.5$, $\beta_2 = 0.999$ as Adam optimizer parameters for both networks. We used grid search to set our hyperparameters empirically of $\mathcal{L}_T$ and $\mathcal{L}_S$ and we set σ to 0.1, λ to 0.5, λ_1, λ_2 and λ_3 are all set to 1.

Comparison Methods and Evaluation. Using a single Tesla V100 GPU (NVIDIA GeForce GTX TITAN with 32 GB memory), we evaluate our framework using a 3-fold cross-validation strategy where we train the model on two folds and test it on the left fold in an iterative manner. Since there is no framework aiming to learn an inter-domain alignment while superresolving brain graphs, we evaluated our proposed model L2S-KDnet with three ablated versions. All comparison methods are TS learning-based frameworks:

1. **Baseline:** we use a TS architecture where we do not include any adversarial regularization.
2. **Baseline+Discriminator:** a variant of the first method where we add a discriminator for distinguishing between the ground truth and predicted HR brain graphs.
3. **L2S-KDnet w/o Local Topology:** it is an ablated version of our L2S-KDnet framework which adopts only a global topology loss function.
4. **L2S-KDnet w/o TD regularization:** in this method, we remove the teacher's decoder from the training process of the student network.

Figure 2-(I) shows the MAE between the ground truth and the predicted HR brain graphs and the MAE between the ground truth and synthesized centrality scores. In addition to the node strength measure, we further evaluate our framework using the eigenvector (EC) [3] and PageRank (PC) [5] centralities measures since they are commonly used for graph analysis [4]. We plot the average results over three folds of the student network as it is the model deployed in

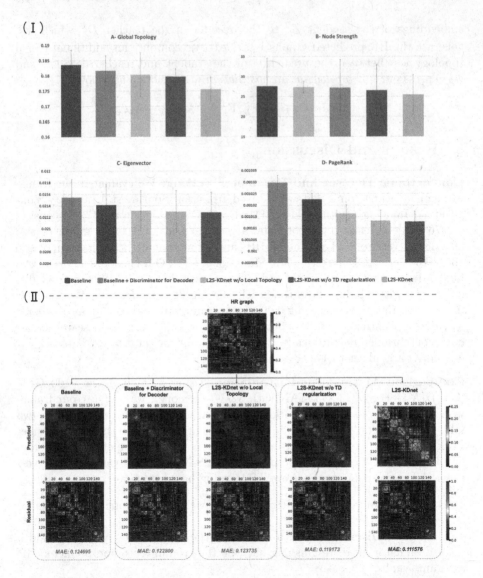

Fig. 2. *Comparison of student model of our L2S-KDnet framework against different ablated methods.* In **(I)** we plot the Mean Absolute Error (MAE) between ground truth and predicted HR brain graphs and the MAE of the node strength, eigenvector and PageRank centrality scores of the predicted and original graphs. In **(II)** we display the residual error computed using the mean absolute error (MAE) metric between the original HR brain graph and the predicted graph of a representative testing subject.

the testing stage. Our L2S-KDnet framework achieved the best prediction performance across all benchmark methods when using global topology and node strength evaluation measures (Fig. 2-A, Fig. 2-B). Interestingly, it outperformed the L2S-KDnet w/o Local Topology method which demonstrates the importance of our *topology-aware adversarial loss* for training the teacher thereby distilling both global and local topological knowledge of the HR brain graphs to train the student network.

While L2S-KDnet ranked first best when evaluated on global topology and node strength, it ranked second-best in MAE between the eigenvector and PageRank centralities as displayed in Fig. 2-C, Fig. 2-D. This shows that the incorporation of the teacher's decoder in the regularization of the student does not improve a lot the preservation of the local topological properties –in particular the eigenvector and PageRank centralities. This is explainable since the teacher's decoder acts as a generator in a GAN architecture, therefore it is exposed to the mode collapse issue [8]. In other words, the decoder of the teacher might produce graphs that mimic a few modes of the original HR brain graphs, thus it solely captures a limited variability of the local topological properties. Designing a solution for circumventing the mode collapse issue of our teacher network is our future avenue. However, we argue that although the TD regularization method did not boost our model performance in preserving complementary local topological properties, our model still outperformed the three remaining competing methods and achieved the best performance when preserving the local topology depicted by the node strength measure. In Fig. 2-(II), we display the residual prediction error computed using MAE between the ground truth and predicted HR brain graphs for a representative subject. It shows that our framework yields a low residual in comparison to its ablated versions. More specifically, such results demonstrate the breadth of our inter-domain alignment-based teacher network which clearly boosts the network superresolution accuracy.

4   Conclusion

We proposed L2S-KDnet the first teacher-student framework designed for brain graph super-resolution based on inter-domain alignment learning. Our key contributions consist in: (i) designing a domain-aware teacher network based on a graph encoder-decoder able to predict a high-resolution brain graph from a low-resolution one, (ii) introducing a *topology-aware adversarial loss function* using a node strength measure to enforce the teacher to preserve topological properties of the ground truth HR graphs, and (iii) proposing a novel way to train the student network which leverages the pretrained teacher decoder to boost the graph super-resolution result. Experiments on a healthy connectomic dataset showed that our proposed L2S-KDnet is able to achieve a low prediction error compared to its variants. In our future work, we plan to generalize our framework to a multiresolution brain graph synthesis task where given a single LR graph we can generate multiple HR graphs at different resolutions. Another interesting future direction is to leverage our framework to synthesize HR brain graphs for unhealthy subjects and then leverage the predicted graphs for boosting early disease diagnosis from low-resolution connectomes.

Acknowledgements. This work was funded by generous grants from the European H2020 Marie Sklodowska-Curie action (grant no. 101003403, http://basira-lab.com/

normnets/) to I.R. and the Scientific and Technological Research Council of Turkey to I.R. under the TUBITAK 2232 Fellowship for Outstanding Researchers (no. 118C288, http://basira-lab.com/reprime/). However, all scientific contributions made in this project are owned and approved solely by the authors. A.B is supported by the same TUBITAK 2232 Fellowship.

References

1. Bassett, D.S., Sporns, O.: Network neuroscience. Nat. Neurosci. **20**(3), 353–364 (2017)
2. Bessadok, A., Mahjoub, M.A., Rekik, I.: Graph neural networks in network neuroscience. arXiv preprint arXiv:2106.03535 (2021)
3. Bonacich, P.: Some unique properties of eigenvector centrality. Soc. Netw. **29**(4), 555–564 (2007)
4. Borgatti, S.P., Everett, M.G.: A graph-theoretic perspective on centrality. Soc. Netw. **28**(4), 466–484 (2006)
5. Brin, S.: The PageRank citation ranking: bringing order to the web. Proc. ASIS **1998**(98), 161–172 (1998)
6. Cengiz, K., Rekik, I.: Predicting high-resolution brain networks using hierarchically embedded and aligned multi-resolution neighborhoods. In: Rekik, I., Adeli, E., Park, S.H. (eds.) PRIME 2019. LNCS, vol. 11843, pp. 115–124. Springer, Cham (2019). https://doi.org/10.1007/978-3-030-32281-6_12
7. Chen, Y., Shi, F., Christodoulou, A.G., Xie, Y., Zhou, Z., Li, D.: Efficient and accurate MRI super-resolution using a generative adversarial network and 3d multi-level densely connected network. In: Frangi, A.F., Schnabel, J.A., Davatzikos, C., Alberola-López, C., Fichtinger, G. (eds.) MICCAI 2018. LNCS, vol. 11070, pp. 91–99. Springer, Cham (2018). https://doi.org/10.1007/978-3-030-00928-1_11
8. Goodfellow, I., et al.: Generative adversarial nets. In: Advances in Neural Information Processing Systems, pp. 2672–2680 (2014)
9. Hinton, G., Vinyals, O., Dean, J.: Distilling the knowledge in a neural network. arXiv preprint arXiv:1503.02531 (2015)
10. Isallari, M., Rekik, I.: GSR-Net: graph super-resolution network for predicting high-resolution from low-resolution functional brain connectomes. In: Liu, M., Yan, P., Lian, C., Cao, X. (eds.) Machine Learning in Medical Imaging, pp. 139–149. Springer, Cham (2020). https://doi.org/10.1007/978-3-030-32692-0
11. Isallari, M., Rekik, I.: Brain graph super-resolution using adversarial graph neural network with application to functional brain connectivity. Med. Image Anal. **71**, 102084 (2021). https://doi.org/10.1016/j.media.2021.102084. https://www.sciencedirect.com/science/article/pii/S1361841521001304
12. Kipf, T.N., Welling, M.: Semi-supervised classification with graph convolutional networks. arXiv preprint arXiv:1609.02907 (2016)
13. Li, W., Andreasen, N.C., Nopoulos, P., Magnotta, V.A.: Automated parcellation of the brain surface generated from magnetic resonance images. Front. Neuroinform. **7**, 23 (2013)
14. Mhiri, I., Khalifa, A.B., Mahjoub, M.A., Rekik, I.: Brain graph super-resolution for boosting neurological disorder diagnosis using unsupervised multi-topology connectional brain template learning. Med. Image Anal. **65**, 101768 (2020). https://doi.org/10.1016/j.media.2020.101768. https://www.sciencedirect.com/science/article/pii/S1361841520301328

15. Pilanci, M., Vural, E.: Domain adaptation on graphs by learning aligned graph bases. IEEE Trans. Knowl. Data Eng. (2020)
16. Pilavci, Y.Y., Guneyi, E.T., Cengiz, C., Vural, E.: Graph domain adaptation with localized graph signal representations. arXiv preprint arXiv:1911.02883 (2019)
17. Qiu, J., Qinglin, Z., Bi, T., Wu, G., Wei, D., Yang, W.: Southwest university longitudinal imaging multimodal (SLIM) brain data repository: a long-term test-retest sample of young healthy adults in Southwest China
18. Qu, L., Zhang, Y., Wang, S., Yap, P.T., Shen, D.: Synthesized 7T MRI from 3T MRI via deep learning in spatial and wavelet domains. Med. Image Anal. **62**, 101663 (2020)
19. Sánchez, I., Vilaplana, V.: Brain MRI super-resolution using 3d generative adversarial networks. arXiv preprint arXiv:1812.11440 (2018)
20. Shen, Y., Gao, M.: Brain tumor segmentation on MRI with missing modalities. In: Chung, A.C.S., Gee, J.C., Yushkevich, P.A., Bao, S. (eds.) IPMI 2019. LNCS, vol. 11492, pp. 417–428. Springer, Cham (2019). https://doi.org/10.1007/978-3-030-20351-1_32
21. Soussia, M., Rekik, I.: 7 years of developing seed techniques for alzheimer's disease diagnosis using brain image and connectivity data largely bypassed prediction for prognosis. In: Rekik, I., Adeli, E., Park, S.H. (eds.) PRIME 2019. LNCS, vol. 11843, pp. 81–93. Springer, Cham (2019). https://doi.org/10.1007/978-3-030-32281-6_9
22. Yang, J., Martinez, B., Bulat, A., Tzimiropoulos, G.: Knowledge distillation via softmax regression representation learning. In: International Conference on Learning Representations (2021). https://openreview.net/forum?id=ZzwDy_wiWv
23. Zhou, B., Lin, X., Eck, B.: Limited angle tomography reconstruction: synthetic reconstruction via unsupervised sinogram adaptation. In: Chung, A.C.S., Gee, J.C., Yushkevich, P.A., Bao, S. (eds.) IPMI 2019. LNCS, vol. 11492, pp. 141–152. Springer, Cham (2019). https://doi.org/10.1007/978-3-030-20351-1_11

Sickle Cell Disease Severity Prediction from Percoll Gradient Images Using Graph Convolutional Networks

Ario Sadafi[1,2], Asya Makhro[3], Leonid Livshits[3], Nassir Navab[2,4], Anna Bogdanova[3], Shadi Albarqouni[2,5], and Carsten Marr[1(✉)]

[1] Institute of Computational Biology, Helmholtz Zentrum München - German Research Center for Environmental Health, Neuherberg, Germany
carsten.marr@helmholtz-muenchen.de
[2] Computer Aided Medical Procedures, Technical University of Munich, Munich, Germany
[3] Red Blood Cell Research Group, Institute of Veterinary Physiology, Vetsuisse Faculty and the Zurich Center for Integrative Human Physiology, University of Zurich, Zurich, Switzerland
[4] Computer Aided Medical Procedures, Johns Hopkins University, Baltimore, USA
[5] Helmholtz AI, Helmholtz Zentrum München - German Research Center for Environmental Health, Neuherberg, Germany

Abstract. Sickle cell disease (SCD) is a severe genetic hemoglobin disorder that results in premature destruction of red blood cells. Assessment of the severity of the disease is a challenging task in clinical routine, since the causes of broad variance in SCD manifestation despite the common genetic cause remain unclear. Identification of biomarkers that would predict the severity grade is of importance for prognosis and assessment of responsiveness of patients to therapy. Detection of the changes in red blood cell (RBC) density by means of separation of Percoll density gradients could be such a marker as it allows to resolve intercellular differences and follow the most damaged dense cells prone to destruction and vasoocclusion. Quantification and interpretation of the images obtained from the distribution of RBCs in Percoll gradients is an important prerequisite for establishment of this approach. Here, we propose a novel approach combining a graph convolutional network, a convolutional neural network, fast Fourier transform, and recursive feature elimination to predict the severity of SCD directly from a Percoll image. Two important but expensive laboratory blood test parameters are used for training the graph convolutional network. To make the model independent from such tests during prediction, these two parameters are estimated by a neural network from the Percoll image directly. On a cohort of 216 subjects, we achieve a prediction performance that is only slightly below an approach where the groundtruth laboratory measurements are used. Our proposed method is the first computational approach for the difficult task of SCD severity prediction. The two-step approach relies solely on inexpensive and simple blood analysis tools and can have a significant impact on the patients' survival in low resource regions where access to medical instruments and doctors is limited.

S. Albarqouni et al. (Eds.): DART 2021/FAIR 2021, LNCS 12968, pp. 216–225, 2021.
https://doi.org/10.1007/978-3-030-87722-4_20

Keywords: Graph convolutional networks · Percoll gradients · Severity prediction · Sickle cell disease

1   Introduction

Sickle cell disease (SCD) is a disorder caused by mutations in position 6 of β-hemoglobin gene (hemoglobin S) with extreme variability at phenotypic level. In some patients the disease manifestation is so mild that they remain asymptomatic most of the time while others die before the age of five from several of the severe complications associated to the SCD [11]. Individuals who have the hemoglobin S variant are naturally protected against malaria, which has a profound influence on the spread of sickle cell disease globally affecting the tropical (African and Asian) countries the most. Many of these countries are not able to support diagnosis and appropriate healthcare for this group of patients leading to a drop in the life expectancy from 45–55 years in high income countries to 90% death rate before the age of 5 in low income countries [12].

Severity monitoring and prediction of the SCD is therefore an important task along with development of new effective and inexpensive therapeutic strategies. Changes in severity allow monitoring of the treatment efficiently and for prediction and prevention of life-threatening complications in short future. To date, there is no practical test based on red blood cells (RBCs) density separation analysis available for prediction of the severity of the disease for a patient.

An important parameter for disease severity assessment is the percentage of hypo- and hyperchromic cells. RBCs with a hemoglobin concentration above $410\,g/l$ are called hyperchromic and characterized by low cellular deformability [4] and increased probability of aggregation of the hemoglobin S which directly associates with advanced severity and poor prognosis for the SCD patients [2,3]. In contrary, Hypochromic RBCs with low hemoglobin content are associated with a lower probability of hemoglobin S aggregation and sickling and thus with mild disease manifestation. Measurement of these parameters using blood smears is laborious and time-consuming, when done manually by skilled personnel, or rather requires expensive medical laboratory equipment, when automated.

The spleen plays an important role in clearing the blood from old, broken, dehydrated or hyperchromic red blood cells (RBCs). A normal and functioning spleen reduces the intravascular hemolysis of damaged cells (where cells rupture in the blood vessels) and prevents vaso-occlusive crisis (where terminally dense sickle cells block circulation of blood vessel leading to painful crisis) and vascular damage in SCD patients [1]. However, fibrosis and progressive atrophy of the spleen resulting finally in necrosis of the organ, known as autosplenectomy, which is often observed in SCD patients with severe disease phenotype. It is a known problem in children with SCD due to repeated splenic vaso-occlusive events in the organ [1]. Measuring spleen size with ultrasound is a common way to evaluate the organ's condition in SCD.

We here propose a computational approach that circumvents expensive lab tests and relies solely on the measurement of spleen size and a Percoll image.

Percoll images are used to assess the density of the cells and particles. After centrifugation, several bands with different thicknesses are formed by RBCs of similar density (see Fig. 1) holding important information about a SCD patient's condition. Back in 1984, Fabry et al. [5] observed a decrease in the dense fraction of Percoll images in SCD patients suffering from painful crisis in 11 patients over 14 painful crisis image. This information can also be computationally analyzed: Sadafi et al. [10] introduced a hybrid approach based on CNNs and features extracted from fast Fourier transform to classify a variety of hereditary hemolytic anemias using Percoll image data.

To predict the severity of the SCD patients, we are proposing an approach based on graph convolutional networks (GCN) to form a population graph [8] on our data. The similarity of the GCN edges is calculated using lab (Percentage of hypo- and hyperchromic RBCs) and clinical data (spleen size). The spleen size is measured using ultrasound. We propose a CNN based approach to have an easy to access and affordable anywhere in the world way to estimate required lab data from the Percoll image.

2   Methodology

Our proposed method, SCD-severity-GCN aims at predicting SCD severity from cheap and easy accessible patient data and consists of the following steps: (i) The abundance of hypo- and hyperchromic cells in the blood sample are predicted based on a Percoll image; (ii) Relevant features are extracted from the Percoll image using a CNN and fast Fourier transform (FFT). (iii) A similarity metric between Percolls based on a patient's spleen size and the predicted abundance of hypo- and hyperchromic cells is calculated to form a population graph. Using GCNs the SCD severity is predicted (Fig. 1).

2.1   Model

Our goal is to have a model f that takes a Percoll image P_i and the spleen size J_i of a patient sample i to return a severity grade S_i:

$$S_i = f(P_i, J_i; \theta) \tag{1}$$

where θ are the model parameters that are learned by training on the dataset.

2.2   Feature Extraction

For primary feature extraction the approach proposed in [10] is employed. There, the extraction of Fourier features from the images has been demonstrated to enhance disease classification performance on Percoll images. Accordingly, we extract features with an AlexNet [7] architecture and combine them with features from FFT (see Fig. 1). We obtain pretrained weights of the model $f_{\text{cnn-fft}}$ and use the activations preceding the final classification layer as features for our GCN approach.

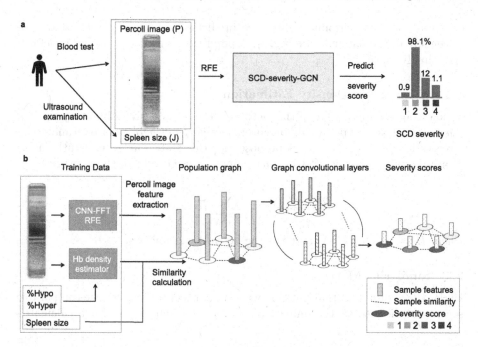

Fig. 1. Overview of the proposed SCD-severity-GCN approach. a) A Percoll image from a conventional blood test and the spleen size obtained by ultrasound examination are passed to our trained SCD-severity-GCN to predict a SCD severity score for the patient. b) The SCD-severity-GCN is trained in the following way: Features from the Percoll image are extracted with a convolutional neural network and a fast Fourier transform (CNN-FFT [10]). Another independently trained CNN estimates the hemoglobin (Hb) density of hypochromic (Hypo) and hyperchromic (Hyper) cells. Together with the spleen size of the patient, a similarity measure is calculated between the nodes of a population graph. After two layers of graph convolutions, a severity score for every sample is predicted.

To reduce feature dimensions, we used recursive feature elimination [6] and a Ridge classifier as suggested by Parisot et al. [8].

$$x_i = \mathrm{RFE}(f_{\mathrm{cnn-fft}})(P_i) \tag{2}$$

where x_i is the feature vector extracted for the Percoll image P_i. Also in our approach this step improved the convergence of the training significantly.

2.3   Graph Convolution Network

One of the most intuitive ways of representing populations and their similarities is through graphs. In our approach, every Percoll image P_i is represented by a vertex $v \in \mathcal{V}$ and the similarity between the Percoll images is modelled by weighted edges $\mathcal{E}$ calculated from the expensive laboratory data (the percentages of hypo- and hyperchromic) which are predicted and cheap clinical data (i.e. spleen size of the patient) (see Fig. 2). A population graph $\mathcal{G} = \{\mathcal{V}, \mathcal{E}\}$ is defined accordingly [8].

2.4 Hemoglobin Density Estimation

To allow for an application of the method without expensive laboratory testing, the percentages of hypo- and hyperchromic cells $\hat{H}$ in the blood are estimated by a regression. A CNN f_{chrome} is proposed for this task. The groundtruth values H are provided for every Percoll image and are used to train the network:

$$\mathcal{L}_{chrome}(\gamma) = \frac{1}{N} \sum_{i=1}^{N} (H_i - \hat{H}_i)^2 \tag{3}$$

where $\hat{H}_i = f_{chrome}(P_i; \gamma)$ and γ is the network parameters.

2.5 Similarity Metric

Under the assumption that patients with similar features experience comparable severity of the disease, the similarity between two samples v and w is calculated via

$$\mathcal{E}(v_i, v_j) = e^{-(||\hat{H}_{v_i} - \hat{H}_{v_j}|| + \lambda[J_{v_i} == J_{v_j}])} \tag{4}$$

where $\hat{H}$ is the vector of estimated percentages of hypo- and hyperchromic cells and J is the spleen size, as above. Iverson brackets yield 1 in case of equality and 0 otherwise. Note that spleen sizes are given as discrete numbers in centimeters (see Fig. 2), obtained in the clinic with a conventional ultrasound device. The coefficient λ is set to weight the importance of spleen and lab measurements.

3 Experiments

3.1 Dataset

Our dataset consists of the 216 samples with Percoll images and laboratory data (% hypo, % hyper) and clinical data (spleen size) obtained from 17 patients diagnosed with SCD, who participated in a clinical trial (NCT03247218) conducted in Emek Medical Center in Afula[1]. The study has been conducted in accordance with local ethics committee guidelines and the Declaration of Helsinki. Blood samples were acquired during pre-planned monthly visits according to the trial protocol. For every visit the patient's health was evaluated using blood analysis, including RBC characteristics and measurement of hemolytic and inflammatory markers, urine analysis and blood pressure measurements. Severity of a patient's condition at each measurement point was estimated using the scoring approach proposed by Sebastiani et al. [11] with minor modifications on disease severity score calculation. Figure 2 shows distribution of severity scores and example samples from the dataset.

---

[1] https://clinicaltrials.gov/ct2/show/NCT03247218.

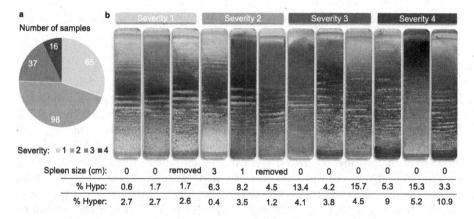

Fig. 2. Dataset overview. a) Pie chart shows distribution of severity values for the patients in the dataset. b) Example images sorted according to their severity scores. Corresponding clinical and laboratory test data for each Percoll is also demonstrated. Spleen size of the patients with autosplenectomy and splenectomy is indicated with removed and 0 respectively.

3.2 Implementation Details

Hemoglobin Density Estimation: A CNN with seven convolutional layers with ReLU activation function and max-pooling is used. After global average pooling and two fully connected layers the output is regulated with a final ReLU. Two dropout layers with a drop rate of 0.5 are used for regularization.

Feature Extraction: The output size from CNN-FFT is 1024, which is reduced with recursive feature elimination (RFE) [6] to 50 features. These features are used as the final feature vector for each Percoll image.

Graph Convolutional Network: A population graph [8] is created based on the defined feature vectors and similarities. We use two hidden layers in the graph and 50 filters in the hidden layers. The dropout rate is set to 0.2. For similarity calculation λ is set to 10.

Training: Both training procedures are carried out on a 10-fold cross validation dataset. The model f_{chrome} estimating hemoglobin density is trained with AMS-Grad variation of Adam optimizer for 100 epochs and a learning rate of 0.0005. The graph convolutional network is trained for 300 epochs using Adam optimizer and a learning rate of 0.01. We use the Tensorflow framework for implementation and training.

Evaluation Metrics: We are reporting root mean square error (RMSE) for the regression task of hemoglobin density estimation. Accuracy, weighted F1-score and area under ROC are reported for the severity grading as well as the area under precision recall curve for every class. Scikit-learn [9] implementation is used for calculation all of the metrics.

Baseline: A linear SVM [9] trained on the feature vectors is used as a baseline for our grading approach.

3.3   Results

The dataset is divided into 10 stratified folds for patient-wise cross validation. All of the models are independently ran on each combination of these folds. Mean and standard deviation is reported for all of the 10 experiments.

First, the values predicted by the Hb density estimation model f_{chrome} based on Percoll images are compared against the actual lab tests. The root mean square error (RMSE) of the percentage for hypochromic cells is 6.5 ± 4.0 and for hyperchromic cells 0.90 ± 0.12. Considering the ranges of the hypo and hyper values, which are $[0.6, 37.5]$ and $[0.2, 10.9]$, respectively, we consider the estimation sufficiently good.

Next, we compare our SCD-severity-GCN approach with the following methods: (i) A linear SVM trained on the x_i features vectors extracted from the Percoll image (SVM), (ii) a linear SVM trained on x_i feature vectors and the cheap clinical ultrasound and newly proposed and time consuming groundtruth lab information (SVM - Lab), (iii) a GCN based on randomized laboratory information (GCN - Rand), and (iv) a GCN using not the estimated, but the actual laboratory information (GCN - Lab) as the upper limit. Table 1 shows that our SCD-severity-GCN approach using estimated Hb densities is close to the GCN that required hard to obtain lab data (GCN - Lab) in terms of accuracy, weighted F1-score and area under ROC. Since the dataset is unbalanced, we are reporting the area under precision recall curve in Fig. 3 for every class and different approaches.

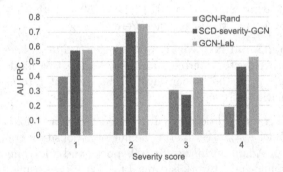

Fig. 3. Area under precision recall curve for different methods per class.

3.4   Ablation Study

GCNs are sensitive to the formulation of the graph adjacency matrix based on the pairwise similarity that is defined between the nodes. Choosing parameters

Table 1. Our proposed SCD-severity-GCN method based on GCN outperforms other approaches. Prediction of hypo- and hyperchromic cells percentage slightly affects the performance while providing independence from expensive cell counters.

	Accuracy	F1 - score	AU ROC
SVM	0.44 ± 0.07	0.28 ± 0.02	–
SVM - Lab	0.39 ± 0.14	0.29 ± 0.08	–
GCN - Rand	0.53 ± 0.05	0.42 ± 0.08	0.53 ± 0.20
SCD-severity-GCN	0.61 ± 0.13	0.53 ± 0.17	0.61 ± 0.25
GCN - Lab	0.65 ± 0.15	0.59 ± 0.19	0.67 ± 0.24

that are biologically significant and easy to obtain is crucial. To evaluate the importance of the different clinical (spleen size) and laboratory (% of hypo- and hyperchromic cells) information used for the formation of our GCN, we designed an ablation study and compare GCNs trained with different combinations of these parameters. As Table 2 shows, the combination based on spleen size and percentages of hypo- and hyperchromic RBCs yields the best result.

Table 2. Combination of different clinical and laboratory measurements results in a slightly different shape for the GCN and thus different performance.

GCN similarity parameters	Accuracy	F1 - score	AU ROC
Spleen	0.45 ± 0.01	0.28 ± 0.01	0.48 ± 0.18
Spleen & Hypo	0.63 ± 0.11	0.56 ± 0.14	0.63 ± 0.23
Spleen & Hyper	0.62 ± 0.16	0.54 ± 0.21	0.60 ± 0.25
Hypo & Hyper	0.55 ± 0.08	0.46 ± 0.12	0.55 ± 0.22
Spleen & Hypo & Hyper	**0.65 ± 0.15**	**0.59 ± 0.19**	**0.67 ± 0.24**

3.5 Discussion

Severity prediction of SCD is a challenging task normally preformed with several clinical and laboratory tests. Here we propose a novel potential severity prediction approach based on RBC density separation (as provided by Percoll gradients) analysis that may amend the currently existing ones. Information obtained solely from Percoll images is not be sufficient for an acceptable classification (see Table 1), even though those features sufficed for successful diagnosis of different anemias [10]. By combining Percoll derived features with complementary clinical and laboratory data and training a GCN with this information, we can achieve an accuracy that is surprisingly high for this challenging clinical task. This is illustrated by the UMAP embedding of feature vectors (Fig. 4a), and GCN outputs with estimated (Fig. 4b) and groundtruth lab information

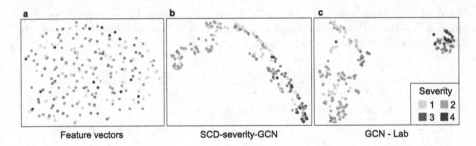

Fig. 4. UMAP embedding of the (a) processed feature vectors, (b) the GCN with predicted ones and (c) the GCN with groundtruth lab information. Clear separation of samples with different severity scores by the proposed GCN is evident.

(Fig. 4c). Samples from different severity classes are nicely disentangled in the UMAP thanks to the GCN approach we utilized. Although clustering using the groundtruth lab information (GCN - Lab) is a lot better, a smooth transition from low to high severity is already evident in the approach that uses estimated Hb density only (SCD-severity-GCN).

4   Conclusion

Sickle cell disease severity prediction is an important task that allows to prevent life-threatening complications, reduce morbidity and mortality and refine the choice of optimal therapeutic strategies [11]. Offering affordable and versatile solutions for improving life quality of the SCD patients is a necessity specially in low resource areas of the planet. Here, we proposed the first computational method requiring only the Percoll gradient image and spleen size obtained from a conventional ultrasound. Analysis of Percoll gradient images with CNNs nicely predicted percentages of hypo- and hyperchromic cells and the proposed GCN predicted SCD severity score with a surprisingly high accuracy. Our approach uses a unique combination of methods, with a GCN at its heart.

Results look very promising and provide a solid ground for future work. Next we will analyze more patients, especially more severe ones as well as pediatric datasets. Our SCD-severity-GCN based on Percoll images requires much smaller volumes of blood compared to common hematological tests (1 ml or less instead of 7–10 ml), which is particularly relevant for kids and patients suffering from severe anemia.

Acknowledgments. Special thanks to Prof. Ariel Koren and Dr. Carina Levin from the Emek Medical Center in Afula who made this work possible. This project has received funding from the European Union's Horizon 2020 research and innovation programme under grant agreement No. 675115—RELEVANCE—H2020-MSCA-ITN-2015/H2020-MSCA-ITN-2015. The work of L.L. was funded by UZH Foundation. C.M. and A.S. have received funding from the European Research Council (ERC) under the European Union's Horizon 2020 research and innovation programme (Grant agreement No. 866411).

References

1. Brousse, V., Buffet, P., Rees, D.: The spleen and sickle cell disease: the sick (led) spleen. Br. J. Haematol. **166**(2), 165–176 (2014)
2. Brugnara, C., Mohandas, N.: Red cell indices in classification and treatment of anemias: from MM Wintrobes's original 1934 classification to the third millennium. Curr. Opin. Hematol. **20**(3), 222–230 (2013)
3. De Franceschi, L.: Pathophisiology of sickle cell disease and new drugs for the treatment. Mediterr. J. Hematol. Infect. Diseases **1**(1) (2009)
4. Deuel, J.W., Lutz, H.U., Misselwitz, B., Goede, J.S.: Asymptomatic elevation of the hyperchromic red blood cell subpopulation is associated with decreased red cell deformability. Ann. Hematol. **91**(9), 1427–1434 (2012)
5. Fabry, M.E., Benjamin, L., Lawrence, C., Nagel, R.L.: An objective sign in painful crisis in sickle cell anemia: the concomitant reduction of high density red cells. Blood **64**(2), 559–563 (1984)
6. Guyon, I., Weston, J., Barnhill, S., Vapnik, V.: Gene selection for cancer classification using support vector machines. Mach. Learn. **46**(1), 389–422 (2002)
7. Krizhevsky, A., Sutskever, I., Hinton, G.E.: ImageNet classification with deep convolutional neural networks. Adv. Neural. Inf. Process. Syst. **25**, 1097–1105 (2012)
8. Parisot, S., et al.: Spectral graph convolutions for population-based disease prediction. In: Descoteaux, M., Maier-Hein, L., Franz, A., Jannin, P., Collins, D.L., Duchesne, S. (eds.) MICCAI 2017. LNCS, vol. 10435, pp. 177–185. Springer, Cham (2017). https://doi.org/10.1007/978-3-319-66179-7_21
9. Pedregosa, F.: Scikit-learn: machine learning in Python. J. Mach. Learn. Res. **12**, 2825–2830 (2011)
10. Sadafi, A., et al.: Fourier transform of percoll gradients boosts CNN classification of hereditary hemolytic anemias. In: 2021 IEEE International Symposium on Biomedical Imaging (ISBI). IEEE (2021)
11. Sebastiani, P., et al.: A network model to predict the risk of death in sickle cell disease. Blood **110**(7), 2727–2735 (2007)
12. Wastnedge, E., et al.: The global burden of sickle cell disease in children under five years of age: a systematic review and meta-analysis. J. Global Health **8**(2), 021103 (2018)

Continual Domain Incremental Learning for Chest X-Ray Classification in Low-Resource Clinical Settings

Shikhar Srivastava[1]([✉])[iD], Mohammad Yaqub[1][iD], Karthik Nandakumar[1][iD], Zongyuan Ge[2,4][iD], and Dwarikanath Mahapatra[3]

[1] Mohamed Bin Zayed University of Artificial Intelligence, Abu Dhabi, UAE
{shikhar.srivastava,mohammad.yaqub,karthik.nandakumar}@mbzuai.ac.ae
[2] Monash University, Melbourne, Australia
zongyuan.ge@monash.edu
[3] Inception Institute of Artificial Intelligence, Abu Dhabi, UAE
dwarikanath.mahapatra@inceptioniai.org
[4] Airdoc Research, Melbourne, Australia

Abstract. Within clinical practise, a shift in the distribution of data collected over time is commonly observed. This occurs generally due to deliberate changes in acquisition hardware but also through natural, unforeseen shifts in the hardware's physical properties like scanner SNR and gradient non-linearities. These domain shifts thus may not be known a-priori, but may cause significant degradation in the diagnostic performance of machine learning models. A deployed diagnostic system must therefore be robust to such unpredictable and continuous domain shifts. However, given the infrastructure and resource constraints pervasive in clinical settings in developing countries, such robustness must be achieved under finite memory and data privacy constraints. In this work, we propose a domain-incremental learning approach that leverages vector quantization to efficiently store and replay hidden representations under limited memory constraints. Our proposed approach validated on well known large-scale public Chest X-ray datasets achieves significant reduction in catastrophic forgetting over previous approaches in medical imaging, while requiring no prior knowledge of domain shift boundaries and a constrained memory. Finally, we also formulate a more natural continual learning setting for medical imaging using a tapered uniform distribution schedule with gradual interleaved domain shifts.

Keywords: Continuous learning · Domain adaptation · Catastrophic forgetting · Chest X-ray classification

1 Introduction

Modern machine learning approaches are highly sensitive to changes in data distribution and suffer from catastrophic forgetting [11]. This refers to the

© Springer Nature Switzerland AG 2021
S. Albarqouni et al. (Eds.): DART 2021/FAIR 2021, LNCS 12968, pp. 226–238, 2021.
https://doi.org/10.1007/978-3-030-87722-4_21

performance degradation of diagnostic models when adapted to shifting data. New data is learnt at the cost of forgetting previously seen data. In real world clinical practice, a shift in the distribution of data collected over time is common, and occurs largely due to continuously changing data acquisition technology [1,10]. Consequently, any deployed diagnostic model may undergo severe degradation in predictive performance over time as the data distribution shifts and is no longer i.i.d, eventually making the model obsolete. Retraining these diagnostic models from scratch however as this degradation occurs is unfortunately not a feasible solution in medical imaging. The significant costs incurred in obtaining expert annotations, practical constraints of low infrastructure and computing resources, and the regulatory constraints surrounding sharing of patient's private data [24] make it unfeasible. Therefore, diagnostic models that can adapt incrementally and mitigate catastrophic forgetting under resource and privacy constraints are inherently valuable in any deployed setting. This is especially true for resource-constrained clinical settings in developing countries. Infrastructure constraints may preclude a complete retraining, and constraints such as communication bandwidth limits may make bulky continuous data transmission unreliable [4]. Combined with the regulatory constraints on data transmission to the cloud [24,28], both model parameters and stored samples may therefore need to reside efficiently on-device in many scenarios. Additionally, in applications such as point-of-care-testing, models must be able to adapt efficiently on deployed mobile devices with finite memory [22]. Another challenge in mitigating catastrophic forgetting in practical settings, is that the domain shift boundaries may not be known in advance. Observed domain shifts may be of a continuous and even unpredictable nature. Manual modifications to the acquisition technology (such as changes in protocols, policies, scanner hardware, settings and diagnostic procedures) are phased out gradually, and natural time-varying properties of the acquisition source (like scanner SNR shifts, gradient non-linearities) can cause *continuous* unforeseen changes to the image characteristics [10,20]. Thus it is desirable to develop approaches with minimal resource requirements that make diagnostic systems robust against these *continuous* domain shifts.

In broader machine learning, this problem is tackled through formulations such as domain adaptation, continual learning and streaming learning. In these settings, approaches are generally aimed at mitigating catastrophic forgetting by scaling [17,26] or regularizing [13,27] the parameters available to the model, or by replaying stored [16,25] or synthetically generated [29] samples from previously observed domains. Recently, there has been some work in medical imaging towards leveraging continual learning approaches to mitigate catastrophic forgetting [1,10,14,24,30]. However, work on approaches that operate under realistic clinical settings have been limited. Existing works have largely been proposed and validated on singular [1,10,30] or short [24] uni-dimensional domain shifts, typically by splitting datasets into smaller chunks which are treated as separate domains, and even by considering synthetic datasets [9]. Most works in incremental domain adaptation also consider a simpler binary-classification task [1,9,24]. Finally, the popular continual learning approaches like LwF [15] and EWC [12] employed in medical imaging require a-priori knowledge of domain shift boundaries which may not be practically available. In this work, we:

- Formulate a more natural continual learning setting for medical imaging using a tapered uniform distribution schedule, that contains continuous domain shifts with interleaved boundaries. Thus, approaches are evaluated in this more realistic setting where no clear domain-shift boundaries exist and no a-priori knowledge of domain shift boundaries is possible.
- Propose a *domain incremental learning* approach for multi-label classification of Chest X-ray images which mitigates catastrophic forgetting under low memory constraints. We leverage vector quantization to efficiently store and replay hidden representations.
- Validate our approach on a sequence of three well-known large-scale Chest X-ray datasets providing rich domain shifts. We show significant reduction in catastrophic forgetting over previous approaches, while requiring no prior knowledge of task boundaries and a finite memory.

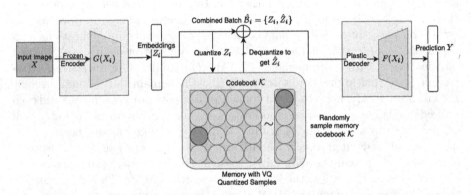

Fig. 1. Proposed method workflow: During continuous training, the encoder generates intermediate features Z that are quantized into the codebook $\mathcal{K}$. Samples from this codebook are then sampled, decoded and added to the training batch $\hat{B}$ for a single forward and backward pass.

2 Related Work

Continual Learning in Medical Imaging. A popular approach used in medical imaging to mitigate catastrophic forgetting [1,9] is Elastic Weight Consolidation (EWC) [12]. EWC uses the fisher information matrix to calculate parameter importances for a given domain. These parameters are then regularized to prevent catastrophic forgetting. LwF [15] is another commonly used approach in medical imaging literature. LwF leverages distillation to regularize the current loss with soft targets taken from a previous version of the model. In recent works, Baweja et al. [1] proposed the application of Elastic Weight Consolidation (EWC) [12] to sequentially learn brain segmentation tasks and mitigate catastrophic forgetting. The work was focused on the class-incremental learning with shifts in the label space, however the approach was evaluated only on a single pair of domains. Karani et al. proposed a novel use of domain-specific batch-normalization for lifelong learning on brain MR segmentation [10]. Entire batch-normalization layers were dedicated to modelling domain difference. However, this required scaling parameters linearly with new tasks.

Another class of approaches replay previously observed samples either by storing or generating them, termed as rehearsal and psuedo-rehearsal respectively [16,25,29]. Within medical imaging analysis, Ravishankar et al. in their recent work propose a pseudo-rehearsal technique and training of task-specific dense layers for pneumothorax classification [24], where activations of network are stored and transformed to mitigate catastrophic forgetting. The technique presumes a-priori knowledge of domain boundaries, linear scaling of parameters with every domain and an unbounded memory to store activations. Hofmanninger et al. [9] present a rehearsal approach where raw samples are retained in-memory based on their degree of uniqueness inferred using the Gram matrix of their activations. They also introduce the idea of continuous training, however their study focuses on in-house, synthetic data with generated domain shifts. Unlike previous rehearsal and psuedo-rehearsal based approaches in medical imaging, we assume a fixed number of parameters, retain no raw samples and assume no a-priori knowledge of domain shift boundaries. We operate in a more natural clinical setting formulated by a proposed sequence of continuous tapered uniform distributions with no clear domain boundaries, and validate on a sequence of well-known large-scale chest X-ray datasets with rich domain differences. Finally, our approach follows recent work in leveraging vector quantization [3,8] to efficiently store and replay hidden representations utilizing an order of magnitude less memory.

3   Method

Continuous Learning Setting. Within the continuous learning setting for our Chest X-ray Classification task, the objective is for the diagnostic system $\mathcal{M}$ to learn to perform multi-label classification on an unending stream $\mathcal{T}$ of images X and labels $Y : \mathcal{T} = \{(X_1, Y_1), (X_2, Y_2), ..., (X_N, Y_N), ...\}$ where any image, label pair X_i, Y_i may be sampled from an infinite set of input distributions or domains $\mathcal{D}$ i.e. $(X_i, Y_i) \sim \mathcal{D}$ where $\mathcal{D} = \{D_1, D_2, ..., D_N, ..\}$. The stream $\mathcal{T}$ observes a continuous shift in the distribution of data it samples from. Generally, the diagnostic model $\mathcal{M}$ is assumed to be trained on a given initial base domain (Fig. 2) and is then continuously adapted on the stream $\mathcal{T}$ in mini-batches of size $\mathcal{B}$ where it must reliably predict $P(Y|X)$. This simulates a typical clinical setting where a diagnostic model trained on an initial domain is deployed to a clinical environment where domain shifts may be observed during its lifetime.

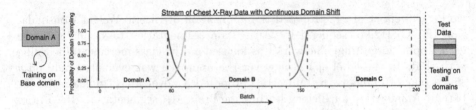

Fig. 2. The continual stream is approximated as a sequence of tapered uniform distributions, where domain shift boundaries are interleaved and not known a-priori. The multi-label classification model is trained initially on Chest X-ray data-domain A, then adapted continuously on domains A to C, and is finally evaluated on the test sets of all domains.

In our work, we incorporate more realistic constraints by formulating the continual stream T such that no distinct boundary exists between domains. In fact samples from consecutive domains may even be interleaved. We enforce this setting by generating the continuous learning stream as a sequence of *tapered uniform distributions* (see Fig. 2). (This is also referred to as a Planck-taper window in scientific literature [19]). Thus, in a stream T where:

$$T = ...(X_{i-2}, Y_{i-2}), (X_{i-1}, Y_{i-1}), (X_i, Y_i), (X_{i+1}, Y_{i+1}), ...$$

it may be possible for samples from consecutive domains to be interleaved i.e. $(X_{i-2}, Y_{i-2}), (X_i, Y_i) \sim D_1$ and $(X_{i-1}, Y_{i-1}), (X_{i+1}, Y_{i+1}) \sim D_2$.

3.1 Proposed Approach

Our proposed approach is a continuous domain-incremental approach that performs multi-label classification of Chest X-ray pathologies by leveraging vector quantization [7] to efficiently replay the network's intermediate representations. The central idea is to efficiently maps the network's intermediate representations to a quantized vector space which could be reused in order to provide robustness against catastrophic forgetting. We assume a highly constrained memory (<5% of sample activations are compressed and stored), fixed number of model parameters and no a-priori knowledge of domain shift boundaries.

Our Chest X-ray multi-label classification model $\mathcal{M}_{\theta,\phi}$ can be visualized as a composite of two convolutional functions where $\mathcal{G}_\phi(X)$ consists of the first L layers of the network NET (referred to as the base model), and $\mathcal{F}_\theta$ consisting of the remaining layers of the network.

$$\mathcal{M} = \mathcal{F}_\theta(\mathcal{G}_\phi(X)); \text{ where } \mathcal{G}_\phi = \text{NET}[: L], \mathcal{F}_\theta = \text{NET}[L :] \tag{1}$$

Our method generates a learnt codebook $\mathcal{K}$ to encode and mitigate the drift in the learnt intermediate representations Z of the convolutional layers. The activations Z from the learner's model are encoded and stored in the codebook

---

Algorithm 1: Vector Quantization for Efficient Continual Chest X-ray Classification

---

1 Train $\mathcal{M} = \mathcal{F}_\theta(\mathcal{G}_\phi(X))$ on training data of base domain
2 Fit Vector Quantization (VQ) model on base domain
3 Freeze $\mathcal{G}_\phi(X)$; Begin continuous adaptation on continual learning stream:
   repeat
4 $X_i, Y_i \longleftarrow$ Fetch Input mini-batch
5 $Z_i \longleftarrow \mathcal{G}(X_i)$; Quantize Z_i using VQ Model and write to buffer
6 $\hat{Z}_i \longleftarrow$ Randomly Sample from Buffer and Decode using VQ Model
7 $\hat{\mathcal{B}}_i \longleftarrow \{Z_i, \hat{Z}_i\}$ with respective labels forms replay-modified batch
8   Generate prediction $P(Y|X) \longleftarrow \mathcal{F}(\hat{\mathcal{B}}_i)$; Calculate cross-entropy loss and perform backward pass
9 **until** *end of continual stream*;

$\mathcal{K}$ and rehearsed along with the continuous training data. Algorithm 1 specifies the workflow of our proposed approach.

Sampling and Writing to the Quantized Memory. The memory is randomly sampled at every input mini-batch X, to generate a mixed batch $\hat{\mathcal{B}}_i$, that is used to do a single forward and backward pass of the learner $\mathcal{F}(\mathcal{G}(X))$. Additionally, during adaptation in the continual stream, samples from every mini-batch are quantized and written to memory. Once the memory size is exceeded, the samples corresponding to the most frequently observed labels are removed first. We found that a random sampling of the written samples in memory works sufficiently with minimal added complexity compared to more sophisticated memory population approaches.

Quantization of Intermediate Representations. We use a vector quantization (VQ) model to quantize the intermediate representations Z_i into a discrete codebook $\mathcal{K}$. The encoder $\mathcal{G}$ encodes input samples into Z_i of shape $H \times W \times L$ where H, W and L are the height, width and channels of the latent representation. In the quantization step, each of the $H \times W$ vectors of L dimensions in Z_i, are partitioned into S sub-vectors of D dimensions (such that $D = L/S$). A vector quantization model is then trained on the S sub-vectors independently to learn codebooks $\mathcal{K} \in \mathcal{R}^{C \times D}$ of size C. Corresponding to their respective codebooks, each of the S sub-vectors are assigned an index, which is finally stored in the memory buffer as an $H \times W \times S$ array.

Mitigating Class Imbalance in Backward Pass: A consequence of the non-iid nature of the continual process is the class imbalance that causes certain pathologies to be over-represented within the learning process. In order to mitigate against this class imbalance in our continual multi-label chest-xray classification scenario, we leverage the protocol in [5], where each label is weighted in the cross-entropy loss according to it's global frequency in the data stream observed so far.

4   Experiments and Results

We conduct our experiments on three of the largest public medical imaging datasets - NIH Chest-X-rays14 [18], PadChest [2], CheXpert [31], with over 150k unique chest X-rays (NIH: 30k images, Padchest: 62k images, CheXpert: 64k images) across various pathologies. In order to replicate domain shifts observed on a large-scale real world clinical settings, we generate a continuous interleaved stream using these datasets, as illustrated in Sect. 3 and Fig. 2.

Datasets. The datasets Chest X-ray14 [18] termed here as 'NIH', PadChest [2] termed as 'PC' and CheXpert [31], vary across a richness of dimensions that affect their visual characteristics and data distribution: including their collection process, acquisition hardware, labelling method, patient population, error/noise ratios, distribution of pathologies among others [2,18,31]. The interested reader is directed to the respective references of the datasets for a full description.

We consider a multi-label prediction task for our diagnostic model by considering the 7 common pathologies among the datasets: {'Cardiomegaly', 'Pneumonia', 'Pneumothorax', 'Consolidation', 'Edema', 'Atelectasis', 'Effusion'}. We follow standard protocols [5,23] for these datasets. All images have been standardized with center cropping to 224×224 size and intensity normalization to within $[-1024, 1024]$. We filter the datasets to include only the 7 common labels, a single posteroanterior (PA) chest view per patient, and remove images with no findings, as per protocol. Our generated continual stream thus has the distribution: {Total Samples: 29904, CheXpert: 9069, NIH: 9730 and Padchest: 11000 images}.

Baselines. We benchmark our method against the most popular continual learning approaches in medical imaging - Elastic Weight Consolidation (EWC) [12] and LwF [15] that has served as a standard benchmark in medical imaging. We additionally consider the Fine-Tuning (FT) and Joint-Training (JT) approaches as reference baselines.

Fine-tuning (FT) & Joint-Training (JT) serve as baseline approaches to indicate the expected lower and upper bound performances respectively. In the Fine-tuning (FT) approach, the diagnostic system is naively fine-tuned continuously without any strategy to account for catastrophic forgetting. In Joint-Training (JT), the diagnostic model is trained jointly on all domains.

Elastic Weight Consolidation (EWC) [12]: In the absence of domain shift boundaries, EWC is known to suffer degradation in performance due to the relative invariance of the batch normalization layers over a representational shift [9]. We therefore follow protocol in [9] to adapt the approach to learn the weight importances for the base domain and then freeze the Batch Normalization (BN) layers. We refer to this modification of EWC or EWC (fBN) interchangeably during the rest of the sections. For the EWC baselines, we experimentally select the regularization hyperparameter $\lambda = 1e6$ after performing a hyperparameter search in the range of $\lambda \in \{1e+1, 1e+2, 1e+3, ..1e+6\}$.

LwF (Learning without Forgetting) [15] - In the absence of domain shift boundaries, the approach is adapted to enforce distillation losses to the network trained on the base dataset. We select the hyperparameters temperature $T = 50.0$ and distillation loss weight $\alpha = 100$ after performing a grid search on the joint space for both $T, \alpha \in \{1e-2, 1e-1, .., 1e+3, 1e+4\}$.

Our Approach. We implement our proposed method as described in Sect. 3.1. The codebook $\mathcal{K}$ of the Vector Quantization (VQ) model is fit on the base domain, and contains $S = 128$ codebooks with size $C = 512$. We experimentally select $L = 4$ for our network such that $\mathcal{G}_\phi = \text{NET}[: 4], \mathcal{F}_\theta = \text{NET}[4 :]$. After training on the base domain, the first four layers of the ResNet101 network are frozen and training proceeds as detailed in Sect. 3.1. We report performance for buffer size of 1000 samples (or 3% of total samples in stream).

Evaluation Metrics. As illustrated in Fig. 2, approaches are validated at regular intervals during the continuous stream, and later tested offline. We calculate multi-label classification performance as the average AUC (Area under ROC curve) metric. While training on the continuous stream, validation on hold-out sets (for each domain) is carried out every 8 mini-batches and the domain specific AUCs are reported. The continuous validation performance is indicative of the reliability of the models against domain shifts during their lifetime.

For the test performances, we adapt standard continual learning metrics [16] to our multi-label Chest X-ray classification task. For a given set of domains T,

$$\text{Backward Transfer (BWT)} = \frac{1}{T-1} \sum_{i=1}^{T-1} R_{T,i} - R_{i,i},$$

$$\text{Forward Transfer (FWT)} = \frac{1}{T-1} \sum_{i=2}^{T} R_{i-1,i} - \hat{b}_i,$$

where $R \in \mathcal{R}^{T \times T}$ is a matrix where each element R_{ij} is the classification AUC of the diagnostic model tested on domain t_j after learning domain t_i. $\hat{b}_i$ is the AUC of the model on domain t_i initially before continual training. The average BWT and FWT metrics quantify the change in a model's performance on past and future domains after training on given domain. Therefore, a negative BWT is indicative of the degree of catastrophic forgetting observed. A positive FWT indicates beneficial transfer learning from learnt domains into future domains has occurred.

Experimental Settings. The experiment is conducted using an Nvidia V100 card. We implement our experiments using PyTorch [21] framework and torchxrayvision [6]. For normalization of our findings with broader literature, we use a ResNet architecture for all baselines. We use the standard Pytorch implementation of the ResNet101 model commonly used in the literature for the Chest X-ray classification tasks. A mini-batch size of 128, the binary cross-entropy loss and the Adam optimizer with weight decay of $1e-5$ is fixed for training the baselines across all experiments. For training on the base domain, a initial learning rate of 0.001 is used till model convergence.

4.1   Results

Validation During Continuous Training. Figure 3 presents the continual performance of the baselines through domain-wise validation AUCs while Fig. 4 shows the pathology-specific AUCs averaged across domains. We note that the performance of LwF does not deviate significantly from the FT approach and infact drops below FT while the NIH domain is observed. We attribute this to the inability of the distillation loss (from the base model trained on CheXpert) to be relevant for adaptation to the NIH domain. We note that the EWC (with fixed Batch Normalization layers) performs with high volatility during domain shifts, but recovers performance eventually. This is attributed to the high regularization penalty of EWC on observing domain shifts, that causes a near complete degradation in performance. In contrast, our approach steadily maintains the initial performance after base training, and although a minute drop in performance is observed after the NIH domain shift, it maintains a reliable, clearly superior retention of performance throughout the continual stream.

Quantitative Results on All Datasets After Continuous Training. Performance at the end of continuous stream is shown in Table 1. LwF once again performed no better than the naive fine-tuned approach on the test metrics. The negative BWT values indicate catastrophic forgetting has occurred. EWC and our approach observe a positive BWT on the test performance, indicating mitigation of catastrophic forgetting. Our approach consistently maintains the

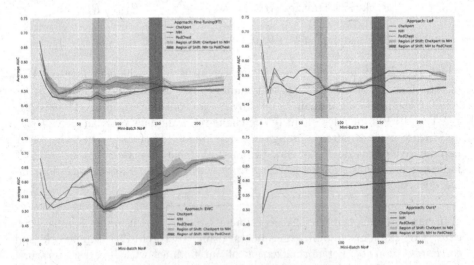

Fig. 3. Validation Performances on the Continual Stream every 8 min-batches. Top Left: Fine-Tuning, Top Right: LwF, Bottom Left: EWC (fBN), Bottom Right: Our Approach. Y-axis indicates the Validation AUC and X-axis indicates the mini-batch number. Shaded lines represent the mean and variance in performance across 3 runs. Shaded bars represent the region of domain shifts.

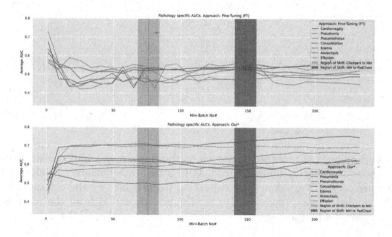

Fig. 4. Pathology-specific Validation Performance on Continual Stream: Fig. shows pathology specific AUCs averaged across the validation sets of all domains. Top: Fine-Tuning approach, Bottom: Ours*.

Table 1. Test performance metrics avg. AUCs, BWT and FWT at the end of the continuous learning stream for three runs on the order CheXpert-NIH-PC.

	CheXpert [31]	Chest-Xrays-8 [18]	Padchest [2]	BWT	FWT
Fine-tuning	0.52 ± 0.00	0.49 ± 0.00	0.52 ± 0.00	-0.003 ± 0.002	-0.102 ± 0.002
LwF [9,12]	0.53 ± 0.01	0.52 ± 0.01	0.61 ± 0.02	-0.016 ± 0.002	-0.097 ± 0.004
EWC [15]	0.62 ± 0.01	0.56 ± 0.00	0.62 ± 0.01	$+0.02 \pm 0.01$	-0.02 ± 0.01
Ours*	0.63 ± 0.01	0.59 ± 0.00	0.67 ± 0.0	$+0.03 \pm 0.01$	$+0.11 \pm 0.00$
Joint training	0.83 ± 0.0	0.81 ± 0.0	0.72 ± 0.0	NA	NA

highest test AUCs, and also exhibits positive forward transfer (FWT) suggesting encountering continual domain shifts infact benefited its final performance.

Memory Buffer Size Requirements. In Fig. 5, a sensitivity analysis of our approach to the memory buffer size is presented. In the left figure, we show test AUC scores across domains for varying buffer sizes in the range of $[0, 3 \times 1e + 4]$ total samples. We observe a minimal change in test performance after a buffer of size 1000 samples or less than 3% of observed samples (in the continual stream of 29k images). Figure 5 (right) shows the continual validation performance w.r.t variations in memory buffer size between [1000, 10000] samples. We note mild sensitivity after the NIH domain shift, which we infer is caused by the same large CheXpert to NIH domain shift expressed across all baselines. Yet our approach shows the greatest robustness to catastrophic forgetting by storing just ∼5% of compressed samples. With the proposed quantization approach, the stored samples are compressed by a ratio of 96:1. Thus, even with the quantization codebooks occupying a finite space, our approach significantly mitigates catastrophic forgetting while having minimal memory requirements.

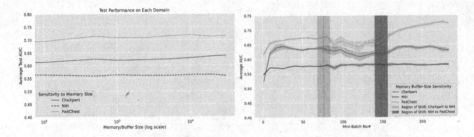

Fig. 5. Performance of proposed approach to varying memory buffer sizes. Figure on the left shows the change in average test AUC across each domain for different buffer sizes. Figure on the right shows the validation performance on the continual stream, with the shaded regions representing the standard deviation from the mean performance represented by the lines.

5   Conclusion

In this work, we furthered a scarcely explored direction of work on mitigating catastrophic forgetting under realistic clinical scenarios with finite resources and the absence of known domain shift boundaries. We formulate a continual learning setting to better model such a scenario. Results show that our proposed approach of leveraging vector quantization to store intermediate representations is effective at mitigating catastrophic forgetting for continual chest X-ray classification under strict memory constraints. We validated our approach on large-scale public medical imaging datasets of chest-xray classification (CheXpert, NIH Chest-xray 8, Padchest) where our method displayed favorable performance over benchmarks, while satisfying all required desiderata. We believe our works offers a viable solution to mitigating the catastrophic effects of domain shift in diverse, low-resource and realistic clinical settings.

References

1. Baweja, C., Glocker, B., Kamnitsas, K.: Towards continual learning in medical imaging. arXiv preprint arXiv:1811.02496 (2018)
2. Bustos, A., Pertusa, A., Salinas, J.M., de la Iglesia-Vayá, M.: PadChest: a large chest x-ray image dataset with multi-label annotated reports. Med. Image Anal. **66**, 101797 (2020)
3. Caccia, L., Belilovsky, E., Caccia, M., Pineau, J.: Online learned continual compression with adaptive quantization modules. In: International Conference on Machine Learning, pp. 1240–1250. PMLR (2020)
4. Chavez, A., Littman-Quinn, R., Ndlovu, K., Kovarik, C.L.: Using TV white space spectrum to practise telemedicine: a promising technology to enhance broadband internet connectivity within healthcare facilities in rural regions of developing countries. J. Telemed. Telecare **22**(4), 260–263 (2016)
5. Cohen, J.P., Hashir, M., Brooks, R., Bertrand, H.: On the limits of cross-domain generalization in automated X-ray prediction. In: Medical Imaging with Deep Learning, pp. 136–155. PMLR (2020)

6. Cohen, J.P., Viviano, J., Morrison, P., Brooks, R., Hashir, M., Bertrand, H.: TorchXRayVision: a library of chest X-ray datasets and models (2020). https://github.com/mlmed/torchxrayvision

7. Gray, R.: Vector quantization. IEEE ASSP Mag. **1**(2), 4–29 (1984). https://doi.org/10.1109/MASSP.1984.1162229

8. Hayes, T.L., Kafle, K., Shrestha, R., Acharya, M., Kanan, C.: REMIND your neural network to prevent catastrophic forgetting. In: Vedaldi, A., Bischof, H., Brox, T., Frahm, J.-M. (eds.) ECCV 2020. LNCS, vol. 12353, pp. 466–483. Springer, Cham (2020). https://doi.org/10.1007/978-3-030-58598-3_28

9. Hofmanninger, J., Perkonigg, M., Brink, J.A., Pianykh, O., Herold, C., Langs, G.: Dynamic memory to alleviate catastrophic forgetting in continuous learning settings. In: Martel, A.L., et al. (eds.) MICCAI 2020. LNCS, vol. 12262, pp. 359–368. Springer, Cham (2020). https://doi.org/10.1007/978-3-030-59713-9_35

10. Karani, N., Chaitanya, K., Baumgartner, C., Konukoglu, E.: A lifelong learning approach to brain MR segmentation across scanners and protocols. In: Frangi, A.F., Schnabel, J.A., Davatzikos, C., Alberola-López, C., Fichtinger, G. (eds.) MICCAI 2018. LNCS, vol. 11070, pp. 476–484. Springer, Cham (2018). https://doi.org/10.1007/978-3-030-00928-1_54

11. Kemker, R., McClure, M., Abitino, A., Hayes, T., Kanan, C.: Measuring catastrophic forgetting in neural networks. In: Proceedings of the AAAI Conference on Artificial Intelligence, vol. 32 (2018)

12. Kirkpatrick, J., et al.: Overcoming catastrophic forgetting in neural networks. Proc. Natl. Acad. Sci. **114**(13), 3521–3526 (2017)

13. Kurle, R., Cseke, B., Klushyn, A., van der Smagt, P., Günnemann, S.: Continual learning with Bayesian neural networks for non-stationary data. In: International Conference on Learning Representations (2019)

14. Lenga, M., Schulz, H., Saalbach, A.: Continual learning for domain adaptation in chest X-ray classification. In: Medical Imaging with Deep Learning, pp. 413–423. PMLR (2020)

15. Li, Z., Hoiem, D.: Learning without forgetting. IEEE Trans. Pattern Anal. Mach. Intell. **40**(12), 2935–2947 (2017)

16. Lopez-Paz, D., Ranzato, M.: Gradient episodic memory for continual learning. In: Advances in Neural Information Processing Systems, pp. 6467–6476 (2017)

17. Mallya, A., Lazebnik, S.: PackNet: adding multiple tasks to a single network by iterative pruning. In: Proceedings of the IEEE Conference on Computer Vision and Pattern Recognition, pp. 7765–7773 (2018)

18. McDermott, M.B., Hsu, T.M.H., Weng, W.H., Ghassemi, M., Szolovits, P.: CheXpert++: approximating the CheXpert labeler for speed, differentiability, and probabilistic output. In: Machine Learning for Healthcare Conference, pp. 913–927. PMLR (2020)

19. McKechan, D., Robinson, C., Sathyaprakash, B.S.: A tapering window for time-domain templates and simulated signals in the detection of gravitational waves from coalescing compact binaries. Class. Quantum Gravity **27**(8), 084020 (2010)

20. Mesri, H.Y., David, S., Viergever, M.A., Leemans, A.: The adverse effect of gradient nonlinearities on diffusion MRI: from voxels to group studies. Neuroimage **205**, 116127 (2020)

21. Paszke, A., et al.: PyTorch: an imperative style, high-performance deep learning library. arXiv preprint arXiv:1912.01703 (2019)

22. Rahman, M.A., Hossain, M.S., Alrajeh, N.A., Gupta, B.: A multimodal, multimedia point-of-care deep learning framework for COVID-19 diagnosis. ACM Trans. Multimedia Comput. Commun. Appl. **17**(1s), 1–24 (2021)

23. Rajpurkar, P., et al.: CheXNet: radiologist-level pneumonia detection on chest X-rays with deep learning. arXiv preprint arXiv:1711.05225 (2017)

24. Ravishankar, H., Venkataramani, R., Anamandra, S., Sudhakar, P., Annangi, P.: Feature transformers: privacy preserving lifelong learners for medical imaging. In: Shen, D., et al. (eds.) MICCAI 2019. LNCS, vol. 11767, pp. 347–355. Springer, Cham (2019). https://doi.org/10.1007/978-3-030-32251-9_38

25. Rebuffi, S.A., Kolesnikov, A., Sperl, G., Lampert, C.H.: iCaRL: incremental classifier and representation learning. In: Proceedings of the IEEE Conference on Computer Vision and Pattern Recognition, pp. 2001–2010 (2017)

26. Rusu, A.A., et al.: Progressive neural networks. arXiv preprint arXiv:1606.04671 (2016)

27. Schwarz, J., et al.: Progress & compress: a scalable framework for continual learning. arXiv preprint arXiv:1805.06370 (2018)

28. Sun, W., Cai, Z., Li, Y., Liu, F., Fang, S., Wang, G.: Security and privacy in the medical Internet of Things: a review. Secur. Commun. Netw. **2018**, 1–9 (2018)

29. van de Ven, G.M., Tolias, A.S.: Generative replay with feedback connections as a general strategy for continual learning. arXiv preprint arXiv:1809.10635 (2018)

30. Venkataramani, R., Ravishankar, H., Anamandra, S.: Towards continuous domain adaptation for medical imaging. In: 2019 IEEE 16th International Symposium on Biomedical Imaging (ISBI 2019), pp. 443–446. IEEE (2019)

31. Wang, X., Peng, Y., Lu, L., Lu, Z., Bagheri, M., Summers, R.M.: ChestX-ray8: hospital-scale chest X-ray database and benchmarks on weakly-supervised classification and localization of common thorax diseases. In: Proceedings of the IEEE Conference on Computer Vision and Pattern Recognition, pp. 2097–2106 (2017)

Deep Learning Based Automatic Detection of Adequately Positioned Mammograms

Vikash Gupta[1]([✉]), Clayton Taylor[2], Sarah Bonnet[3], Luciano M. Prevedello[2],
Jeffrey Hawley[2], Richard D. White[1], Mona G. Flores[4],
and Barbaros Selnur Erdal[1]

[1] Department of Radiology, Mayo Clinic, Jacksonville, FL, USA
gupta.vikash@mayo.edu
[2] Department of Radiology, The Ohio State University, Columbus, OH, USA
[3] Department of Radiology, West Virginia School of Medicine,
Morgantown, WV, USA
[4] Medical AI, NVIDIA Inc., Santa Clara, CA, USA

Abstract. Screening mammograms are a routine imaging exam performed to detect breast cancer in its early stages to reduce morbidity and mortality attributed to this disease. In the United States alone, approximately 40 million mammograms are performed each year. In order to maximize the efficacy of breast cancer screening programs proper mammographic positioning is paramount. Proper positioning ensures adequate visualization of breast tissue and is necessary for breast cancer detection. Therefore, breast imaging radiologists must assess each mammogram for the adequacy of positioning before providing a final interpretation of the exam. In this paper we propose a method that mimics and automates the decision-making process performed by breast imaging radiologists to identify adequately positioned mammograms. Our objective is to improve the quality of mammographic positioning and performance, and ultimately reduce repeat visits for patients whose imaging is technically inadequate. If the method is not able to identify the adequately positioned mammogram, the patient will be scanned again. This AI driven method will be useful in reducing the cost and anxiety associated with mammogram scanning. The method can be very useful in serving the needs in developing countries, where mammogram scanning is not considered a routine procedure due to increased cost to the patient. In addition, the methodology can also be used for training the mammogram techs by providing actionable feedback during scan. The proposed method has a true positive rate of 91.4% for detecting a correctly positioned mediolateral oblique view. For detecting a correctly positioned craniocaudal view, the true positive rate is 95.11%.

Keywords: Mammogram · Mammography positioning · Breast cancer · Deep learning · MQSA

V. Gupta and C. Taylor—Co-First Authors.

© Springer Nature Switzerland AG 2021
S. Albarqouni et al. (Eds.): DART 2021/FAIR 2021, LNCS 12968, pp. 239–250, 2021.
https://doi.org/10.1007/978-3-030-87722-4_22

1 Introduction

Breast cancer is the most common malignancy in women worldwide, accounting for an estimated 25% of all cancers in women, and is the leading cause of cancer-related mortality [9]. Mammography-based screening programs have been shown to result in atleast 30% [11] decrease in breast cancer mortality. Screening programs have been widely adopted; in the United States alone, approximately 40 million mammograms are performed annually [5]. Nevertheless, the efficacy of screening mammography relies on proper technique. Poor breast positioning in mammography can result in undetected malignancy, as well as repeat examinations with added costs [1–3,8,12,13]. It has been found to be the most common technical cause of undetected breast cancers [4,12].

The typical screening mammography workflow is standardized and involves a single imaging encounter with the acquisition of standard mammographic views in two positions, cranial-caudal (CC) and mediolateral-oblique (MLO) for each breast. Once the patient leaves the imaging facility, the images are reviewed offline by the interpreting physician in batches with results subsequently communicated to the patient. The adequacy of breast positioning and technique is assessed by the interpreting physician during the course of reviewing each mammographic image. If the study does not meet the necessary quality standards, the interpreting physician will request that the necessary mammographic images are repeated. This process is inefficient for both the patient and the facility, prevents the interpreting physician from providing immediate constructive feedback to the technologist, and delays interpretation and subsequent patient results; the associated costs are manifested in expenses (e.g. patient return travel, repeat imaging related), time (e.g. patient away from work or family), and emotional toll (e.g. patient anxiety during result delay).

The adequacy of tissue included on the MLO view is assessed by ascertaining that the pectoralis muscle (PEC) extends to or below a line drawn from the nipple perpendicular to the course of the PEC, known as the posterior nipple line (PNL). If the PEC does not extend to the PNL, there is a possibility of breast tissue exclusion from the mammogram. The CC view is assessed by looking for visible fibers of the PEC (visible only in approximately 15% of mammograms), or more often by extending a line from the nipple perpendicular to the chest wall, also referred to as the posterior nipple line (PNL) (Fig. 1). If the distance of the PNL on the CC view is within 1 cm of the PNL of the MLO view, then the CC view is deemed to include enough posterior tissue. While much attention has been placed on deep learning and artificial intelligence automated cancerous lesion detection methods, there has been a minimal emphasis on novel strategies to improve quality and positioning.

Deep learning and artificial intelligence have produced great results in the field of medical imaging and can aid radiologists and clinicians in decision-making processes. Transfer learning has been successfully used to detect bone fracture on radiographs [7,14,16] and coronary stenosis in cardiac computed tomography angiography images [6]. All of these tasks were classification based tasks. However, in the context of the present paper, we use transfer learning for regression.

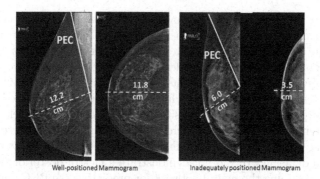

Fig. 1. Relationship between the posterior nipple line (PNL) on mediolateral oblique (MLO) and craniocaudal (CC) views for well-positioned and inadequately positioned mammograms. **Left:** the difference between the PNL lengths in the MLO and CC views should be less than 1 cm. **Right:** if the difference between the PNL lengths in the MLO and CC view is more than 1 cm, this indicates that the CC view may be inadequately positioned leading to the exclusion of breast tissue.

2   Data

This retrospective study was approved by the institutional review board, with a waiver of the requirement to obtain informed consent. Screening mammograms performed between January 2012 and July 2017 at an academic tertiary referral center were searched in our PACS system for keywords indicating issues with positioning. The exclusion criteria, includes age under 30 years; prior history of breast conservation surgery; breast piercings; and breast implants. Using these criteria, a total of 327 patients were collected. Out of which 194 patients were inadequately and 133 patients were adequately positioned. As 194 patients were inadequately positioned it contained multiple acquisitions; it comprised of 508 MLO and 379 CC views. The correctly positioned 133 patients comprises of 266 MLO and CC views.

Our final goal is to evaluate the position of the mammogram. However, in doing so first, we need to evaluate our deep learning algorithm for PEC and PNL predictions. Also, we will use a separate algorithm for detecting the radiopaque marker (BB), which is used for predicting the PNL on the CC view of the image. Only after achieving these tasks we will use the prediction results to evaluate the positioning of mammograms. In the following sections, we will treat each of these problems as a separate problem and report results in each case.

2.1   Data Labeling

The PEC and PNL lines are marked on both MLO and CC views using LabelMe [15]. The resident radiologist involved in the study drew straight lines demarcating the pectoral muscle line on the MLO view. The PNL lines were drawn on both MLO and CC view. In case of MLO, the PNL is approximately perpendicular to the PEC line. However, in the case of the CC view, the PNL line is approximately a horizontal line starting at the BB marker and reaching the end of the image either on the left or on the right, depending on whether the left or the right breast is under consideration. The reviewer only drew these lines on the inadequately positioned patients. Out of the 508 labeled MLO images, the reviewer identified 66 MLO images, which were specifically identified at the time of initial interpretation as poorly positioned only because of a lack of posterior tissue. The remaining 442 images included labeled images that were technically adequate and those that were technically inadequate but exhibited more than one technical deficiency or did not demonstrate a lack of posterior tissue. These were used for training the neural network for predicting the PEC and PNL. The 442 MLOs were divided into training, test, and validation sets, which comprised of 277, 71, and 94 images, respectively. The remaining 66 poorly positioned images were kept for testing the positioning algorithm performance according to Mammography Quality Standards Act (MQSA) standards. We have applied the following transformations to the images: random flips about the x and y axes as well as random rotations about the center of the image. A random flip about the vertical axis creates an invariance in terms of the left or right breast. In order to determine whether the positioning was adequate, both the MLO and the CC views were evaluated as described in the previous section. Our proposed method comprises the following steps, as mentioned in Algorithm 1. We would like to point out again that, the adequacy of MLO view can be determined independently; the adequacy of the CC view is dependent on the adequacy of the MLO view. If we cannot predict the MLO positioning, we cannot comment about the CC positioning.

3   Predicting the PEC and PNL on the MLO View

A line is defined by the Cartesian Coordinates (x, y) of its two end-points. For predicting the PEC and PNL lines, we need to predict these four points. Separate neural networks were trained for each of the lines. Predicting the coordinates is essentially a regression task. For the MLO view, we would like to identify both the line delineating the pectoral muscle from the breast tissue (PEC) as well as the posterior nipple line (PNL) as shown in Fig. 3. For the MLO view, if the PEC and PNL lines intersect within the bounds of the image, we conclude that the MLO view was acquired correctly. The annotations made by the resident radiologist is used as training data.

We chose InceptionV3 [10] as the base network. Inception-V3 is a neural network designed for image classification whose last layer contains 1000 fully connected nodes. The last layer outputs the softmax probability for each of the classes. We replaced the last layer with a single output node. The output node outputs a vector with four elements corresponding to the two ordinates (x and y) of the two output points. We initialize the network with the pre-trained weights from a trained model using the ImageNet dataset. In addition, instead of using the cross-entropy as the cost function, we used the logarithm of hyperbolic cosine (logcosh) as the cost function to be optimized. The cost function is defined as

---

Algorithm 1: Algorithm to compute decision making on mammogram positioning

---

Input: MLO view, CC view of the mammogram
Output: Decision on mammogram positioning
Predict the PEC and PNL lines on the MLO view.
Check,

1. If PEC and PNL lines intersect within the image boundaries, then MLO is correctly positioned and **accept the view**.
2. Else, the MLO is poorly positioned and **reject the view**.

Predict the BB location on the CC view.
Draw the PNL lines on the CC view.
Calculate the length of PNL lines on CC (d_{cc}) and MLO (d_{MLO}) views.
Calculate $d_{diff} = |d_{CC} - d_{MLO}|$.
Check,

1. If $d_{diff} < 1$ cm, **accept the view**.
2. Else, the CC is poorly positioned and **reject the view**.

Exam complete.

$$L(y^{true}, y^{pred}) = \sum_{i=1}^{n} log(cosh(y_i^{true} - y_i^{pred}))$$

where L is the loss value, y^{true} and y^{pred} are the true and the prediction values respectively. The Log-Cosh loss takes on the behavior of squared loss when the cost function is small and acts as absolute loss when the cost function is large. This behavior is similar to Huber loss, but the cost function has the advantage of being twice differentiable everywhere.

The average size of a mammogram is approximately 3500 pixels in width and 4000 pixels in height. The images were resampled to 250×250 pixels. In addition, the image intensities were normalized to a range $(0, 1)$. The annotations were also resampled along with the image. For training, we used a batch size of 12 images and a learning rate of 0.0001. The training process terminated after 150 epochs. During the training process, we monitored the validation loss and save the model which achieves the lowest validation loss.

4   Detecting the BB (Nipple) and the PNL on CC View

As we know that the shape of the BB marker is circular, we can use the Hough Circle detector [17]. The output of the Hough-Circle transform is a set of center-points and the radii of the detected circles. As shown in Fig. 2, the circle detector will yield multiple potential BB candidates (shown in green). We can use the fact that the BB has a uniform intensity for filtering out the false-positive candidates for BB detection. This filtering process is achieved using the 9-way connectivity rule. We check if all the 9 pixels connected to the center point of the circle has

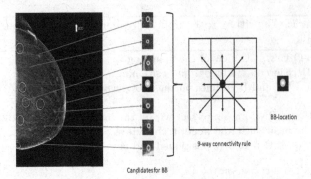

Fig. 2. BB detection: a pipeline showing the algorithm for BB detection. The image on the left shows all the possible candidates for BB location. Each of the candidates for BB is then subjected to a 9-way connectivity rule, that is, all the pixels in the vicinity of the center point should have the same intensity as that of the center. Only the BB qualifies this test. (Color figure online)

the same intensity values. Only the candidate passing this connectivity rule is the BB marker. The BB detection process is depicted in Fig. 2. Once the BB is detected, we draw a horizontal line to either the left or right extremity of the image, depending on whether it is the left or the right breast. Then the length of the line segment is calculated, which gives the PNL length on the CC view of the breast.

5   Results

We will present the results in four stages as (a) prediction of PEC and PNL lines on MLO, (b) predicting the adequacy of MLO, (c) predicting the PNL on CC view and (d) predicting the adequacy of the MLO/CC pair. Results in each of these problems should be considered independent of each other.

5.1   Predicting PEC and PNL Lines

Figure 3 shows a comparison between the ground truth labels and the predicted labels on the downsampled images. In the next step, the images the predictions on the downsampled images are resampled to the original resolution of the image using linear interpolation techniques. In Fig. 4, the plots show the distribution of errors in the end-points for the PEC and PNL lines across the test set. Further, in Fig. 4, we show a comparison between the lengths of the predicted posterior nipple line and the ones drawn by the radiologists.

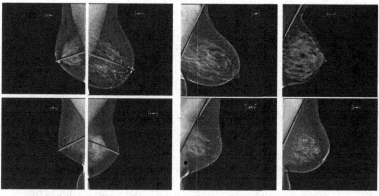

Posterior Nipple Line (PNL) prediction   Pectoralis muscle (PEC) prediction

Fig. 3. Predicting the pectoral muscle (PEC) and the posterior nipple line (PNL) on the downsampled images. The blue line shows the ground truth and the yellow line shows the prediction results. (Color figure online)

5.2 Predicting the Adequacy of MLO

After predicting the PNL and PEC lines on an MLO image, we can calculate the point of intersection of these two lines. As per the MQSA standards, the point of intersection should lie within the bounds of the image for the MLO view to be acceptable. We compared 266 adequately positioned MLOs (from 133 normal patients) to 66 inadequately positioned. The results of our proposed algorithm is shown using the confusion matrix in Fig. 5. From the confusion matrix, one can infer that our proposed method is very accurate in detecting the correctly positioned MLO. The true positive rate is 0.9053. The technique is not particularly useful in detecting the poorly positioned MLOs (True negative rate = 0.45). However, we cannot conclude that this method is not a good alternative for detecting poorly positioned MLOs. There are several reasons to support this claim. First, the number of poorly positioned MLO images in our dataset is small compared to the number of normal images. Hence, it is difficult to make a statistically relevant claim. As the positioning algorithm is dependent on the guidelines proposed by MQSA, the errors are also dependent on the accuracy of PEC and PNL detection. Further, as our neural network was trained on normal MLO views only, the PEC/PNL detection accuracy is relatively lower in the case of poorly positioned MLOs. We can improve on this aspect of the proposed work by collecting more poorly-positioned images and using some part of them in training our model.

5.3 Predicting the Positioning of CC View of the Mammogram

The positioning on the craniocaudal view is judged based on its corresponding MLO view. For the CC view, we first need to predict the BB, as discussed in Sect. 4. Our proposed algorithm was able to detect the correct BB location in

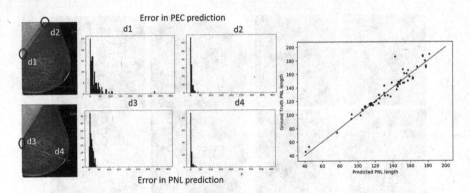

Fig. 4. Errors on the end-points while predicting the pectoral line (PEC) and the posterior nipple line (PNL). The errors are measured based on the end-points of PEC and PNL lines. In the adjoining histograms in the middle, the error distribution for all the four points in the test set is shown. Blue represents the predicted line, whereas green represents the ground-truth. On the right, we show the agreement between the ground truth and prediction for the PNL lengths. (Color figure online)

Predicted positioning

	N = 332	Adequate	Inadequate	Total
Ground Truth	Adequate	243	23	266
	Inadequate	36	30	66

Fig. 5. Confusion matrix showing the efficacy of prediction of MLO positioning based on the intersection of PEC and PNL.

84.5% of cases (225 out of 266). On the CC view, we predicted the PNL line as mentioned in Sect. 4, and calculated the length of PNL length (d_{CC}). If a BB is not correctly detected, we cannot make any prediction about the PNL length on the CC view.

5.4 Predicting the Adequacy of the MLO/CC Pair

We calculated the absolute difference between d_{MLO} and d_{CC}. If the difference lies within 1 cm, the MLO and CC pair of views are considered to be adequately positioned. To detect the efficacy of the 1 cm rule as suggested by the MQSA standards, we only included the CC images, where the correct BB was predicted. Out of 225 adequately positioned MLO/CC pairs, we were able to accurately predict the adequate positioning for 214 MLO/CC pair and failed to do so for the remaining 11 MLO/CC pair. Thus, our true positive rate is 0.95 and the false negative rate is 0.05.

5.5 Generating an Automated Report on the Positioning of the Breast: Real-World Application

So far, we have discussed the efficacy of the individual components of the algorithm. In practice, we generate a text report along with images where the predicted PEC and PNL lines are drawn. The goal of this report is to provide enough information to the technologist about the positioning of the breast. If the technologist sees that the predicted PEC and PNLs are not correct, he/she can disregard the text report. Here we will discuss two automatically generated reports for two different patients.

In Fig. 6, we show the report generated for a patient with four scans. The point of intersection of PEC and PNL lines on both LMLO and RMLO images is shown along with their corresponding lengths. Similarly. PNL lengths are calculated on CC, and the difference between the two lengths is reported.

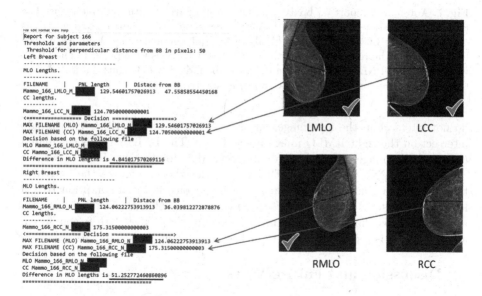

Fig. 6. Automated report on breast positioning for a patient. The PEC and PNL lines intersect on both LMLO and RMLO views. The difference between the lengths of the posterior nipple line is reported.

In Fig. 7, we show the report generation for a patient where the proposed algorithm failed to make any decisions. As the results are produced along with the images, the technologist can see that for the LMLO view, the PEC and PNL are not intersecting within the image bounds. So, he/she can use their judgment on the positioning of the image. Similarly, our algorithm failed to detect the BB on the LCC view of the image. Thus, no decision was made on the positioning of the left breast. However, in the case of the right breast, multiple RMLO and RCC views were available. Out of the two RMLO views, the PEC and PNL

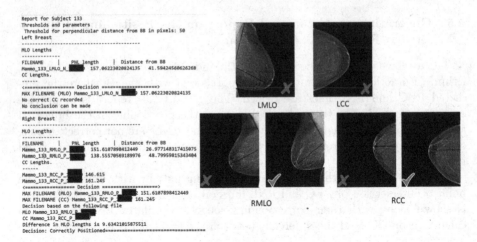

Fig. 7. Automated report on breast positioning of a patient with multiple scans. The point of intersection on the LMLO view is outside the image bounds and no BB is detected on the LCC view. Thus the algorithm did not make any decisions on the left breast. For the right breast, multiple MLO and CC views were present. The MLO, where PEC and PNL intersect within the image bounds (RMLO-right) is selected. The CC view with the longest PNL (RCC-right) is selected.

do not intersect in the left image, and so is not considered. However, they do intersect on the right RMLO image in the figure. The PNL length is calculated based on the image in which they intersect. Also, for the two RCC views, we calculate the PNL lengths in both but consider the one where the length of the PNL is maximum. The reason for such a choice is the assumption that a longer PNL ensures more breast tissue to be present. Thus the RCC image with a longer PNL is chosen. Finally, the difference in the PNL length is computed, and a decision is reported for the right breast positioning.

6   Discussion and Future Work

In this paper, we have presented a deep learning-based method for determining if the breast is well-positioned on a mammographic image. The proposed method is based on the MQSA standards followed for breast scan positioning. We showed that by combining techniques from deep learning and feature based machine learning algorithms, it is possible to accurately predict if the breast is adequately positioned under the scanner. We showed that our proposed method has a high true positive rate (TPR or sensitivity) for detecting the adequately positioned mammograms. However, the true negative rate (TNR or specificity) for detecting the inadequately positioned mammograms is low. We think that the specificity can be improved if we include more inadequately positioned mammograms in the training data.

As far as we know, this is one of first papers delving into automated quality control for breast imaging. One of the major contributions of the paper is to breakdown the mammogram positioning problem into a deep learning driven regression problem. We used heuristics to compute the measurement guidelines provided by the MQSA. In the end, we produce an automated report which shows the respective measurements along with the predictions while the patient is present in the imaging center. These reports will likely reduce the time spent by the technologist, determining the quality of images. In the future, we can improve the method to identify the inadequately positioned mammograms with high accuracy. Having a small number of inadequately positioned mammograms limits our ability to report the accuracy of the algorithm on the same at this moment. In the future, we can evaluate the efficacy of the framework for identifying inadequately positioned mammograms on a sufficiently large dataset.

Acknowledgements. The authors would like to thank Dr. Ratnesh Kumar for his invaluable support and insights during the course of this project. We will also like to acknowledge that the data collection and algorithm development was performed in the Department of Radiology at The Ohio State University.

References

1. Andersson, I., et al.: Breast tomosynthesis and digital mammography: a comparison of breast cancer visibility and BIRADS classification in a population of cancers with subtle mammographic findings. Eur. Radiol. **18**(12), 2817–2825 (2008). https://doi.org/10.1007/s00330-008-1076-9
2. Bae, M.S., et al.: Breast cancer detected with screening US: reasons for nondetection at mammography. Radiology **270**(2), 369–377 (2014)
3. Bird, R.E., Wallace, T.W., Yankaskas, B.C.: Analysis of cancers missed at screening mammography. Radiology **184**(3), 613–617 (1992)
4. Birdwell, R.L., Ikeda, D.M., O'Shaughnessy, K.F., Sickles, E.A.: Mammographic characteristics of 115 missed cancers later detected with screening mammography and the potential utility of computer-aided detection. Radiology **219**(1), 192–202 (2001)
5. United States Food and Drug Administration: MQSA national statistics
6. Gupta, V., et al.: Performance of a deep neural network algorithm based on a small medical image dataset: incremental impact of 3D-to-2D reformation combined with novel data augmentation, photometric conversion, or transfer learning. J. Digit. Imaging **33**, 431–438 (2020). https://doi.org/10.1007/s10278-019-00267-3
7. Gupta, V., et al.: Using transfer learning and class activation maps supporting detection and localization of femoral fractures on anteroposterior radiographs. In: 2020 IEEE 17th International Symposium on Biomedical Imaging (ISBI), pp. 1526–1529. IEEE (2020)
8. Skaane, P., et al.: Follow-up and final results of the Oslo I Study comparing screen-film mammography and full-field digital mammography with soft-copy reading. Acta Radiol. **46**(7), 679–689 (2005)
9. American Cancer Society: Global Cancer Facts & Figures, 3rd edn (2015)
10. Szegedy, C., Vanhoucke, V., Ioffe, S., Shlens, J., Wojna, Z.: Rethinking the inception architecture for computer vision. In: Proceedings of the IEEE Conference on Computer Vision and Pattern Recognition, pp. 2818–2826 (2016)

11. Tabár, L., et al.: Swedish two-county trial: impact of mammographic screening on breast cancer mortality during 3 decades. Radiology **260**(3), 658–663 (2011)

12. Taplin, S.H., Rutter, C.M., Finder, C., Mandelson, M.T., Houn, F., White, E.: Screening mammography: clinical image quality and the risk of interval breast cancer. Am. J. Roentgenol. **178**(4), 797–803 (2002)

13. Uematsu, T.: Ultrasonographic findings of missed breast cancer: pitfalls and pearls. Breast Cancer **21**(1), 10–19 (2013). https://doi.org/10.1007/s12282-013-0498-7

14. Varma, M., et al.: Automated abnormality detection in lower extremity radiographs using deep learning. Nat. Mach. Intell. **1**(12), 578–583 (2019)

15. Wada, K.: Labelme: Image Polygonal Annotation with Python (2016). https://github.com/wkentaro/labelme

16. Yu, J., et al.: Detection and localisation of hip fractures on anteroposterior radiographs with artificial intelligence: proof of concept. Clin. Radiol. **75**(3), 237-e1 (2020)

17. Yuen, H., Princen, J., Illingworth, J., Kittler, J.: Comparative study of Hough transform methods for circle finding. Image Vis. Comput. **8**(1), 71–77 (1990)

Can Non-specialists Provide High Quality Gold Standard Labels in Challenging Modalities?

Samuel Budd[1]([envelope])([ORCID]), Thomas Day[2,3], John Simpson[2,3], Karen Lloyd[2],
Jacqueline Matthew[2,3], Emily Skelton[2,3,4], Reza Razavi[2,3],
and Bernhard Kainz[1,5]([ORCID])

[1] Department of Computing, BioMedIA, Imperial College London, London, UK
samuel.budd13@imperial.ac.uk
[2] King's College London, London, UK
[3] Guy's and St Thomas' NHS Foundation Trust, London, UK
[4] School of Health Sciences, City, University of London, London, UK
[5] Friedrich–Alexander University Erlangen–Nürnberg, Erlangen, Germany

Abstract. Probably yes.—Supervised Deep Learning dominates performance scores for many computer vision tasks and defines the state-of-the-art. However, medical image analysis lags behind natural image applications. One of the many reasons is the lack of well annotated medical image data available to researchers. One of the first things researchers are told is that we require significant expertise to reliably and accurately interpret and label such data. We see significant inter- and intra-observer variability between expert annotations of medical images. Still, it is a widely held assumption that novice annotators are unable to provide useful annotations for use by clinical Deep Learning models. In this work we challenge this assumption and examine the implications of using a minimally trained novice labelling workforce to acquire annotations for a complex medical image dataset. We study the time and cost implications of using novice annotators, the raw performance of novice annotators compared to gold-standard expert annotators, and the downstream effects on a trained Deep Learning segmentation model's performance for detecting a specific congenital heart disease (hypoplastic left heart syndrome) in fetal ultrasound imaging.

Keywords: Expert · Novice · Labels · Annotations

1 Introduction

It is commonly believed that domain experts are the only reliable source for annotating medical image data. This assumption has resulted in a dearth of

Electronic supplementary material The online version of this chapter (https://doi.org/10.1007/978-3-030-87722-4_23) contains supplementary material, which is available to authorized users.

© Springer Nature Switzerland AG 2021
S. Albarqouni et al. (Eds.): DART 2021/FAIR 2021, LNCS 12968, pp. 251–262, 2021.
https://doi.org/10.1007/978-3-030-87722-4_23

annotated medical image datasets due to the time and high costs associated with expert labelling time. In this work we challenge this assumption and employ novice annotators to perform a complex multi-class fetal cardiac ultrasound (US) segmentation task.

A core goal of medical image analysis is to free up experts' time for more challenging tasks and time with patients. Our current view is that expert annotation efforts that aid in the development of models, will save expert time in the long term. However we hypothesise that in many cases, this annotation effort can be performed by novice annotators at a lower cost, saving both resources and experts' time, with minimal impact on the performance of automated downstream models.

Segmentation is widely regarded as among the most labour intensive medical image analysis tasks, requiring pixel-level labels to enable supervised learning methods to learn complex segmentation tasks. In this study we use a multi-class fetal cardiac US segmentation task as our initial test case, as this task is challenging in both anatomy and modality (noisy, heterogeneous and often contains artefacts). This makes the task of annotating fetal US images challenging for both experts and novices, and an ideal test case for comparing the efficacy of novice annotations. Segmentation of the fetal heart from '4-Chamber view' images provides quantitative biomarkers that can be used for the diagnosis of Hypoplastic Left Heart Syndrome (HLHS). As such we include in our dataset several HLHS cases. The presence of pathology within our dataset makes this annotation task even more challenging, and enables us to compare the performance of novice and expert annotations on a segmentation-informed diagnostic classification task.

We provide evidence that the reliability of novice annotators is greater than expected and that this approach might be a viable option for annotation of medical image datasets in the future.

Related Work: Significant work has been done to mitigate for a lack of well-annotated medical imaging data. Learning from fewer labels, unsupervised learning and active learning are all valuable contributions in this and their benefits go beyond our setting. Advances in these fields can only benefit from the increasing sizes of annotated medical image datasets, and as such we do not challenge these approaches. More tightly related to our work are methods for learning from crowd-sourced noisy labels, where annotations of varying quality are acquired [3,4,10,13,14,16].

In [12] it is shown that novice annotators are comparable to expert annotators for a series of natural language annotation tasks, and that only a small number of novice annotations are necessary to equal the performance of expert annotators. In [5] it is shown that novice annotators are able to effectively prune non-informative text from training data for sentiment classifiers to improve classification performance of trained models.

In [9] it is shown that crowd-sourcing many noisy labels for heavily class imbalanced text classification datasets is expensive and the usual benefits of redundant labelling seen in crowd-sourcing scenarios is lesser in imbalanced settings. [9] provide techniques for discarding redundant instances such that annotations can be acquired in a cost-effective way over a five-way majority vote aggregation.

In [15] the authors assess the effects of aggregating progressively more labels per instance on model performance for mitotic figure detection from histologic images. They show that high accuracy can be achieved with a single annotation per image, and improved by aggregating three annotations per image, while aggregating beyond three annotations per image results in only minor very minor performance increases.

In [8] criteria are proposed by which the suitability of a text sentiment classification task for crowdsourcing can be assessed (1. Noise level, 2. Inherent Ambiguity and 3. Informativeness to the model). Models trained on expert and novice annotations are compared. By considering the three proposed criteria, it is shown that comparable model performance can be achieved using expert or novice annotations.

For a 3d segmentation correction task, there is evidence for little to no difference between novices and expert performance (engineers with domain knowledge, medical students, and radiologists) in the ability to detect and correct errors made by a segmentation algorithm [7], although novice annotators need significantly more time per annotation.

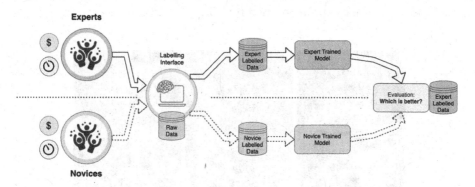

Fig. 1. Graphical overview of our process: we assess the upstream and downstream impacts of using novice annotations in place of expert annotations on a challenging medical image segmentation task. Each set of annotations is acquired, pre-processed and used to train models in the same way. We evaluate both expert and novice models against an expert annotated test set.

Contribution: We assess the upstream and downstream impacts of training medical image multi-class segmentation models, and downstream classification models on noisy labels from novice annotators compared against gold-standard labels from expert annotators.

We show that novice annotators are capable of performing complex medical image annotation tasks to a high standard, and that variability between novices and experts is comparable to that amongst experts themselves. We show that models trained on novice labels are comparable to those trained on expert labels for multi-class segmentation and downstream classification.

We analyse the time and costs associated with using expert vs. novice labels to show that using novice annotations is more resource efficient, and that the major parameter governing model performance is dataset size, rather than label quality in this setting. This will enable clinical and translational researchers to develop a greater understanding of the trade-offs associated with acquiring medical image annotations with respect to cost, time and supervised learning method performance.

2   Method

Annotation Labels Collection: The current paradigm for collecting annotations for medical image data is to present experts with un-annotated data in an annotation interface that allows them to delineate structures of interest in every image (Fig. 1). Once complete, the annotations and input can be exported for use. In this work we employ novice annotators to perform the same task using the same annotation tools on the same data to provide us with novice annotated for later use, as shown in the bottom half on Fig. 1. We use the Labelbox web-based interface as our annotation tool [1].

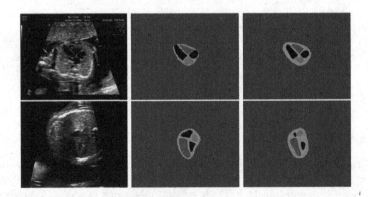

Fig. 2. Example US images and manual segmentations of anatomical areas. Top row: Healthy image, expert manual label and novice manual label (left to right). Bottom row: HLHS image, expert manual label and novice manual label (left to right)

Segmentation Model: From a single US image of the '4-Chamber Heart View' (4CH view) acquired during fetal screening, we train a model to delineate 5 anatomical areas: *'Whole Heart'* (WH), *'Left Ventricle'* (LV), *'Right Ventricle'* (RV), *'Left Atrium'* (LA) and *'Right Atrium'* (RA) (Fig. 2).

We use the UNet architecture as our segmentation network [11], known to perform well for US segmentation. We train using dropout [6], for a fixed number of epochs, then select the best performing model on the validation set. Random horizontal and vertical flipping, cropping, translation, rotation and scaling is applied during training.

Classification Model: We extract numerical features from $\hat{y}_i^{seg}$ (manual or automated segmentation) in order to classify HLHS vs. healthy patients from interpretable features $f = \{f_0, f_1, ..., f_N\}$ where $f_i = r_{ab} = A_a/A_b$ if $a \neq b$ and r_{ba} is not in f already. Here r_{ab} is the ratio between two quantities and consider r_{ab} and r_{ba} to contain equivalent information and exclude the latter from f. A_a is the count of pixels belonging to class a in $\hat{y}_i^{seg}$ which acts as an estimate to the area.

We apply an L2 regularised, class weight balanced logistic regression classifier implementation to classify the extracted segmentation area ratio features as healthy vs. HLHS as in [2].

Statistical Analysis: Here we pose the questions answered in this paper and outline our approach to answering them. For tests of statistically significant difference between distributions we use a two-tailed Z-test for the null hypothesis of identical means: $Z = \frac{\hat{X} - \mu_0}{s}$ where Z is our test statistic, μ is our population mean and s is our population standard deviation.

Q1: Are novice annotations as similar to experts as expert annotations are to other experts? We answer this question by computing the average DICE similarity coefficient between novice and expert annotations, and between pairs of expert annotations. We calculate the Dice score for each class separately, and test for statistically significant difference between the two sets.

Q2: How different are automated segmentations trained on experts annotations to automated segmentations trained on novice annotations? We show evidence by computing the average DICE similarity coefficient between novice and expert trained model predictions and an expert annotated test set. We calculate the DICE score for each class separately, and test for statistically significant difference between the two sets.

Q3: How different are classification predictions trained on either manual, or model based segmentations from novices compared to experts? We train a classifier using training data from manual expert and manual novice annotations, as well as expert model and novice model predictions. We test each classifier on our expert test set and compare key performance metrics to assess the discriminative powers of novice vs. expert based segmentations.

Q4: In resource limited scenarios are expert or novice annotations more cost effective to attain the same model performance? We observe the time/cost/quality trade-off by measuring the DICE scores obtained by models trained using novice and expert data on progressively more labels (50 to 1000 labels) using a UNet with 200 epochs. We use DICE scores on the test set as a measure of prediction quality, and use time taken and estimated financial cost to acquire each annotation, to plot the time vs. cost vs. performance of our models for both experts and novice annotations.

3   Experiments and Results

Data and Pre-processing

- Raw images: We use a private and ethics/IP-restricted, de-identified dataset of 2380 4CH US images, with 1000 for training, 380 for validation, 1000 for testing acquired on Toshiba Aplio i700, i800 and Philips EPIQ V7 G devices.
- Expert segmentations: A fetal cardiologist and three expert sonographers delineated the images using Labelbox [1]. Multiple expert annotations for 319 images were acquired and used to calculate expert-expert annotation similarity. A single annotation for each image is used for training.
- Non-expert segmentations: A novice workforce with no experience annotating medical US data was employed to delineate the images using Labelbox [1], this workforce was provided with the instruction pdf included in the Supplementary Material. Three novice annotations for every image in the training set were acquired, each further calculation made used novice annotations was performed three times and the results averaged.
- Time: Experts annotated images in an average time of 127 s per image, and Novices annotated images in an average time of 253 s per image.
- Cost: Experts costs were set at $60 per labelling hour, and Novices cost $6 per labelling hour.

During analysis 10 cases were found to have two hearts visible (split screen view), resulting in zero DICE agreement amongst experts and experts and novices, having annotating different sides of the image. These cases have been removed. A significant proportion of the worst performing remaining cases are a result of mislabelling of left/right atriums and ventricles resulting in very low DICE scores for those cases.

Q1: In Table 1 the average DICE scores for expert-expert and novice-expert segmentations show that no statistical difference is found between the variability of annotators on three out of five annotated classes. This shows that novice annotators are better at annotating complex medical data than is assumed and the variability between experts and novices is similar to that amongst experts for these three classes. Figure 3 highlights the similarity in DICE distributions between novice-novice and expert-novice annotations, indicating it may not be possible to avoid variation in annotations even when using experts annotations alone.

Q2: Average DICE scores and segmented class sizes for expert trained vs. novice trained models show there is no statistical difference in the performance of the models on three out of five classes (Tables 2, and 4 (Supplementary Material)). Expert models average higher DICE scores on all but one class, and one reason for this better performance is that both models are tested against an expert

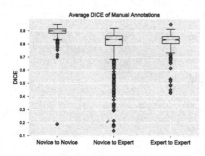
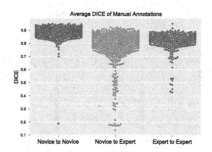

Fig. 3. Distributions of DICE similarity scores between raw labels for Novice to Novice labels, Novice to Expert labels and Expert to Expert labels. Left: Box and whisker plots. Right: Swarm plots.

Table 1. Mean DICE scores of manual annotations performed by Experts compared with DICE scores of manual annotation performed by Novices. Statistically significant (95%) results shown in bold.

DICE	LV	RV	LA	RA	WH
Expert to expert	0.807	0.787	0.764	0.808	0.887
Novice to expert	0.778	0.761	0.757	0.806	0.894
p-value	**0.009**	**0.005**	0.551	0.866	0.359

annotated test set. Models trained on novice annotations can perform almost equally well as those trained on expert annotations for multi-class US segmentations problems. We see a significant difference between average class sizes predicted by the two models, most noticeably in the right ventricle class (RV), however their overall similarity is highlighted in Figs. 4 and Figs. 7–8 in the Supplementary Material where both DICE and sizes appear very similar across all classes.

Q3: The results of HLHS classification methods trained on manual and automated expert and novice segmentations show that novice trained models attain very similar results to those trained by experts, in both manual and model cases (Table 3). We see a slight improvement from expert annotations in Precision and F1 scores but the overall performance is remarkably similar. This result again shows the viability of acquiring a significant proportion of medical image annotations from non-experts during annotation efforts. Figure 5 highlights that in some scenarios novice manual annotations may out-perform expert annotations on some metrics. Both ROC curves and Precision-Recall Curves for experts and novices follow very similar trajectories demonstrating the similarity in their performance for classification.

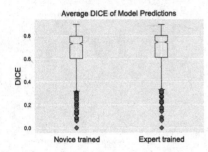
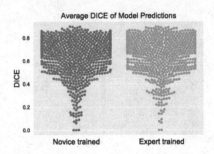

Fig. 4. Distributions of segmentation predictions DICE scores against the expert test set

Table 2. Class average DICE scores of model predictions and class average sizes in pixels of model predictions, comparing models trained using expert annotations and models trained using novice annotations. Statistically significant (95%) results shown in bold.

DICE	LV	RV	LA	RA	WH
Expert model	0.721	0.707	0.663	0.749	0.617
Novice model	0.708	0.679	0.652	0.731	0.634
p-value	0.174	**0.003**	0.321	0.071	**0.001**

Q4: Figure 6a shows the consistent increase of both expert and novice trained models as the size of the dataset increases, demonstrating that collecting initial annotations from novices may well suffice to achieve a good accuracy in many tasks. We calculate the cost per image for both novices and experts using the average cost of an hour of labelling work and the average time each annotation took to create. Figure 6b shows how when time is the priority, then expert annotators achieve higher quality models in a shorter time-span, however this comes at a much greater financial cost. If cost is the priority then novice annotators achieve higher quality models at a much smaller financial cost, however the same number of annotations take longer to acquire from novices than from experts. We can see from this that the dominant driving force of improving model quality is dataset size, regardless of whether annotations come from experts or novices, indicating that to train high performing models in a resource efficient way that novice annotations are a useful mechanism by which this can be achieved.

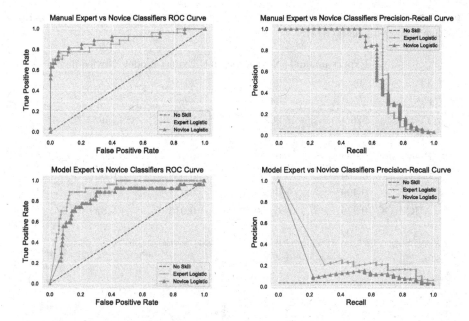

Fig. 5. Top row: Classification performance for manual annotations predicted classifications. Bottom row: Classification performance for model predicted classifications. Left to right: ROC curves and Precision-Recall curves.

4   Discussion

We have assessed the upstream and downstream effects of acquiring complex medical image segmentation annotations from novices compared to experts. We have found that raw novice annotations are of remarkable quality, and that novice trained models show only a minor performance decrease compared expert trained models. Our results highlight that annotations performed by novices are of great utility for complex tasks such as segmentation and classification. A time and cost analysis for using limited resources more efficiently is provided, guiding practitioners in acquiring annotations to give the best performing models under their constraints. Through future studies on other complex tasks, we aim to develop protocols through which confidence can be given that novice annotations are sufficient in many use cases.

Additional combination of crowd-sourcing from novice labels with models incorporating measures of annotator skill and merging of multiple annotations show great promise in enabling highly accurate models to be developed on a wide variety of tasks for which expert annotated data has been infeasible to acquire at a large enough scale.

We note that we are unsure of how representative our Labelbox workforce is of the wider novice annotator community. Through our engagement with Labelbox they were made aware of our intentions with the annotated data and it is our hope that no special measures were taken to improve the quality of annotations

Table 3. Classification results: Precision, Recall and F1 scores are reported for the positive prediction class (HLHS)

	Expert manual	Novice manual	Expert model	Novice model
TP	19	20	24	22
FP	47	64	126	224
TN	926	909	847	749
FN	8	7	3	4
Precision	0.288	0.242	0.16	0.091
Recall	0.704	0.753	0.889	0.827
F1	0.409	0.367	0.271	0.165
AUC-ROC	0.879	0.900	0.915	0.829

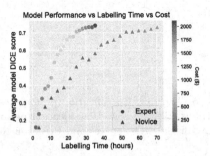

(a) DICE scores as we increase the training dataset size from 50 to 1000 images

(b) Time, Cost and DICE scores as we increase the training dataset size from 50 to 1000 images

Fig. 6. Analysis of the Time/Cost/Model performance trade-off.

beyond that of the wider novice annotator community. Similarly, when comparing costs of annotating large datasets, we must consider the ethical implications of employing low-cost workers to perform these tasks - while the low cost makes using workforce services appealing, care must be taken to ensure that workers are paid fairly and under suitable working conditions. Limited information given regarding the locations and working conditions of annotation workforces creates difficultly in making this judgement. Additional consideration must be given to data privacy when using external labelling services, as regulation surrounding data storage and transfer must be adhered to ensure patient data remains protected.

5   Conclusion

We have demonstrated that novice annotators are capable of performing complex medical image segmentation tasks to a high standard, with a comparable variability to experts as experts show to themselves. We have shown that training models with novice annotations is both resource efficient and can give comparable models in terms of prediction performance against expert annotations for both segmentation and downstream classification tasks. We foresee that in combination with existing methods that better handle noisy annotations, and active learning methods selectively choosing the most informative annotations to acquire next, that novice annotations will play a vital role in developing high-performing models at a fraction of the cost of using expert annotations.

References

1. Labelbox (2021). https://labelbox.com. Accessed 27 Feb 2021
2. Budd, S., et al.: Detecting Hypo-plastic Left Heart Syndrome in Fetal Ultrasound via Disease-specific Atlas Maps, July 2021. https://arxiv.org/abs/2107.02643v1
3. Chang, J.C., Amershi, S., Kamar, E.: Revolt: collaborative crowdsourcing for labeling machine learning datasets. In: Proceedings of the 2017 CHI Conference on Human Factors in Computing Systems, CHI 2017, pp. 2334–2346. Association for Computing Machinery, New York (2017). https://doi.org/10.1145/3025453.3026044
4. Cheplygina, V., Perez-Rovira, A., Kuo, W., Tiddens, H.A.W.M., de Bruijne, M.: Early experiences with crowdsourcing airway annotations in chest CT. In: Carneiro, G., et al. (eds.) LABELS/DLMIA -2016. LNCS, vol. 10008, pp. 209–218. Springer, Cham (2016). https://doi.org/10.1007/978-3-319-46976-8_22
5. Fang, J., Price, B., Price, L.: Pruning non-informative text through non-expert annotations to improve aspect-level sentiment classification. In: Proceedings of the 2nd Workshop on The People's Web Meets NLP: Collaboratively Constructed Semantic Resources, pp. 37–45. Coling 2010 Organizing Committee, Beijing, August 2010. https://www.aclweb.org/anthology/W10-3505
6. Gal, Y., Ghahramani, Z.: Dropout as a Bayesian approximation: representing model uncertainty in deep learning. In: ICLR 2016, pp. 1050–1059 (2016)
7. Heim, E., et al.: Large-scale medical image annotation with crowd-powered algorithms. J. Med. Imaging 5(03), 1 (2018). https://doi.org/10.1117/1.jmi.5.3.034002
8. Hsueh, P.Y., Melville, P., Sindhwani, V.: Data quality from crowdsourcing: a study of annotation selection criteria. In: Proceedings of the NAACL HLT 2009 Workshop on Active Learning for Natural Language Processing, HLT 2009, pp. 27–35. Association for Computational Linguistics (2009)
9. Jamison, E., Gurevych, I.: Needle in a haystack: reducing the costs of annotating rare-class instances in imbalanced datasets. In: Proceedings of the 28th Pacific Asia Conference on Language, Information and Computing, pp. 244–253. Department of Linguistics, Chulalongkorn University, Phuket, December 2014. https://www.aclweb.org/anthology/Y14-1030
10. Rodrigues, F., Pereira, F.C.: Deep learning from crowds. https://arxiv.org/pdf/1709.01779v2.pdf
11. Ronneberger, O., Fischer, P., Brox, T.: U-Net: convolutional networks for biomedical image segmentation. In: Navab, N., Hornegger, J., Wells, W.M., Frangi, A.F. (eds.) MICCAI 2015. LNCS, vol. 9351, pp. 234–241. Springer, Cham (2015). https://doi.org/10.1007/978-3-319-24574-4_28

12. Snow, R., O'Connor, B., Jurafsky, D., Ng, A.: Cheap and fast - but is it good? Evaluating non-expert annotations for natural language tasks. In: Proceedings of the 2008 Conference on Empirical Methods in Natural Language Processing, pp. 254–263. Association for Computational Linguistics, Honolulu, October 2008. https://www.aclweb.org/anthology/D08-1027

13. Tajbakhsh, N., et al.: Embracing imperfect datasets: a review of deep learning solutions for medical image segmentation. Med. Image Anal. **63**, 101693 (2020). https://doi.org/10.1016/j.media.2020.101693

14. Tinati, R., Luczak-Roesch, M., Simperl, E., Hall, W.: An investigation of player motivations in Eyewire, a gamified citizen science project. Comput. Hum. Behav. **73**, 527–540 (2017). https://doi.org/10.1016/j.chb.2016.12.074

15. Wilm, F., et al.: How many annotators do we need? A study on the influence of inter-observer variability on the reliability of automatic mitotic figure assessment, December 2020. http://arxiv.org/abs/2012.02495

16. Yu, S., et al.: Robustness study of noisy annotation in deep learning based medical image segmentation. Phys. Med. Biol. **65**(17), 175007 (2020). https://doi.org/10.1088/1361-6560/ab99e5

Author Index

Printed in the United States
by Baker & Taylor Publisher Services